Molecular Sensors and Nanodevices

Molecular Sensors and Nanodevices

Principles, Designs and Applications in Biomedical Engineering

John X. J. Zhang

Kazunori Hoshino

AMSTERDAM • BOSTON • HEIDELBERG • LONDON
NEW YORK • OXFORD • PARIS • SAN DIEGO
SAN FRANCISCO • SINGAPORE • SYDNEY • TOKYO
William Andrew is an imprint of Elsevier

William Andrew is an imprint of Elsevier.
225 Wyman Street, Waltham, MA 02451, USA
The Boulevard, Langford Lane, Kidlington, Oxford OX5 1GB, UK

Library of Congress Cataloging-in-Publication Data
A catalog record for this book is available from the Library of Congress

British Library Cataloguing-in-Publication Data
A catalogue record for this book is available from the British Library

ISBN: 978-1-4557-7631-3

For information on all William Andrew publications
Visit our Web site at http://store.elsevier.com/

14 15 16 17 18   10 9 8 7 6 5 4 3 2 1

Printed in the United States of America

Working together
to grow libraries in
developing countries

www.elsevier.com • www.bookaid.org

To Ting, Michael and Lauren

for the love, fun and bringing out the teacher in me

— **John X.J. Zhang**

Contents

About the Authors

John X.J. Zhang

Xiaojing (John) Zhang is an Associate Professor in the Department of Biomedical Engineering at the University of Texas of Austin (UT Austin). He received his Ph.D. from Stanford University, California, and was a Research Scientist at Massachusetts Institute of Technology (MIT) before joining the faculty at UT Austin.

Dr. Zhang's research focuses on integrating Micro-electro-mechanical Systems (MEMS), nano-materials, micro-imaging and biosensors to provide innovative solutions to critical healthcare issues. Dr. Zhang has actively engaged in teaching new concepts and methods in these emerging biomedical engineering frontiers and has demonstrated innovation and excellence in engineering education. The materials presented in this textbook are based on his lecture notes while teaching a popular course on molecular sensors and nanodevices at UT Austin for the past 8 years.

He has a track record for developing both core and emerging engineering curriculum along with developing well-funded research programs with fellow students. Among his numerous awards, Dr. Zhang received the Wallace H. Coulter Foundation Early Career Award in Biomedical Engineering, NSF CAREER award, DARPA Young Faculty Award among many others. To recognize his accomplishment in research and education, Dr. Zhang was selected to attend the prestigious US National Academy of Engineering, Frontiers of Engineering (NAE FOE) program in 2011, the NAE Frontiers of Engineering Education (NAE FOEE) program in 2012, and subsequently China-America Frontiers of Engineering Symposium (CAFOE) program in 2013.

As an active member in his professional community, Dr. Zhang has served on numerous international conference organizing committees and editorial boards. He is an editor for *ASME/IEEE Journal of Microelectromechanical Systems (JMEMS)* and an associate editor *Biomedical Microdevices*.

Kazunori Hoshino

Kazunori Hoshino received his PhD degree in mechanical engineering from the University of Tokyo, Tokyo, Japan, in 2000. He worked for the University of Tokyo from 2003 to

2006 as a lecturer in the Department of Mechano-Informatics, School of Information Science and Technology, where he conducted several government funded project as the principal investigator. In 2006, he joined the University of Texas at Austin, where he works as a Senior Research Associate in the Department of Biomedical Engineering.

His research interests are (1) NEMS/MEMS-based nanophotonic sensing and imaging systems and (2) microfluidic detection, imaging, and analysis of cancer cells. He has more than 100 peer reviewed publications in top international journals and conferences in the field of sensors, micro-electro-mechanical system (MEMS), and micro total analytical systems (μTAS). He is the inventor of 6 issued US patents and 12 issued Japanese patents. The courses he has taught at the University of Tokyo and the University of Texas at Austin include: molecular sensors and nanodevices, intelligent micro-electro-mechanical systems, digital control systems, and engineering mathematics.

Preface

Whether you are an undergraduate or graduate student, researchers, industrial professions or medical practitioners, you will find a body of enjoyable and useful information within the covers of this book. The comprehensive coverage of fundamental concepts, practical approach and use of case studies makes this book essential reading for a wide range of scientists and engineers using and evaluating molecular sensors and nanodevices across a variety of research settings and industry sectors. With detailed descriptions of principles and the end of chapter problems, it serves as an ideal textbook in engineering, biomedicine and interdisciplinary fields related to biomedical engineering.

In this book, we define the major categories of traditional and state of the art molecular sensors, and discuss details of the background, functions, structures and the sensing requirements towards the applications. The advent of the nanotechnologies has provided unlimited possibilities for molecular sensors design and applications, but the fast moving frontiers have also made it a daunting task for learners to capture the essential design principles and methods behind the implementations. This book aims to show a clear outline and detailed analysis on the technology of molecular sensors in the context of contemporary and future nanotechnology in biomedicine. The book is composed of the following seven chapters. **Chapter 1: Introduction to molecular sensors.** We will introduce fundamental concepts of the molecular sensor, the capture and recognition mechanism, and the sensor signal transduction. First, we define molecular sensors and describe their applications and performances. We will explain how recognition agents such as DNAs and antibodies work in a sensing system. We will also outline transduction technologies which will be detailed in chapter 4, 5 and 6. Finally, we will introduce the interesting use of animals as molecular sensors. There are many well-known sensing applications which utilize animals as detectors, processors, and outputs. Studies of animal-based sensors not only show us the sensing mechanism, but also provide vital insights into the future of man-made molecular sensors and their sensing requirements. **Chapter 2: Introduction to Nano/Microfabrication.** The key technology that has enabled realization of modern intelligent devices is nano/microfabrication. Introduction of nano/microfabrication has expanded the possibilities for more versatile, miniaturized molecular sensors. We start the chapter from discussion on the benefits of system miniaturization by introducing the effect of size and shape in molecular sensing. We will then describe basic silicon microfabrication techniques including film deposition, photolithography, and wet and dry etching. We discuss design theories and fabrication techniques of a microelectrostatic actuator as a basic model element of microelectromehcanical systems (MEMS). We then introduce techniques of "soft lithography," where several novel materials including polymers, proteins, DNAs and other nanomaterial are implemented into molecular sensors. **Chapter 3: Microfluidics basics and total analytical systems.** Handling of carrier and analyte fluids is critical in many molecular sensing systems. Microfluidics enables control of small volumes in the range of nano- to microliters. The reduction of analyte volume can dramatically improve the reaction rate of conventional assays which are otherwise costly to automate and require extensive human intervention to complete. Smaller immunoassay platforms also use less power and allow sample preparation and analysis steps to be coupled together in an automatic streamlined process. We will start with fundamental theories of diffusion and fluid dynamics that govern the reaction processes in molecular sensors. It is followed by introduction of important applications of microfluidic bioassays, including immunoassay, cell separation, and DNA amplification and analysis. Emphasis is put on the authors' main areas of expertise, detection and analysis of cancer cells. **Chapter 4: Electrical transducers: electrochemical sensors and FET-based sensors.** We describe the fundamental principles and important applications of electrical transduction-based molecular sensors. The first part is a description of electrochemical measurements, followed by an introduction of important applications such as ion and gas sensors. The second part introduces the basics of semiconductors. We explain how semiconductor materials can be used as fundamental elements of integrated circuits, followed by an introduction of silicon field effect transistor (FET)-based molecular sensors. We also introduce FET molecular sensors based on novel nanomaterials such as organic thin films, polymers, carbon nanotubes, and graphene. **Chapter 5: Optical transducers: optical molecular sensors and optical spectroscopy.** We describe molecular sensors based on optical transduction. We explain the basics of electromagnetic waves which begin with Maxwell's equations. We then proceed to principles of optical guiding, which is the basis of many optical molecular sensors such as surface plasmon resonance (SPR) sensors and waveguide sensors. Another important part of the chapter is the introduction of optical spectroscopy. Optical microscopy and spectroscopy are cornerstones of modern life sciences. We describe theories and practices of optical absorption, scattering and fluorescence, along with details of important microscopy techniques. **Chapter 6: Mechanical transducers: cantilevers, acoustic wave sensors and thermal sensors.** We describe sensors that utilize mechanical structures for transduction. Binding events are detected using two general strategies: by detecting mechanical deflections induced by molecular binding or by detecting added mass as the change in resonant frequency. We start from sensors based on a mechanical cantilever, which is the fundamental building element of many mechanical structures. Structural analysis of a cantilever is included in the theoretical study. We then introduce acoustic sensors that rely on measurement of mechanical waves for detection of molecular binding.

Based on the piezoelectric effect, acoustic sensors transduce change in mechanical oscillations into measureable electrical signals. At the end of the chapter we introduce principles and applications of thermal sensors for biomedical applications. **Chapter 7: Implantable sensors.** We describe practical applications of miniaturized implantable sensors. The continual breakthroughs in device theory, design and fabrication methods have enabled engineers to implant smaller and smaller sensors into the human body. Sensing applications for human implanted biochips have been an increasingly popular field of research. First, we introduce the principles and the design of implantable pacemakers, which is one of the most important practical examples among existing implantable systems. Then we introduce recent studies that incorporate nano/ microfabrication in an attempt in create highly integrated multifunctional implantable systems. Such systems include microelectodes for electrochemical analysis, pressure sensors for blood monitoring, and techniques for energy harvesting. Finally, we discuss issues related to materials, designs, and use of implantable systems.

Throughout the book, some of the key features include:

- Reviews of state-of-the-art molecular sensors and nanotechnologies
- Explains principles of sensors and fundamental theories with homework problems at the end of each chapter to facilitate learning
- Demystifies the vertical integration from nanomaterials to devices design
- Covers practical applications the recent progress in state-of-the-art sensor technologies.
- Includes case studies of important commercial products
- Covers the critical issues of implantability, biocompatibility and the regulatory framework

On more personal notes, the inspiration for this book arose from many years of teaching experience of a popular course on molecular sensors and nanodevices at UT Austin. A rich set of lecture notes were developed, but the desire to have a textbook has grown stronger year by year. It can be clearly felt across the lecture room to enlighten and instill a greater appreciation of the fundamentals about a seemingly diverse subject and to inspire students to explore its fascinating, and almost infinite, applications. The mission of our writing is therefore to provide a comprehensive learning experience of the current and emerging technologies of molecular sensing, explaining the principles of molecular sensor design and assessing the sensor types currently available. We describe the role of nano/micro fabrication techniques in molecular sensors, including MEMS, BioMEMS, MicroTAS, and Lab on a Chip technologies. The miniaturization of versatile molecular sensors opens up a new design paradigm and a range of novel biotechnologies, which is illustrated through case studies of groundbreaking applications in the life sciences and elsewhere. The book is also aimed at a broader audience of engineering professionals, life scientists and students working in the multidisciplinary area of biomedical engineering. It explains essential principles of electrical, chemical, optical and mechanical engineering as well as biomedical science, intended for readers with a variety of scientific backgrounds. In addition, since each chapter contains discussions on practical applications and limitations of the sensors, it will be valuable for medical professionals and researchers. An online tutorial developed by the authors provides learning reinforcement for students and professionals alike.

Through these pages, we hope the readers will also appreciate just how profound the influence of these molecular sensors and small scale systems to our lives. The book also explores some of the most exciting frontiers at the interfaces of modern science, engineering and medicine, and the opportunities they present to young generation of students and engineers for future careers. New generation of functional microsystems can be designed to provide a variety of small scale sensing, imaging and manipulation techniques to the fundamental building blocks of materials. There is so much yet to be discovered and invented by the future leaders in science and technology, who choose to aim their minds and chisels towards exploring the almost infinite number of yet undiscovered land of molecular sensors and nanodevices.

Acknowledgement

This manuscript would not have come into shape without the great inspiration and help from many colleagues, who shared their experiences and insights with us throughout the writing of this book.

We would like a special note of appreciation to go to the students in the BME 354 Molecular Sensors and Nanodevices class at UT Austin, who have contributed though their active participation in our lectures over the years. They help expand the technical content through the interactive learning, timely presentations on recent publications and the identifications of emerging commercial systems based on molecular sensors and nanotechnologies. Cumulatively, the effort provided technical information that greatly strengthened the breadth and depth of our book.

The past and current members of the Zhang research group helped refining the content and providing detailed suggestions. Follow the sequence of the book chapters, Elaine Ng helped refine chapter 1, 2, 3 and 7, Kaarthik Rajendran helped refine chapter 1 and 3. Yu-Yen Huang and Gauri Bhave helped on the section of microfabrication in cleanroom facilities. Peng Chen provided summary of the methods on cancer cell separation. Dr. Dajing Chen and Nick Triesault reviewed the technical writing of chapter 4. Taewoo Ha and Youngkyu Lee provided the comments on the optics and surface plasmons in chapter 5. Dr. Tushar Sharma and Sahil Naik shared their knowledge on the theories and practices of implantable sensors in chapter 7. In addition, Dr. Ashwini Gopal and Dr. Eiji Iwase helped review the content of the book to further polish the content of our book. Our appreciation also goes to the colleagues whose names we may miss here. Together, we hope this book will reflect your contributions towards better serve the need of our community.

Our particular appreciation also goes to the editorial team at Elsevier. We enjoyed the productive collaborations, which brought a seemingly endless writing process to a successful end. In particular, we thank Wayne Yuhasz, Frank Hellwig and Matthew Deans for their expertise and experience in scientific publishing.

Finally, we see this book as the beginning of a journey for learning. While much of the materials are time-independent, we expect that frequent updates of certain areas such as applications will be desirable. Also, materials that could not be completely covered in the book because of space limitations, such as additional appendix, emerging concepts and methods, and end of chapter problems, should be of interest to some readers.

CHAPTER 1

Introduction to Molecular Sensors

Chapter Outline

1.1 Introduction

Sensors act as interfaces that receive and translate information across physical, chemical, and biological domains. Molecular sensors measure the physical, chemical, and biological quantities at small scales, such as concentration of ions or proteins, existence of toxic molecules, and genetic information from cells, with a sensitivity and specificity at molecular scales. The purpose of this book is to provide comprehensive coverage of current and emerging technologies of molecular sensing, explaining the principles of molecular sensor design and assessing the sensor types currently available for different applications. We will discuss the theories and roles of molecular sensors in light of the development of new materials and fabrication technologies. We also discuss the role of nano/microfabrication techniques in designing and creating molecular sensors. The miniaturization of versatile molecular sensors opens up a new range of nanoscale sensing technology.

Key sensing components or the design of many molecular sensors are based on those found in living organisms. Living organisms have wonderful machinery that contains specialized sensors designed to detect a broad range of molecules, such as toxins and nutrients in the external environment, biomolecular interactions, metabolic activities, glucose levels, and hormones for the internal environment. These natural systems are the original form of "molecular sensors", which combine biomolecular recognitions with special forms of reporters so that the presence of guests can be "felt" by the organism. These original molecular sensors are the key biotechnologies that enable life to be sustainable, robust and adaptive. Some of the sensor technologies today may not offer the level of sophistication and form factor found easily in nature, but it is improving rapidly with the convergence of information technology, breakthrough of physical design, and advanced manufacturing. These revolutions change the way people live every day. The requirement for molecular sensors in which molecules interact with an analyte to produce a detectable change, is becoming more demanding. Molecular sensors have been used in a large variety of applications. Modern molecular sensors are essential parts of many emerging medical diagnostic devices, which play significant roles in global health, reducing healthcare costs in developed countries, and in increasing the response of the world to pandemics or bioterrorism.

Prominent areas of molecular sensor research are the detection of biomolecules for disease diagnosis, the detection of volatile components for air pollutant characterization, and the detection of chemical analytes for evaluation of physiological activity. Molecular sensors are used in diagnostics, such as glucose monitoring in diabetes patients, body fluids screening for disease detection, and in measurement of other physical parameters such as temperature and pressure. They are used in biomedical R&D in drug discovery and evaluation, and protein engineering. Sensors are used in environmental applications such as water quality detection, air pollution monitoring, detection of pathogens, and gas monitoring. Sensors are also used in food safety, such as the detection of contaminants in drinking water, allergens, and the determination of drug residues. Recent trends in both scientific research and business development indicate a growing interest in molecular sensors with the ultimate goal of developing a broad range of inexpensive, batch-fabricated, high-performance biomedical microdevices, and intelligent platforms, which are easily interfaced with digital electronics for biomedical diagnostics and therapy.

We define the functions and structures of traditional and state of the art molecular sensors, and discuss details of applied areas, including the background and sensing requirements for the application. For each type of sensor we will describe fundamental concepts, theories and practical implementation methods.

1.2 Principles of Molecular Sensors

We start from the definition of molecular sensors, and will introduce fundamental concepts of the molecular sensor, the capture and recognition mechanism, and the sensor signal transduction, followed by description of applications and performances of molecular sensors.

1.2.1 Definition of Molecular Sensors

Sensors are devices that detect, record, and indicate a physical or chemical property, with potential capability for further processing. In this book, we mostly describe **molecular sensors**, which are designed to quantify the amount of molecules associated with gas, ions, proteins, DNA, and live cells. Although the terms **chemical sensors** or **biosensors** are sometimes used almost interchangeably, molecular sensors are considered chemical sensors that are able to detect molecules of interest, or **analytes**, which are often present at very low concentrations. Some types of physical sensors, such as temperature sensors, may be considered molecular sensors when the measurement is highly related to the concentration or the activities of molecules.

A sensor can be broken down into three elements: (1) detector, (2) transducer and (3) processor. In the case of a molecular sensor, they can be renamed as the following three elements:

1. Capture and recognition
2. Transduction
3. Measurement and analysis.

Figure 1.1 shows the conceptual sketch of the three elements.

Capture and Recognition

Capture and recognition refers to binding of target molecules. Binding of molecules induces the **sensing effect**, or the physical or chemical changes, due to a detection event. Such a signal triggering molecule binding to a sensing site is called a **ligand**. In most cases, the selectivity to a particular type of molecule is achieved at this stage, although it may be implemented in any other stage. The desired characteristics for this stage include:

• Observation of the sensing effect with only the target molecule (selectivity)

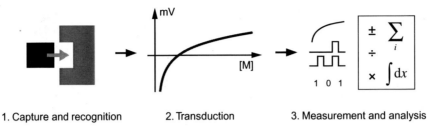

1. Capture and recognition   2. Transduction   3. Measurement and analysis

Figure 1.1: The Three Elements of a Molecular Sensor.

- Obtaining a large sensing effect with a small number of molecules (sensitivity)
- The sensing effect is a good indication of the amount of captured molecules (resolution)
- The sensing effect is not saturated with a large number of molecules (dynamic range).

Transduction

Transduction refers to the conversion of the detection event into a measurable signal. In most cases, the signal is an electrical signal of either voltage or current. A wide variety of transduction methods could be used for molecular sensing. Most transducers are physical sensors themselves that convert the sensing effect of molecular binding into measurable electrical signals. High sensitivity, good resolution, and a large dynamic range are desirable characteristics for transducers.

Measurement and Analysis

The signal from the transducer has to be measured or recorded, and interpreted as the quantity of molecular activities. In many cases, measurements may be simple and straightforward. However, measurements of biological quantities, such as detection of smells or tastes and finding diseases are combinations of nonlinear reactions of biological systems. High levels of signal processing may be needed to retrieve meaningful data in such cases. In modern sensing systems, this part is mostly done by digital processors, or computers. The focus of our book will mostly be on the first two parts of molecular sensors. Analysis or interpretation of data will be discussed as needed for each application.

1.2.2 Applications of Molecular Sensors

The application areas of molecular sensors include clinical testing, biomedical studies, environmental monitoring, and food allergen testing. Box 1.1 summarizes the application areas of molecular sensors.

Box 1.1 Application Areas of Molecular Sensors

- Biomedical, diagnostic (in vivo)
  - Glucose monitoring in diabetes patients
  - Body fluids screening for disease detection
  - Other physical parameters that are related to activities of molecules (temperature, voltage, pressure, light intensity)
- Biomedical (ex vivo)
  - Blood screening
  - Drug discovery and evaluation
  - Protein engineering in biosensors
- Environmental and safety
  - River water (detection of pesticides, heavy metal ions)
  - Air pollution (gas, particulate matter)
  - Detection of pathogens
  - Explosive detection
  - Gas monitoring
- Food related
  - Drinking water
  - Allergens (egg, wheat, milk, tree nuts, shellfish, soy)
  - Determination of drug residues (antibiotics, growth promoters)

1.2.3 Model of a Molecular Sensor

Along with detailed studies on the theories behind molecular sensors, one important aim of this book is to introduce the implementation methods that have enabled the integration of molecular sensors. Here, we describe the actual composition and structure of sensors using a model molecular sensor.

A simplified model of a molecular sensor is shown in Fig. 1.2. When we discuss the structure of a molecular sensor, we mostly focus on the first two elements of the molecular sensor, namely capture and recognition, and transduction.

Capture and Recognition

Elements that enable capture and recognition are usually molecules that bind a specific analyte. Such molecules include enzymes, receptors, antibodies, nucleic acid, and specialized polymers. Other types of recognition elements include porous membranes that select molecules based on pore size. Processes used for capture and recognition include antigen-antibody binding, DNA hybridization, adsorption, covalent attachment, and cross-linking. Captured molecules change physical properties of the sensing site to induce the sensing effect; they change physical quantities such as electrical resistance and capacitance, optical index and absorption coefficient, or mechanical stiffness and

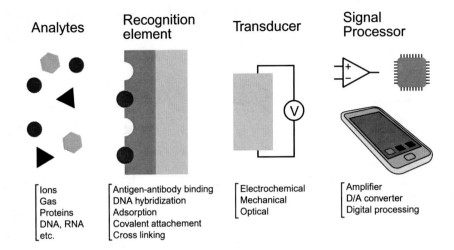

Figure 1.2: Structure of a Molecular Sensor.

displacement. They may undergo a chemical reaction such as reduction−oxidation (redox) reaction.

Transduction

The sensing effect is transduced into electrical signals by a **transducer**. There are a wide variety of transducers. Transducers and sensors are often exchangeable terminologies for engineers, since transducers alone could be used as a sensor to measure physical quantities. We categorize transducers used in molecular sensors into three types based on the phenomena utilized for signal conversion, namely: electrical, optical, and mechanical transducers. Transduction can also be realized through thermal and magnetic/electromagnetic domain measurements. Details of the transduction principles and implementation methods for the three types of transducers are described in Chapters 4, 5, and 6, respectively.

Measurement and Analysis

Electrical signals are usually voltages or currents. They are amplified, converted into digital signals by analog−digital (AD) converters, and analyzed by digital processors.

1.2.4 Example of Molecular Sensor 1: Immunosensor Based on Field Effect Transistor

Figure 1.3 shows the structure of a typical molecular sensor based on Field Effect Transistor (FET). The details of theories and implementation will be described in the following chapters. Here, we use this example to illustrate the concept behind the structure of general molecular sensors.

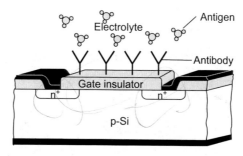

Figure 1.3: Example of a Typical FET-Based Molecular Sensor.

Capture and Recognition Elements

Antibodies (Ab) attached on the surface of the gate insulator serve as recognition elements. Antibodies selectively bind antigen (Ag) molecules. Binding of antigen molecules changes the electrical charge around the surface of the gate insulator. Details of antibodies are described in Section 1.3.1.

Recognition Element: Antibodies
Antigen molecules (=analyte) are recognized as surface charges (=sensing effect)

Transducer

The two electrodes, n+ regions, and the gate insulator integrated on the silicon body (p-Si) compose a device called a **field effect transistor (FET)**, which changes its conductivity based on surface charges around the gate insulator. The amount of antigen molecules is thus converted to **conductivity**. Details of the electrical characteristics of FET-based molecular sensors are described in Chapter 4.

Transducer: Field Effect Transistor
Changes in surface charges (=sensing effect) is transduced to conductivity (=signal)

Measurement and Analysis

Conductivity, which is the ratio between current and voltage, can be easily found by measuring currents at different applied voltages. Since the sensor is made of silicon in this example, it is also possible to implement all the processing circuitry including amplifiers, noise filters, AD converters, and digital processing circuitry.

> **Measurement: Voltage-current (V-I) Measurement**
> Conductivity(=signal) is measured and analyzed

1.2.5 Example of Molecular Sensor 2: Animal Olfactory System

The animal olfactory system is a perfect example of a sophisticated and powerful molecular sensing system, capable of detecting single molecules from a diverse and complex mixture See Section 1.6 for interesting uses of animals as molecular sensors. It can be modeled or understood as a type of molecular sensor as shown in Box 1.2.

> **Box 1.2 Animal Olfactory System as a Molecular Sensor**
>
> | | |
> |---|---|
> | Capture and recognition | Olfactory membrane |
> | Transduction | Nerve cells |
> | Measurement and analysis | Brain |

From the soothing smell of coffee to the pungent smell of body odor, the olfactory system serves to provide us with information of our surroundings, including odor intensity and concentration, which we paraphrase in terms such as "strong", "stinky", "fresh", "good", or "bad".

The human olfactory system is capable of detecting atomic-sized molecules called odorants at extremely low concentrations. Odorants are usually small organic molecules less than 400 Da in size. Odorants can vary in size, shape, functional groups, and charge, from various alcohols to ketones and aromatic ring structures. Odorant sensing begins in the nasal cavity where millions of olfactory receptor cells, or neurons, act as the sensory signaling cells. Olfactory receptor neurons, covered with cilia, are located on the olfactory epithelium, a small patch of tissue in the back of the nasal cavity (see Figure 1.4). The cilia, in turn, are riddled with many odorant-specific receptors. Each odorant molecule has its own unique receptor to bind to in a ligand-receptor binding, or interaction. You can therefore imagine the large number of receptors (about 1000 different types) [1] that are present and play a role in the olfactory system. The presence of cilia on the olfactory receptor neurons is therefore critical in increasing surface area on which the receptors are located. Upon binding of an odorant to a receptor on an olfactory receptor neuron, the binding effect is transduced via the olfactory nerve from a chemical stimulus into an electrical signal that is ultimately sent to the brain. Thus, the olfactory serves as a transducer, connecting the brain to the external air. The electric signal travels along the

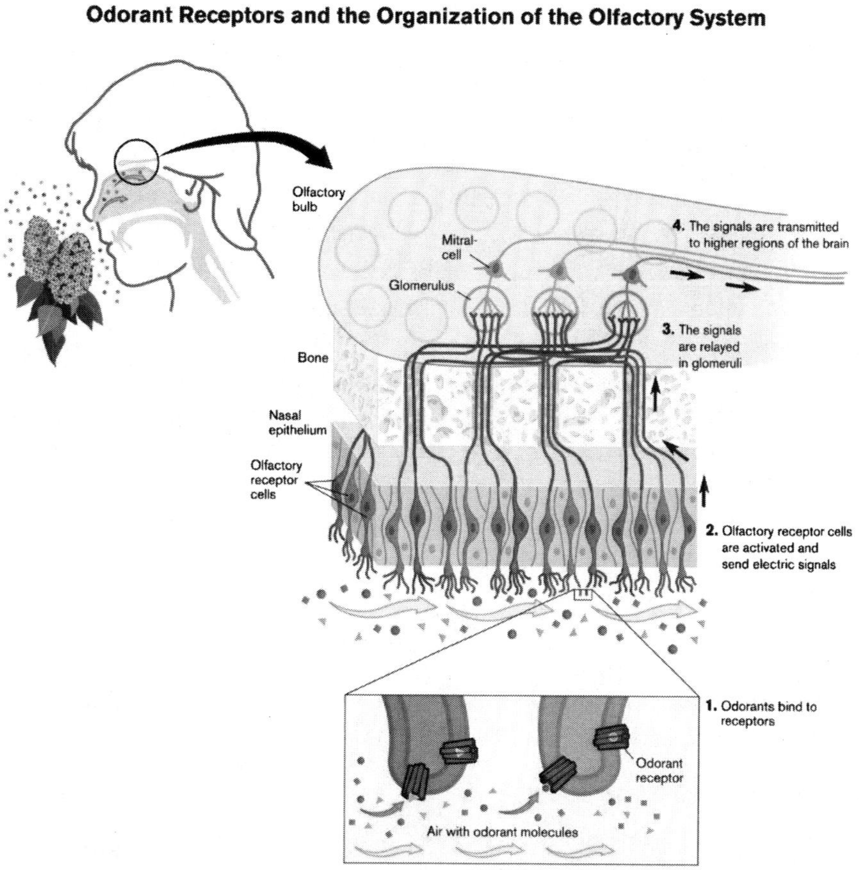

Odorant Receptors and the Organization of the Olfactory System

Figure 1.4: The Human Olfactory System.

Press Release: The 2004 The Nobel Prize in Physiology or Medicine, www.Nobelprize.org. [Online]. Available: http://www.nobelprize.org/nobel_prizes/medicine/laureates/2004/press.html (accessed 22 July 2013). © The Nobel Committee for Physiology or Medicine.

olfactory nerve and eventually reaches the olfactory bulb in the limbic system of the brain where it is interpreted, or processed. The information is further relayed to the olfactory cortex of the brain where it is further processed as part of the central nervous system [2].

Olfaction begins when odorants bind to olfactory receptor neurons and are detected. The olfactory nerve transduces the chemical signal into an electrical signal and sends the signal to the olfactory bulb. The olfactory bulb processes the electrical signal into information that is relayed and further processed in the brain for information about the individual's surrounding environment.

The presence of ligand-specific receptors on the olfactory receptor neurons makes for an extremely precise, selective, and sensitive molecular sensor. As mentioned before, each ligand is paired with a corresponding receptor unique to the ligand. The uniqueness of the ligand−receptor binding site is characterized by various attributes such as steric, structural, and chemical difference [1,3]. So sensitive is our olfactory system that we are able to detect and amplify the minute subtle differences across the different odorants. For example, the carvone molecule has two mirror image enantiomers, R-(−)-carvone and S-(+)-carvone (see Figure 1.5). This subtle difference results in a drastic difference in odor. R-(−)-carvone smells like spearmint while S-(+)-carvone smells like caraway [4]. The two enantiomers are perceived as two unique smells because they bind to two different receptors. The receptors unique to R-(−)-carvone do not and cannot bind to S-(+)-carvone, and vice versa, due to steric differences. Humans can discriminate between two odorants differing, on average, in concentrations of 19%, and as low as 7% [5]. It is important to note though, that factors such as the environment and repeated exposure to the odor can add "noise" to or reduce an individual's detection threshold of an odor signal.

Odor intensity is the perceived strength of odor sensation and can be measured relative to odor concentration via a series of sample dilutions and modeled in an equation known as the Weber−Fechner law:

$$I = a \log(c) + b \tag{1.1}$$

where I = the perceived psychological intensity, a = the Weber−Fechner coefficient, c = the chemical concentration, and b = the intercept constant [6].

Surprisingly, the olfactory system is one that shares strikingly similar characteristics across a broad range of organisms, including vertebrates, mollusks, arthropods, and nematodes. These organisms possess the same three critical components of the olfactory sensory system as described earlier in humans. In all the aforementioned organisms, odorants are detected by olfactory receptor neurons, transduced from a chemical to electrical signal via the olfactory nerve, and finally the signal is processed in the brain, more specifically at the olfactory bulb which is part of the organism's central nervous system. The only difference between organisms of varying phyla is the location of the components. For example, vertebrate olfactory receptor neurons are found on the olfactory epithelium while insect

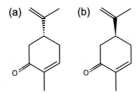

Figure 1.5: The Carvone Molecule in Two Enantiomer Forms: (a) R-(−)-carvone and (b) S-(+)-carvone.

olfactory receptor neurons are found in clusters inside cuticle-covered sensillae along appendages, such as antennae [7].

The high sensitivity of living cells have motivated interesting approaches that utilize living cell sensitive to analyte molecules as a part of molecular sensor. They can be categorized into three types [8]. In the first type, the sensing effect is changes in the mechanical contact between cells and between cells and substrates which are measured as electrical conductivity [9]. The second type uses an additional chemical sensor to measure metabolic products delivered from cultured cells as the sensing effect [10,11]. In the third approach the direct electrical response of cells or a network of neural cells is measured [12–14]. There is another approach that even uses a part of the animal as a sensing element. Examples include moth antennae-circuit hybrids for the ultrasensitive detection of moth sex pheromones [15].

1.3 Capture and Recognition Elements in Molecular Sensors

To capture the molecule of interest, molecular sensors employ a variety of capture and recognition mechanisms. For biological molecules, we can use proteins (antibodies, receptors), nucleotides (DNA, RNA, aptamers), synthetic polymers that simulate a binding site, and phages. For the molecule entrapment , we can use: membrane entrapment, such as ion exchange membranes; physical adsorption, such as van der Waals forces, hydrophobic forces, hydrogen bonds, and ionic forces, among many others; porous matrix entrapment, such as porous silica, silicon, among many others; covalent bonding, such as gold, silicon, carbon surfaces, among many others. Here we describe two important examples: antibodies, and DNA.

1.3.1 Antibody–Antigen Binding

Antibody Overview

The immune system produces proteins called antibodies or immunoglobulins (Ig) that recognize and bind to antigens, specific molecules with a particular shape and charge. Once the antibody and antigen bind, antibodies either tag or neutralize the antigen. Antibodies are 10–20 nm in diameter and 2–3 nm in length [16]. Antibodies are ligands with affinity to many targets including pathogens [17,18], malignant cells [19], food allergens [20], toxins [21], and explosives [22]. For each person, over 100 million different antibodies, each with a particular antigen binding site, are produced by B cells, members of a group of white blood cells classified as lymphocytes. The shape of an immunogen is identical to that of the B cell that synthesized the antigen. Igs make up about 20% of the total protein in plasma by weight and can be classified into five categories in mammals: IgM, IgG, IgA, IgD, and IgE.

Antibodies are produced by immunizing host animals, such as chickens, goats, horses, mice, rats, and donkeys, with an antigen. When an antigen is injected into an animal, a mixture of antibodies to the antigen is generated since there are many different B cells in an animal. Such antibodies that are secreted by different B-cell lineages are called polyclonal antibodies. In contrast, antibodies made by clones of an isolated unique parent cell are called monoclonal antibodies. It is possible to produce monoclonal antibodies for almost any substance.

As shown in Fig. 1.6, each antibody is roughly a Y-shaped molecule consisting of four polypeptide chains, two identical light (L) chains, and two identical heavy (H) chains. Antigens bind to the fragment antigen binding (Fab) region, the variable region of the antibody while the constant stem region, the fragment crystallizable (Fc) region, of the Y-shaped molecule interacts with receptors and molecules on cells. The constant region does not vary much while the variable region varies depending on the antigen and is responsible for providing an antibody with specificity, as represented in Fig. 1.7. H chains are about 50 kDa, L chains are about 25 kDa, making an antibody the size of about 150 kDa. An antibody can simultaneously bind to two identical antigens because each

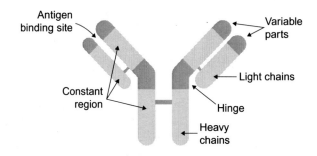

Figure 1.6: Antibody Structures.

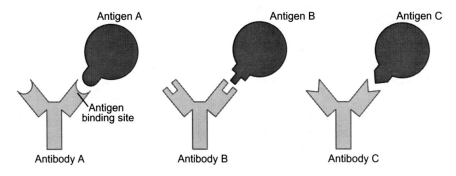

Figure 1.7: Antigen–Antibody Binding.

antibody has two available antigen binding sites. The light chains consist of about 220 amino acids while the heavy chains contain about 440 amino acids and the four chains are held by noncovalent and covalent (disulfide) bonds. Antigen binding sites are situated between the ends of heavy and light chains on two identical halves. Antibodies recognize and bind to a region on the antigen known as an epitope, a 5–8 amino acid long chain. The binding between the antigen and antibody is held by non-covalent forces including hydrogen bonds, hydrophobic van der Waals forces and ionic interactions and as a result, the binding of antibody to antigen is a reversible process.

Antibody—Antigen Binding

Antibody—antigen binding is one of the most selective capture mechanisms utilized in determining concentrations of biomarkers in biological fluids (see Fig. 1.7). A test employing antibody—antigen binding is called an immunoassay. The selectivity and sensitivity of immunoassays makes it an attractive solution for detecting pathological antigens that are often present in nanomolar concentrations in physiological conditions. Traditional immunoassays including agglutination, nephelometry, and western blotting have been invaluable tools for diagnosing pathologies such as human immunodeficiency virus (HIV), or for creating quality control tests for presence of food allergens. However, traditional immunoassays are not feasible in point of care diagnostics in developing countries because of the extensive infrastructure necessary to carry out these tests. Enzyme-linked immunosorbent assay (ELISA) is a test used to quantify or qualify an analyte colorimetrically. Antibodies, labeled with an enzyme, are immobilized on a microliter plate. The substrate (analyte) is introduced and reacts with the enzyme, producing a colored product that can be measured. ELISA is a multistep procedure requiring absorption, blocking, incubation, washing, secondary incubation, secondary washing, and result analysis, which demands extensive human intervention [23]. In an environment where trained laboratory technicians and equipment are a scarcity, these lifesaving tests would be too costly. High throughput automatic assays are desirable for these points of care environments because of their inherent low cost due to its manufacturability and their low infrastructure requirement.

Immunoassays

Immunoassays can be categorized into two types: homogeneous and heterogeneous. In homogeneous immunoassays, the antibody—antigen binding event occurs in solution and leads directly to a measurable change in signal that can be measured and quantified. Homogeneous immunoassays require a separate step to remove antigen—antibody complexes from free antigen in the solution. Separate fluorescent-labeled antibodies can also be added to amplify the signal and improve sensitivity. Most assays used for molecular sensors are heterogeneous immunoassays, where antibodies are immobilized on a solid substrate. Heterogeneous immunoassays can be further subdivided into competitive and noncompetitive types. In competitive immunoassays, the antigen competes with a labeled antigen of known concentration and binding affinity. The signal generated in this kind of immunoassay becomes weaker as the concentration of the measured substance increases due to increased competition with the labeled antigen. Noncompetitive immunoassay involves washing the analyte over an antibody coated surface, allowing it to bind, and then adding another labeled antibody that binds with the antigen—antibody complex. This kind of test is called a sandwich assay, and the signal is proportional to the concentration of antigen. Typical formats of heterogeneous immunoassay used for molecular sensors are shown in Fig. 1.8. They are shown with situations before (upper part) and after equilibration (lower part).

Summary of Antibodies

- Antibodies bind to both small and large molecules including proteins (pathological antigens, food antigens), cells, viruses, microbial toxins, explosives.
- Antibodies are produced by immunizing a host animal
- Antibodies are Y-shaped proteins.
- Antibodies are 10—20 nm in diameter and 2—3 nm in height, 150 kDa in size.
- The antigen binding site is located at the variable (V) region
- Each antibody has two identical antigen-binding sites
- Antibody—antigen binding is a reversible process.

1.3.2 DNA as a Recognition Element

Discovery of DNA

DNA, or deoxyribonucleic acid, is the hereditary molecule of life. The story of DNA begins in 1869, when Swiss physiological chemist Friedrich Miescher isolated the molecule from the nuclei of leukocytes, terming the compound "nuclein" [25]. For many years, scientists thought that proteins, because of their complexity and wider diversity of forms, were the molecule of life. In 1944, Oswald Avery and his associates at Rockefeller University demonstrated that DNA, not protein, was the genetic material [26]. In the 1950s, Erwin Chargaff realized that the total amount of purines (adenine (A) and guanine (G)) was nearly equivalent to the total amount of pyrimidines (cytosine (C) and thymine (T)) [27]. Chargaff's discovery was pivotal for another group of researchers, James Watson, an American scientist, and Francis Crick, a British researcher, to ascertain the double helical nature of DNA in 1953, shown in Fig. 1.9 [28].

DNA Structure and Characteristics

DNA is composed of two polynucleotide chains that are twisted into a helix 2 nm in diameter. Russian biochemist Phoebus Levene proposed the polynucleotide structure of

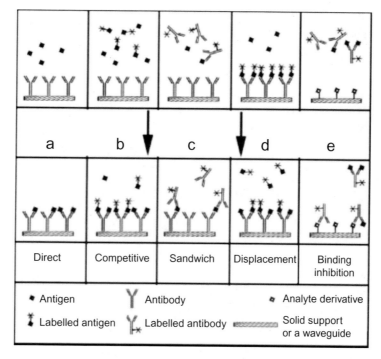

Figure 1.8: Typical Formats of Immunoassay Used for Sensing Applications [24].

nucleic acids in 1919 [29]. Each nucleotide is composed of a nitrogenous base, a pentose sugar molecule (deoxyribose for DNA and ribose for RNA), and a phosphate group. Purines and pyrimidines are the two groups of nitrogenous bases. Both types are heterocyclic aromatic organic compounds. A purine is pyrimidine fused to a five-membered imidazole ring. The nucleic acids adenine (A) and guanine (G) are purines and cytosine (C) and thymine (T) (and uracil (U) in RNA) are pyrimidines. Because of the nature of chemical pairing, the base A on one strand usually forms hydrogen bonds with the base T on another strand, and likewise C with G.

DNA length is described in a unit termed kb (kilobase) or kpb (kilobase-pair). Each base pair is 3.4 Angstroms apart and because each base pair is rotated 36 degrees with respect to the previous pair, the helix repeats every 34 Angstroms. The average size of a protein-coding gene is 30 kbp and the average mass of a base pair is 650 Da. The longest gene in the human genome is Titin (80 781 bp) while the longest piece of synthetic DNA is the *Mycoplasma genitalium* bacterial genome (582 970 bp) [30]. Because of the negatively charged phosphate ions in the backbone, DNA has an overall negative charge.

The sugar and phosphate units are linked alternately forming a sugar–phosphate backbone, which defines the direction of the nucleic acid. The two chains run antiparallel to each

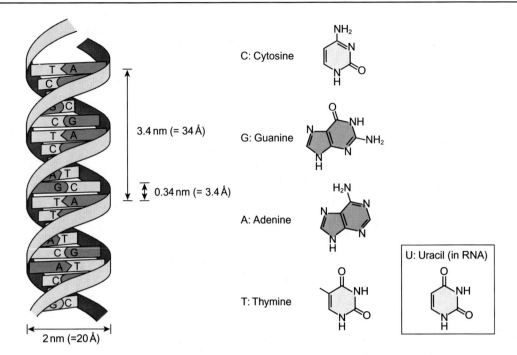

Figure 1.9: DNA Is a Double Helix.

other: one strand runs from the 5′ end to the 3′ end while the other strand runs from the 3′ end to the 5′ end. For each nucleotide, the sugar is the 3′ end, and the phosphate is the 5′ end. The phosphate group attached to the 5′ carbon of the sugar on one nucleotide forms a phosphodiester bond with the hydroxyl group attached to the 3′ carbon of the sugar on the next nucleotide. Figure 1.10 summarizes the overall structure and orientation of DNA molecules and the bonding between complementary groups.

Each strand of DNA can serve as the template for assembling a complementary sequence during replication. Every time a human cell divides, nearly 3.2 billion base pairs must be copied. Interestingly, although Watson–Crick base pairing is highly selective in a human cell, the wrong nucleotide is inserted approximately every 100 000 nucleotides.

RNA Function

While DNA is the hereditary molecule, Ribonucleic acid (RNA) is a family of molecules that transcribes genetic information from DNA and is eventually translated into proteins produced in the cell. Unlike DNA, RNA is single stranded, contains ribose sugar, and the complementary base for adenine is uracil (U). During transcription, a segment of DNA information is transcribed to messenger RNA (mRNA). mRNA specifies the amino acid

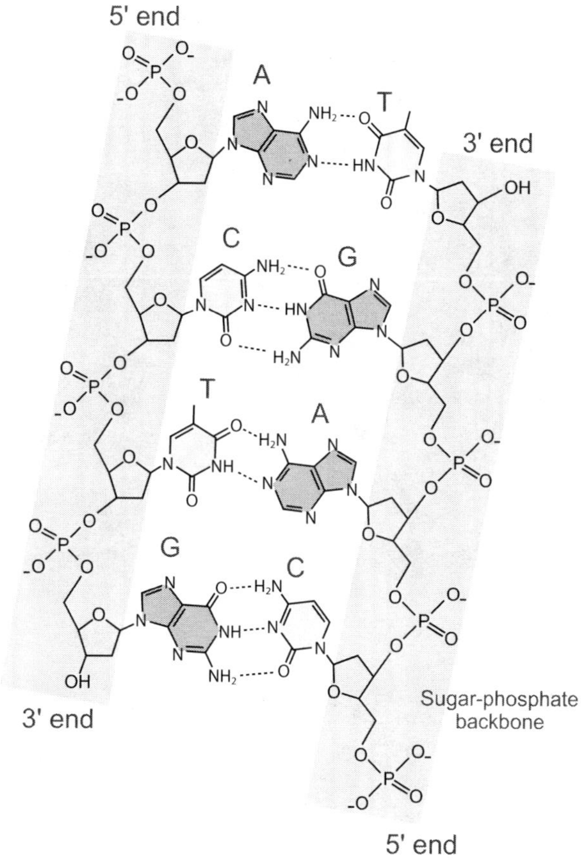

Figure 1.10: Base Pairing in DNA.

sequence of a polypeptide through codons, a group of three nucleotides that identify a single amino acid. After modification in the nucleus, mRNA is transported into the cytoplasm, where it is read by ribosomes. Amino acids are carried to the ribosome by Transfer RNA (tRNA) which transfers the amino acid to a growing polypeptide chain if the tRNA's anticodon can form a base pair with the mRNA codon. The synthesis of a polypeptide is termed translation. Transcription, which occurs in the nucleus, and translation, which occurs in the cytosol, are illustrated in Fig. 1.11.

DNA Hybridization

In hybridization, two complementary strands of DNA from different sources base-pair to form hybrid DNA. Hybridization is applied in many laboratory techniques, including polymerase chain reaction (PCR) and Southern blotting. Hybridization is also used in DNA

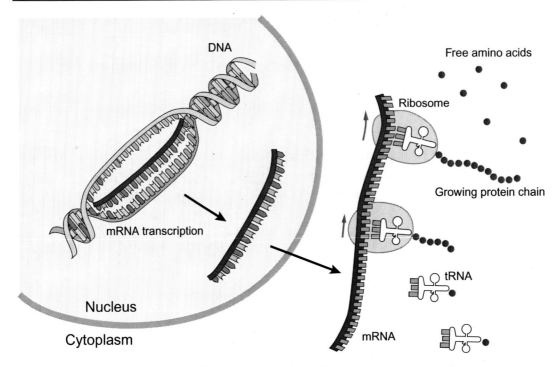

Figure 1.11: mRNA to Protein.
Courtesy of image by Jane Wang.

probes, in which patient sample DNA is hybridized with specific nucleotide sequences for disease diagnosis.

Oligonucleotides

Oligonucleotides, short, single-stranded nucleic acid polymers, can be synthesized using DNA/RNA synthesizers for industrial and medical applications. The nucleic acid chains are generally 13–25 nucleotides in length but can be synthesized up to 200 bases [31]. Because oligos can hybridize with their respective complementary nucleotides, oligos are used in diagnostic tests for genetic diseases and infectious diseases, in research for finding innovating drugs, and for improving agricultural products. Customized oligos are available from companies such as Integrated DNA Technologies.

Nucleic Acid Sensors

DNA/RNA detection can be applied for gene analysis, forensics, medical research and clinical diagnostics, food quality control, environmental analysis, and detection of biological warfare agents. Single-stranded DNA (ssDNA) is directly immobilized on a sensor surface and hybridized with the complementary DNA sequence as shown in

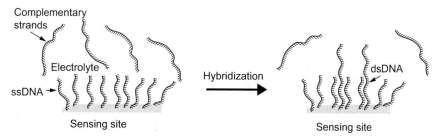

Figure 1.12: DNA as a Recognition Element.

Fig. 1.12. The hybridization (as low as nanomolar range) produces signal changes around the sensing site which can be transduced electronically, optically, mechanically, or electrochemically.

Summary of DNA as a Recognition Element

DNA encodes the genetic material to make proteins. DNA is not protein
Double stranded DNA is 2 nm in width
Helix of DNA repeats every 34 Angstroms
DNA is negatively charged

1.3.3 Aptamers

Aptamers are nucleic acids that specifically bind to target molecules (e.g. proteins, peptides, amino acids, drugs, metal ions, and whole cells) with high affinity and specificity [32,33]. Aptamers include oligonucleotides like RNA and DNA and small peptide molecules. Compared to antibodies, aptamers have several advantages as a selection marker. With the equilibrium dissociation constant in the low nanomolar to picomolar range, aptamers have binding affinities comparable to antibodies. In contrast, aptamers are more easily modified and synthesized because of their small size, which also allows them to access concealed protein epitopes [34]. Furthermore, aptamers are versatile and can be chemically modified with reporters, functional groups, signal moieties, and other payloads. Once selected, aptamers can be mass produced using in vitro techniques. Aptamers have been applied to therapeutically inhibit a protein function or pathway (e.g. Macugen), modified to signal the presence of an analyte for disease detection, and functionalized with a drug for targeted delivery.

Aptamer Selection Process

Aptamers are selected from a large collection (10^{12}–10^{14}) of DNAs or RNAs in a selection process termed "systematic evolution of ligands by exponential enrichment" (SELEX)

[34−37]. A library of random sequences of nucleotides is generated. The nucleotides are generally between 30−40 base pairs in length and the ends have a common primer binding site. Sequences that don't bind strongly to the target are washed out and those that do bind are amplified by polymerase chain reaction (PCR) to create a new library (see Chapter 3, Section 3.3.3 for details of PCR). Multiple rounds of washing are performed and the aptamers are characterized through biological assays.

Example Process: Bead Based Selection

An overview of the in vitro selection process is provided in Fig. 1.13. A target (e.g. His-tagged protein) is immobilized on magnetic or polystyrene beads and any target that does not chemically bind to the beads is washed away. A pool of nucleic acids is introduced and incubated with the immobilized target for 30 minutes or less. Unbounded or weakly bounded species are removed by washing and bound nucleic acids are eluted. The new pool is isolated (contaminants are removed using sodium dodecyl sulfate−polyacrylamide gel electrophoresis (SDS-PAGE)) and amplified via PCR. Prior to reimmobilizing the target, a negative selection is performed between the pool and the beads to remove any non-specific binders. In early rounds, the washing is less stringent to allow weaker interactions. After several successive washing (e.g. six rounds), the nucleic acid is quantified and the binding strength is characterized with a ligand binding assay. Emergent aptamers are sequenced with Sanger sequencing and the structures are visualized for potential motifs (see Chapter 3, Section 3.3.3 for Sanger sequencing.).

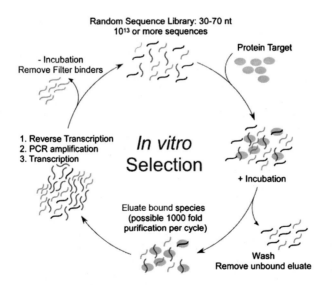

Random Sequence Library: 30-70 nt
10^{13} or more sequences

Protein Target

- Incubation
Remove Filter binders

1. Reverse Transcription
2. PCR amplification
3. Transcription

In vitro Selection

+ Incubation

Eluate bound species
(possible 1000 fold
purification per cycle)

Wash
Remove unbound eluate

Figure 1.13: In vitro Selection of Aptamers.
© Brad Hall, Aptamer Stream, http://aptamer.cns.utexas.edu/Aptamer_Selection/Selection.html

1.4 Transduction Mechanisms

Through signal transductions, sensors convert the detection event into a measurable signal. Biological, chemical or physical quantities are converted into a detectable signal, mostly an electrical signal. We categorize molecular sensors into three categories based on the transduction principle, namely electrical transduction, mechanical transduction, and optical transduction.

1.4.1 Electrical Transduction

In the electrical domain, transduction can be accomplished through impedance, potentiometric, voltammetric, amperometric, and calorimetric measurements. In Fig. 1.14, the resistance between the two terminals at the ends of a nanometer scale polymer tube changes according to the amount of biotin-DNA molecules. The transducer of this sensor is the polymer nanotube transistor. Details of electrical transduction techniques along with basics of semiconductors are described in Chapter 4.

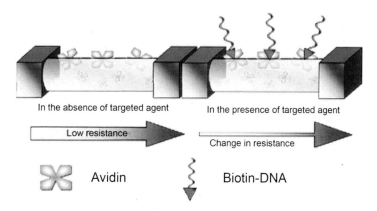

In the absence of targeted agent | In the presence of targeted agent

Low resistance | Change in resistance

Avidin | Biotin-DNA

Figure 1.14: Electrical Transduction. The Transducer of This Sensor Is the Polymer Nanotube Transistor.
From [38].

Optical Transduction

In the optical domain, transduction can be accomplished through absorption, fluorescence, reflection, surface plasmon resonance spectroscopy, and light scattering measurements. Details of optical transduction will be described in Chapter 5.

An example of optical transduction is shown in Fig. 1.15, where the intensity of the surface plasmon wave (SPW), which is propagating along the air−metal interface, changes depending on the amount of captured molecules. The transducer of this sensor is the optical system composed of the light source, the metal waveguide and the photodetector.

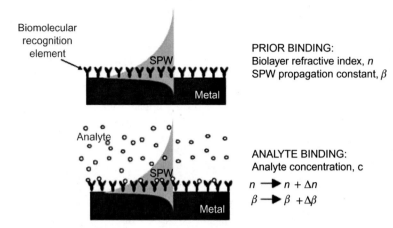

Figure 1.15: Optical Transduction. The Transducer of This Sensor Is the Optical System Composed of the Light Source, the Metal Waveguide and the Photodetector.
From [39].

Mechanical Transduction

In the mechanical domain, binding of molecules is detected as changes in mechanical properties. Details will be described in Chapter 6. Figure 1.16 shows an example where binding of target DNAs results in a mechanical deflection Δx of the cantilever. Transduction can be accomplished through piezoresistive or piezoelectric measurements or combination of electrical and optical measurements. The transducer in this sensor is the mechanical cantilever.

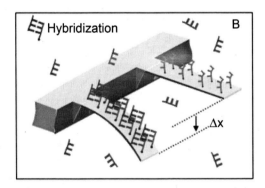

Figure 1.16: Mechanical Transduction [40]. The Transducer in This Sensor Is the Mechanical Cantilever.

1.4.2 Sensitivity of a Transducer

Here we will discuss the sensitivity of a transducer, which convert sensing effects into electrical signals. The sensitivity of a transducer is usually limited by the noise that accompanies the output signals. We will describe the relationship between the sensitivity, responsivity and noise. Let us consider a model device which transduces a change in a physical quantity M into voltage V.

Responsivity

We first define the **responsivity** of a transducer as the mean ratio of the signal output to the signal input. If an average voltage change of ΔV is observed for an average change ΔM in the quantity, the responsivity Rsp of the sensor becomes:

$$Rsp = \frac{\Delta V}{\Delta M} \tag{1.2}$$

Sensitivity is the minimum input level required to produce an output that overcomes the threshold, which is typically defined with the noise level and a specified signal-to-noise ratio. It is often chosen just above the noise level.

Noise in a Sensing System

Noise is expressed in a form of power spectrum, where the **power spectral density** (PSD) of noise is given as a function of frequency. PSD is the mean square voltage variance per hertz of bandwidth:

$$\overline{v_n^2} \quad (V^2/\text{Hz}) \tag{1.3}$$

It is often convenient to use an **amplitude spectral density** (ASD), which is the square root of the PSD. ASD is given in a form of

$$\sqrt{\overline{v_n^2}} \quad \left(V/\sqrt{\text{Hz}}\right) \tag{1.4}$$

Since noise is given as a density, a bandwidth Δf has to be specified In order to find the voltage value of noise v_n in root mean square (RMS).

$$v_n = \sqrt{\overline{v_n^2}} \cdot \sqrt{\Delta f} \quad (V_{\text{RMS}}) \tag{1.5}$$

As Eq 1.5 indicates, the wider the spectrum, the more noise we will observe. When $\overline{v_n^2}$ is a function of frequency, the total noise v_n from f_{low} to f_{high} has to be as follows to be exact:

$$v_n = \sqrt{\int_{f_{\text{low}}}^{f_{\text{high}}} \overline{v_n^2} \cdot df} \quad (V_{\text{RMS}}) \tag{1.6}$$

However, in many cases Eq 1.5 with the value of v_n at $f = f_{\text{high}}$ usually works as a good approximation.

Sensitivity

Now we can discuss the sensitivity of a sensor. The output signal ΔV_{min} has to be at least larger than the noise level:

$$\Delta V_{min} = \Delta M_{min} \cdot Rsp > \sqrt{\overline{v_n^2}} \cdot \sqrt{\Delta f} \qquad (1.7)$$

Where ΔM_{min} is the minimum measurable change. From (Eq 1.7),

$$\Delta M_{min} > \sqrt{\overline{v_n^2}} \cdot Rsp^{-1} \cdot \sqrt{\Delta f} \qquad (1.8)$$

The value of ΔM_{min} gives the sensitivity with a specific bandwidth Δf. The term $\sqrt{\overline{v_n^2}} \cdot Rsp^{-1}$ is also referred to as the sensitivity, because it gives a more general form of minimum measurable quantity. See problem 1.14 for discussions with an example sensor. Here we specifically described a sensor that produce output in voltage. The same discussion is true to transducers that uses output signal of any quantity. Good examples can be found in the discussion on oscillating cantilever as a mass sensor in Chapter 6, Section 6.2.2.

Thermal Noise

Thermal noise in a resistor is given in the following way:

$$\overline{v_n^2} = 4kRT \quad (V^2/\text{Hz}) \qquad (1.9)$$

where k (J/K) is Boltzmann's constant, R (Ω) is the resistance, and T (K) is temperature.

Equation 1.9 can be rewritten in a form of ASD as:

$$\sqrt{\overline{v_n^2}} = \sqrt{4kRT} \quad (V/\sqrt{\text{Hz}}) \qquad (1.10)$$

The RMS of the noise becomes

$$v_n = \sqrt{\overline{v_n^2}} \cdot \sqrt{\Delta f} = \sqrt{4kRT\Delta f} \quad (V_{\text{RMS}}) \qquad (1.11)$$

For example, a resistor of 50 Ω at 1 Hz bandwidth has a thermal noise of 1 nV at room temperature.

Example 1

Consider a sensor which transduces a change in mass into a change in voltage. If an average voltage change of $\Delta V = 5$ mV is observed for an average additional mass of $\Delta M = 10$ mg, the responsivity Rsp of the sensor becomes:

$$Rsp = \Delta V/\Delta M = 0.5 \ V/g$$

Example 2

A force sensor has a responsivity of 1 μV/10 pN. The amplitude spectral density of the noise is 40 nV/$\sqrt{\text{Hz}}$ at 1−2 kHz. The sensitivity of the sensor at this bandwidth of becomes:

$$(40 \times 10^{-9}\ V/\sqrt{\text{Hz}}) \cdot (10 \times 10^{-12}\ N)/(1 \times 10^{-6}\ V) = 0.4\ \text{pN}/\sqrt{\text{Hz}}$$

The minimum measurable force at the bandwidth of 1−2 kHz is:

$$0.4 \times 10^{-12}\ N/\sqrt{\text{Hz}} \cdot \sqrt{1000\ \text{Hz}} = 12\ \text{pN}$$

Example 3

Figure 1.17 is a noise spectrum of silicon piezoresistive cantilevers (see Chapter 6 for details) shown in the units of V/$\sqrt{\text{Hz}}$ in [41]. It shows a clear 1/f component which is called 1/f noise or pink noise. The total noise is found from the integration of Eq 1.6, and is calculated to be is 1.14 mV for the bandwidth from 10 Hz to 1 kHz.

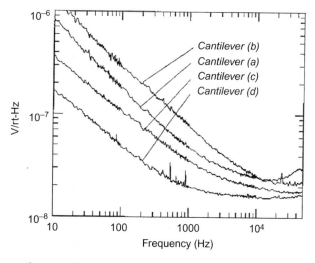

Figure 1.17: Noise Spectra of Silicon Cantilevers.
From [41].

1.5 Performance of Molecular Sensors

Molecular sensors quantify the presence of an analyte down to the single molecule level. Sensor performance factors should be discussed based on the nature of measurement:

- What molecules should be measured in what environment?

- What is the required sensitivity range? Sub-millimolar, or femtomolar?
- What types of substances should be discriminated from target analyte? (What could be "false positive" events?)
- How accurate the measurement should be?

We will look at some general requirements and how molecular sensors can play an important role in meeting these requirements. The following are a few requirements that should be considered when discussing the performance of molecular sensors:

Sensitivity: Signal should be sensitive to small changes of analyte concentration.
Specificity: Detection must be specific to one kind of molecule.
Response time: Response must be faster than the expected changes of analyte concentration.
Dynamic range: Amount of measureable molecules must match the measured environment.
Lifetime: Shelf time and operation duration must be long enough for specific applications.

For an example of blood glucose sensing (see Chapter 4 for details). Blood sugar levels should be measured five to ten times a day. The glucose monitor should be sensitive in the range of $20-500$ mg/dL (~ 130 mM) with a resolution of ~ 1 mg/dL ($\sim 50\,\mu$M). These requirements should be considered when deciding the specifications and discussing the performance of glucose sensors.

Sensitivity of molecular analysis is decided by method used in the measurements. In addition to conventional electrochemical sensing as commonly used for blood glucose sensing, DNA-based diagnostic tools and immunoassays are two fruitful areas. There are many recent reviews that outline progress in these areas [42−44]. Typical concentration ranges measured by different types of bioassays [45] are shown in Fig. 1.18. In this book, we will introduce several types of sensors and discuss their sensitivities. The chart in Fig. 1.18 provides a guideline to compare different types of sensing methods.

1.6 Animals as Molecular Sensors

The principles, design, fabrication, and application of engineered molecular sensors will be described in detail in Chapters 2 to 7. In the remaining part of this chapter, we will take a look at nature's natural biosensors by examining the olfactory systems of animals and insects. By studying the sensory systems of animals and insects found in nature, we can learn to design sensor platforms that will eventually have the selectivity and sensitivity comparable to that of animals and insects.

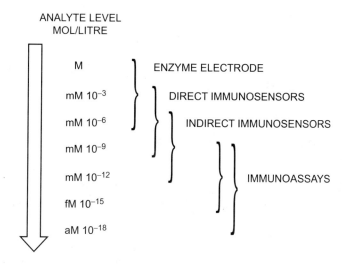

ANALYTE LEVEL
MOL/LITRE

M — ENZYME ELECTRODE

mM 10^{-3} — DIRECT IMMUNOSENSORS

mM 10^{-6} — INDIRECT IMMUNOSENSORS

mM 10^{-9}

mM 10^{-12} — IMMUNOASSAYS

fM 10^{-15}

aM 10^{-18}

Figure 1.18: Typical Concentration Ranges Measured by Bioassays.
From [45].

Canines have been used as molecular detectors for almost as long as they have been domesticated. The earliest instances of humans using dogs for their superior olfactory senses date back to when canines were first used as hunting dogs. It is well known that dogs and pigs have been used to find truffles in Europe. For many tasks, even nowadays, it is difficult to create an artificial sensor as sensitive and as versatile as animals. They still play a number of important roles in real world sensing applications. Such application areas include detection of explosives, drugs, gas leaks, pests, and even cancers within patients [46]. The use of animals capitalize on the fact that compounds of interest, such as cancers and explosives, emit characteristic volatile compounds that may be detected by animals with acute olfactory perceptions. Animals are portable, so they are readily available in nearly all locations regardless of industrial development, making them truly universal. They are amenable to training, and are inexpensive to train and maintain. These two characteristics allow animals to be "tailored" to the application needed. Moreover, animals combine the elements of detection, processing, and output into one "biological implementation." This minimizes bias from signal transduction.

Though various man-made detection modalities exist for the screening of various compounds of interest, such modalities are not necessarily amenable to widespread implementation. Specifically, developing countries may not have the necessary infrastructure nor trained technicians to support the sophisticated nature of these devices. For example, rural countries or regions may not have the technological infrastructure in place to support advanced screening equipment. Moreover, these devices may be expensive, lack high-throughput, are inefficient, and are for one-time use only, further barring them from widespread implementation.

Such interest has thrust "animal molecular sensors" as a practical solution. Certain animals exhibit heightened sensory perceptions, and when exploited, make ideal candidates for *direct animal sensors* in the detection of targets of interest. Animals have already been investigated as sensors in myriad applications, including explosives detection [46,47], drug detection [48] , and cancer diagnosis [49–53].

1.6.1 Sensitivity of Animal Olfactory Systems

Canine Olfactory System

It is already known that canines have olfactory systems that are more sophisticated and more sensitive than that of humans. This fact is largely due to the 220 million olfactory receptors that typical dogs have in their noses compared to a mere 5 million in a human [54]. Due to this large number of receptors, canines have developed an extra olfactory chamber alongside smaller olfactory systems composed of olfactory epithelium containing oral receptors that are different from those found in the nasal cavity. This vomeronasal chamber, which is used when a dog sniffs, can fill with air but will not flush out upon exhalation, demonstrating that the organ evolved out of specialization. All of these factors exponentially increase the efficiency and sensitivity of the olfactory system.

The sensitivity of canine olfactory systems to cocaine was quantified using cocaine hydrochloride and methyl benzoate, a product of cocaine that may become present in vapor [55]. The dogs were tested for both minimum threshold needed and the types of responses given at each concentration of the chemicals. They were first trained through backward chaining in an experimental chamber that had two levers, indicating either a positive or negative response. The chemicals were delivered via vapor by an olfactometer, ranging in value from 960–2 parts per billion (ppb) for methyl benzoate and 0.80–0.002 ppb for cocaine hydrochloride. Multiple trials were run, with control (clean air), two non-targets, and the two target chemicals. The threshold values recorded for cocaine were as low as 0.02 ppb, with 50% of the trials testing correctly at this data point. For methyl benzoate, the threshold was calculated to be 5 ppb, with a steep drop in accuracy under 50 ppb [55]. Values of 9.6% and 6.3% were registered as the average false positive percentage for cocaine and methyl benzoate, respectively. While each dog did have varying curves, the overall trends and thresholds conformed nicely.

Walker et al. aimed to take a more natural approach to testing olfactory sensitivity by using PVC tubes that emitted the target molecule among several other blank tubes [56]. Their approach proved successful as they tested sensitivities as small as 1.14 parts per trillion (ppt) in their Schnauzer and 1.9 ppt for a Rottweiler. These values are orders of magnitude more precise than all other tests and the group felt it was due heavily to their procedures.

Their work is also exhibits an impressive amount of stability in results, with the same values being calculated over several trials.

Insect Olfactory System

Insects smell using a different class of organs: antennae. Molecules are collected by the insect's olfactory sensillae in their antennae and mouthparts [57]. Specific proteins in antennae bind the odor molecules and are collected to a membrane-bound receptor on a nerve cell. Because insects use their sense of smell to communicate, their sensitivity can be extremely high compared to other animals. Antennae of male silkworm moths (*Bombyx mori*) are highly sensitive and selective to sex pheromone (10E,12Z)-hexadeca-10,12-dien-1-ol, or bombykol. Electrophysiological recordings indicated that the sensitivity limit of male antennae for bombykol is just a single molecule, which is the theoretical detection limit for molecular sensing [58]. Another study showed that worker honeybees can discriminate between mixtures of the same two compounds (un- and dodecanoic acids; tri- and pentacosane) with different proportions [59].

1.6.2 Applications of Animal Molecular Sensors

Explosive Detection

Canine Detection of Explosives

Modern explosives are becoming increasingly more complex, but are still composed of the same basic chemicals such as RDX (cyclotrimethylenetrintramine), 2,4,5-TNT (2,4,6-trinitrotoluene), 2,4-DNT (2,4-dinitrotoluene), NG (nitroglycerin), and NC (nitrocellulose). Though these chemicals are the main components of explosives, some may be odorless or have an extremely low vapor pressure and are not present in normal atmospheric conditions. Canines' usefulness as explosive detectors has grown exponentially in the past century, and much research has been done in canines concerning the detection of volatile chemicals [46].

In order for the dog to be useful for explosives detection, a 6−8 week training program with a handler is needed. The training process usually consists of several stages of operant conditioning, with an initial regime of showing the dog where the sample is, followed by increasing levels of independence in detecting the needed chemical. During the operant testing phase, it is important to vary mixtures of these compounds to account for the differing types of bombs that are in use. Negative controls and other common items are also used to raise the sensitivities of the trained canines. A majority of canine sensing and recognition training follows a similar process as previously described in which the dogs are trained to acutely sense specific molecules in minute quantities through some form of conditioning. Such training has been applied to detecting dogs for disease, polymers, cocaine, and explosives.

There is an issue of how the molecules present are affected by the general environment and how the movement of these particles affects the noses of canines. Models of odors are fairly difficult to create and are still very unreliable. Major steps taken to combat this problem involve testing canines in as many situations with differing variables as possible. Recent comparisons have been made between newly lab trained sniffer dogs and canines traditionally used by the police, with results suggesting that for specific purposes (i.e. drug versus explosives), training dogs will adapt to a specific class of odors. Different tests highlighting the versatility and advantages provided by using dogs as explosive vapor detectors (EVD) were also conducted.

For the security of public places such as airports, or stadiums and arenas, EVD instruments are commonly used. They are typically used by security or government departments in order to find explosives before they can be used for undesirable results. The only other place that uses dogs as detectors is local police departments, which train their dogs carefully under their own programs. Compared to developed and man-made devices such as EVD and solid-phase microextraction (SPME), canines have a much higher selectivity and are more adaptable to different scenarios that explosives are present in. Current options for EVDs are fairly bulky and require quite a bit of processing time due to the nature of their analysis methods. Dogs are many times faster, usually needing only a cursory sniff of the suspected container or area. Devices also require a more exact sample besides small amounts of vapor from the exact source, often needing a direct swab or screening through a diagnostic device. Comparisons of down time between both methods come into consideration as well, with dogs being limited by their handlers and the general amount of time until they become tired (~ 8 hours) while machines have a possibility of running 24 hours, though this is a fairly unreasonable estimate. The biggest argument against the use of machines would be cost, for both the initial investment as well as maintenance costs. EVD equipment costs up to ten times the amount that would be spent on a canine and handler team. While it can be argued that dog units are more affected by unforeseen circumstances such as sickness or any sort of physical ailment, machines are actually more likely to give inaccurate results due to electrical or mechanical problems.

Pouched Rats for the Detection of Landmines

A Belgian company Anti-Persoonsmijnen Ontmijnende Product Ontwikkeling (APOPO) has begun training of African pouched rats (*Cricetomys gambianus*) for land mine-detection in Mozambique [60,61]. They are small enough not to activate mines, and are native to sub-Saharan Africa, and resistant to local parasites and diseases. The animas searched 93 400 m^2 of land, and found 41 mines and 54 other explosive devices in 2009. They made an average of

0.33 false positive detections for every 100 m^2 searched, which is below the threshold determined by International Mine Action Standards for accrediting mine-detection animals [61].

Honeybees for the Detection of Landmines

Though dogs have been heavily utilized by the government in drug and explosives screening, this detection modality is not without inherent limitations. Dogs are handler intensive, and require housing, food, and veterinarian visits. All of these factors require specific training and accumulate costs. Therefore, there has been significant interest in utilizing insects as molecular detectors.

The ability of honeybees to discriminate scents has been well-characterized [62,63]. Honey bees exhibit a proboscis extension response (PER) that acts as a positive indication signal, which may be correlated with detection events once honeybees are trained [62]. An example of PER utilization is the training of bees to detect the presence of Mediterranean fruit fly infested Valencia oranges [64]. Moreover, honeybees are small, and therefore do not require as much space or food as larger animals. These characteristics of honeybees make them an ideal candidate for molecular sensing. While the Defense Advanced Research Projects Agency (DARPA) has placed emphasis on incorporating insects such as beetles and moths into military operations, these efforts have focused on utilizing these insects as surveillance equipment carriers. Other efforts have capitalized on the inherent acute sensory perceptions of insects for the detection of specific compounds of interest. An interesting application is the use of honeybees in detecting landmines [62], an effort initiated by DARPA in collaboration with Sandia Laboratories.

Approximately 120 million landmines are buried throughout the world, cutting off access to critical water sources and vast areas of valuable farmland. Landmines kill or injure 2000 people each month, and yet more landmines are being placed than are being recovered and decommissioned [62]. Moreover, thousands of tons of leftover unexploded ordnances from World War I and II, as well as other conflicts, remain buried. Conventional explosives detection schemes are limited in application – detectors require either close proximity to, or direct contact with, the target, placing the user at risk. The detectors are often cumbersome, expensive, and require extensive training to operate. Honeybees offer a far more efficient and inexpensive detection means. Bees are highly amenable to conditioning and training, and may be directed to suspected targets exhibiting a specific odor with high sensitivities. TNT is used as the main charge in approximately 90% of all landmines [62]. As TNT is released into the environment by the landmines, it is subjected to biotransformations, such as plant metabolism and the action of fungi and soil microbes, resulting in byproducts. One such byproduct is 2,4-DNT, which has one of the highest vapor pressures of all of the byproducts formed and is typically found within the vicinity of a landmine. Accordingly, 2,4-DNT provides a good indicator for landmines, and was capitalized as the compound of interest in these studies.

Bees are trained in a manner similar to training dogs — the bees are conditioned to associate a specific target odor with a reward or food. Because bees are natural foragers, they will actively seek out the target odor within the foraging area, and will be attracted to areas with the odor while ignoring areas without the odor. While dogs may take at least 3 months to train, an entire bee colony may be set out at a site, the bees conditioned, and the area searched within a day [62]. Training takes approximately 2 h to accomplish. Moreover, one beehive may rapidly establish training protocols. Only one hive is required to train multiple hives; as one hive is trained, bees from that hive will recruit and train other hives within the area. In this manner, the beehive acts as the functional unit of this system, and becomes "self-training." The bees also exhibit a remarkable sense of discrimination. Given a set of samples, blanks (sugar-water) and areas contaminated with either 2,4-DNT, RDX, or 2,4,6-TNT, the bees were capable of only targeting feeders containing DNT while largely ignoring the TNT- and RDX-contaminated targets. For field testing, experiments focused on distances from the hive and different concentrations of chemicals in order to assess the effects of distance and chemical concentration on the performance of the bees. Hive-mounted cameras monitored bee flight activity. The honeybees were reported to have a probability of detection of DNT targets of 98.7%, with a probability of failure to detect a DNT target around 1.1–1.2%. Bees exhibited sensitivies in the sub-parts-per-trillion or lower range, around 4.48–13.4 ppb of 2,4-DNT in the field. Multiple beehives may be trained concurrently with different compounds for a multi-compound detection scheme.

This detection scheme effectively addresses the limitations of current explosives detection methods. The use of bees eliminates the need for an animal handler, bulky equipment, and does not require placing people in high-risk situations. While such research is still being conducted by Defense Advanced Research Projects Agency (DARPA), companies have also begun to capitalize on the advantages posed by honeybees. The company Inscentinel is employing sniffer honeybees to detect explosives and drugs. However, one of the inherent drawbacks of this system is fear — people are naturally afraid of honeybees. Such phobia may limit widespread commercial appeal of such honeybee systems.

Disease Detection

Canines for Cancer Detection

Although there have been significant advances in the development of highly-specific cancer drug delivery systems and the characterization of cancer-specific biomarkers, the standard of care for cancer treatment remains highly dependent on early detection of cancer. There is a general trend of improved patient outcomes and prognosis with early detection. However, current detection methods are expensive, not sensitive enough, invasive, and may actually introduce health risks to the patient. These detection methods are not readily accessible to the general populace who may not be able to afford such procedures, and are not pragmatic in developing countries that do not possess the facilities requisite for such procedures. There is therefore a need for an inexpensive, efficient, and more facile means of cancer detection.

Dogs as cancer sensors were first triggered by the story of a canine that licked and barked at a particular mole of a patient. The mole was later checked and confirmed for melanoma. Various cancers such as lung cancer, bladder cancer, and melanomas present cancer-specific biomarkers present within the patients' breath, urine, and skin [65]. Dogs, which have an acute sense of smell, have been demonstrated to be able to detect these biomarkers, and have potential as a diagnostic tool. Groups have tested dogs sniffing different types of samples based on the specific type of cancer being targeted including urine samples for bladder and prostate cancer, breath samples for lung and chest diseases, and even direct skin smelling in the case of melanoma. These are all easy to obtain samples compared to tissue biopsies and also open up the possibility of a streamlined process in which the canines analyze several sets of tests at once.

In general, tumors produce volatile, tumor-specific organic compounds, or biomarkers, that are released into the atmosphere [52,66]. In bladder cancer, these biomarkers are released into the urine, and may have characteristic odors that are detectable by dogs [49,51,52]. Operant clicker training was utilized to train the dogs. In the bladder cancer studies, the reported success rate was 41% (95% confidence intervals 23−58%), compared to an expected 14% by chance [51,52]. Canines used for lung cancer detection are trained in almost a similar manner as bladder cancer detecting dogs. These dogs were then presented with breath samples, in the form of sampling tubes with either polypropylene or carboxen-polydimethylsiloxane-coated fibers [50]. Lung cancer seemed to fit more within the limits of canine smell detection. After significant training through the three phase system, canines were able to identify samples positive for lung cancer. In the lung cancer studies, the reported sensitivities and specificities were 80−90%. These results seem to suggest that there is a molecular basis as to how dogs may detect cancer-positive samples. However, these relatively low accuracy rates make dog detection of cancer currently unviable in the stringent requirements of the medical field.

Limitations of dog detection of cancer bar it from widespread medical implementation. Evidence regarding the viability of dogs to detect cancer was mostly anecdotal [49,67]. The relatively low reported success rates are not practical in the medical field, and may lead to misdiagnoses that would harm patients. Especially for cancer diagnosis, high sensitivities and specificities (> 99%) are required to accurately diagnose patients and begin therapy. While these reported success rates are low, they are significantly greater than if the dogs had selected the positive samples correctly by chance. This suggests dog detection of cancer is a promising diagnostic tool that, when optimized, presents a new and more patient-compliant venue of cancer detection. In terms of current medical applications, these canine detectors may be used as a broad, preliminary means of cancer screening. Screening can be routinely performed at ease, and patients whose samples are indicated as positive may be further evaluated by their doctors.

Pouched Rats for the Detection of Tuberculosis

Pouched rats were used for the diagnosis of tuberculosis (TB) [68]. While tuberculosis is relatively rare in developed countries, it remains a predominant threat in sub-Saharan Africa, South East Asia, Latin America, and the Caribbean, with 90 million cases recorded between 1990 and 1999 [68]. The cheapest and most affordable manner of TB detection is smear microscopy, which suffers from low detection rates (between 20 and 40% in sub-Saharan Africa) and requires trained microscopists [68]. Moreover, the high prevalence of HIV in the region has led to a high number of false negatives. The drawbacks of smear microscopy have prompted interest in training African giant pouched rats (*C. gambianus*) as a potentially faster and more accurate TB diagnostic tool through the olfactory detection of TB-positive sputum.

Broadly, these rats may be used as a preliminary screening agent in areas with high TB prevalence and probability of infection, such as slums, prisons, and schools, and are ideal candidates for use in sub-Saharan Africa [68]. The use of rats has been demonstrated to have high throughput; one rat can screen 140 samples in 40 min, with 1680 samples per day. Compared to the current standard of sputum microscopy (40 samples per day), which is susceptible to human error, and it is evident that rat screening of samples is a potentially cheaper and faster means of mass-TB screening.

The study demonstrated that TB detection by rats may be potentially faster than smear microscopy, and is at least as sensitive. That rats were even able to detect TB-positive samples in smear-negative patients, which suggests that *C. gambianus* is capable of detecting TB during its incipient stages. However, more work will have to be performed to assess the accuracy of TB detection of HIV-positive patients, since HIV is highly prevalent in sub-Saharan Africa and produces more smear-negative results [68].

For the culture-confirmed sputum samples, the rats exhibited a combined sensitivity of 86.5% and a specificity of 89.1%. The investigators found that using multiple rats for the same samples increased the sensitivity, and decided to increase the number of rats used from 2 to 16. The group of 16 rats consistently scored above 72% detection and less than 8.1% false-positive. Further refinement may be accomplished by selecting the best-performing rats. For the microscopy-confirmed sputum samples, the rats averaged a sensitivity of 87.9% and a specificity of 95.8%.

Overall, the use of pouched rats in sub-Saharan Africa is well suited to the environment. TB is highly prevalent in this region, and pouched rats are native to this region. Moreover, pouched rats are amenable to training, and exhibit an acute sense of smell. All of these factors make a cogent appeal for *C. gambianus* as a cost-effective, accurate, and high-throughput means of TB detection. Though rats are not currently being implemented as a diagnostic tool, these initial results are highly promising. More work will have to be performed to optimize the workflow for operational situations.

Other Applications

Canine Detection of Pirated DVDs

An unlikely organization that has recently called on the help of sniffer dogs is the Motion Picture Association of America (MPAA), an organization focused on preventing copyright infringement and the distribution of illegally made movies.

It was found that dogs could quicky narrow down on the detection element in DVDs, polycarbonate. The material is a thermoplastic polymer that is used widely in electrical hardware as well as in the construction industry. Two trained dogs have assisted law enforcement agencies in Malaysia, Germany, and England in seizing over 1.9 million pirated movies. The MPAA originally paid $18 000 for the dogs, but feel they have more than made that up with these successful raids [69]. One obvious disadvantage is that they are trained to only detect the polycarbonate in the DVDs, and therefore, cannot distinguish between fake and real DVDs. Nonetheless, this success has inspired many other organizations to look into dogs for unique purposes similar to the MPAA's uses.

Canine Detection of Bed Bugs

Bed bugs (*Cimex lectularius*) are nocturnal feeders and hide during the day, making them difficult to visually detect. This makes early bed bug detection and containment particularly difficult, since many pest control operators will not apply insecticide without visual confirmation. Moreover, many people have delayed reactions, or no reactions, to bed bug bites, making correlations to infestations difficult. These difficulties in early bed bug detection often mean that bed bug infestations go unnoticed until populations are overwhelming. Early containment of bed bug infestations is particularly desired, leading to more successful outcomes, less likelihood of spreading, and cheaper costs. Therefore, a means of bed bug detection other than sight would be valuable in the detection of small and early infestations [70].

Dogs are utilized to detect for the presence of live bed bugs and viable bed bug eggs [70]. The investigators conducted four experiments to determine the ability of canines to detect bed bugs when trained with live adult bed bugs: (A) to determine if trained dogs were able to differentiate bed bugs from other general household pests, (B) to determine if dogs could be trained to discriminate live bed bugs and viable eggs from other bed bug materials, (C) to determine if trained dogs could locate hidden bed bugs in hotel rooms in a controlled experiment, and (D) to assess if a bed bug pseudoscent could be recognized as live bed bugs by trained dogs.

The results obtained for this study indicate favorable outcomes. For experiment (A), positive indications averaged $\sim 98\%$, with no false positive indications. For experiment (B), dogs were able to indicate live bed bugs 100% of the time, which was significantly more accurate than indication of viable bed bug eggs (90%, $p < 0.0001$ by two-way ANOVA). The dogs had an average of 3% false positive indication on bed bug feces, and no indications on cast skins, dead bed bugs, and the control. The overall indication rate was reported to be 95%, which was significantly higher than the false positive rate ($p < 0.0001$). For experiment (C), dogs averaged 98% accuracy, with no false positives. For experiment (D), dogs exhibited positive indications with the pentane extract, with low to nonexistent indications on the other extracts.

The results of this study show promising prospects for bed bug-detecting dogs. Moreover, the hotel room experiment exhibited the high sensitivity of dogs, and the commercial applicability of these dogs. Economically, these dogs, once trained, represent a facile, inexpensive, and reusable means of bed bug detection. Especially in the hotel industry, such "technology" would substantially cut down the costs associated with lawsuits, cleaning, and pesticide treatment.

1.6.3 Discussion on Animals as Molecular Sensors

Efforts described in this section have been motivated by the need to create streamlined and uniform sensors in a high-throughput manner. In spite of the demonstrated attractive advantages, there are several obvious drawbacks. While animals are amenable to training, they are still biological organisms, and are subjected to genetic variation that may affect detection of certain biological compounds. Moreover, animals require extensive handler interface that prohibits training in a high-throughput fashion. Accordingly, these artificial animal-based sensors were created based on biological principles to address these limitations.

The studies we introduced here suggest that animals are still more excellent molecular sensors in many ways than the artificially designed molecular sensors we will describe in the following chapters. They also suggest important ideas about key potential application areas where inexpensive and sensitive "artificial" sensor should be developed and employed in future.

1.7 Conclusion

Traditional methods of small molecule detection involve techniques such as chromatography, ELISA, and DNA/RNA microarrays. However, these methods can be slow and are usually limited in the number and variety of molecules they can detect. In the following chapters, we will discuss new techniques involving smaller scales of detection, which both enhance the sensitivity of the sensor and allow for multiplexed assays that can screen for multiple chemicals simultaneously. Original solutions to this problem have taken the form of microscale sensors, chemically sensitive field effect transistors, and techniques based on traditional silicon fabrication. Improvements in this field can be found by integrating existing nanotechnology into the currently existing fabrication methods.

Problems

P1.1 Molecular Sensor

Provide one idea of a molecular sensor related to the content discussed in this chapter coming across your mind. You need to back-up your idea with a brief statement showing why you believe it is feasible.

P1.2 Molecular Sensor

What are the three key elements of a sensor? Distinguish between chemical sensors, physical sensors, biosensors and molecular sensors, give one example for each. What characteristics are required for a "useful" molecular sensor?

P1.3 Recognition Element

What are requirements for recognition elements?

P1.4 Basics of Molecular Sensing

Answer the following questions:

pH: What is an acid?
Forces: Which is stronger, electrostatic forces or van-der-Waals forces? Why?
Molecules: What is a covalent bond?
Biochemistry: What is a ligand?
Protein: How many amino acids exist?
DNA: What is hybridization?
Gene: What is an intron?
Antibody: What is a Fab fragment?
Enzyme: What does an enzyme do?
Cells: What constitutes a cell membrane?

P1.5 Antibodies

Describe how we can produce antibodies for a new antigen molecule. Discuss the difference between polyclonal and monoclonal antibodies.

P1.6 Immunosensing

Design an antibody-based biosensor. What are the three elements of the sensor? How does it work to provide us with information? What are the advantages and disadvantages of such a sensor?

P1.7 DNA Biosensor

Design a DNA-based biosensor. What are the three elements of the sensor? How does it work to provide us with information? What are the advantages and disadvantages of such a sensor?

P1.8 DNA Basics

What is the base sequence of the DNA strand that would be complementary to the following single-stranded DNA molecule?

5′ GAAGACATGC 3′

P1.9 DNA Basics

What is the base sequence of the RNA strand that would be complementary to the following single-stranded DNA molecule?

5′ ATCGATGCTAGC 3′

P1.10 DNA Basics

The percent of cytosine in a double-stranded DNA is 19. What is the percent of adenine in that DNA?

P1.11 DNA Basics

A nucleic acid was analyzed and found to contain 20% A, 30% T, 36% G, 14% C. Based on the nucleotide content, what kind of nucleic acid does this represent?

P1.12 Thermal Noise

What is the value of RMS noise voltage for a 1 kΩ resistor at room temperature and a 10 kHz bandwidth?

P1.13 Thermal Noise, Responsivity and Sensitivity

Let us consider a 1 MΩ resistor as a transducer that transduce a current I into a voltage V.

1. What is the responsivity of the resistor?
2. What is the amplitude spectral density of the thermal noise?
3. If the thermal noise is the limiting factor, what is the sensitivity?

P1.14 Sensitivity of a Force Sensor

1. If an average voltage change of $\Delta V = 10\ \mu V$ is observed for an increase force of $\Delta F = 120\ pF$, what is the responsivity of the sensor?
2. The noise was measured to be 40 nV/$\sqrt{Hz}$. What is the total noise from 1 Hz to 1 kHz?
3. What is the value of minimum measurable force in the bandwidth of 1 Hz to 1 kHz?

P1.15 Animals as Molecular Sensors

1. Name examples of animals used as molecular sensors other than the ones introduced in this chapter.
2. Discuss why animals are still used for those applications. Are there any alternative artificial devices?

References

[1] Llorens J. The physiology of taste and smell: how and why we sense flavors. Water Science and Technology: A Journal of the International Association on Water Pollution Research 2004;49:1.
[2] Gaillard I, Rouquier S, Giorgi D. Olfactory receptors. Cellular and Molecular Life Sciences CMLS 2004;61:456–69.
[3] Sicard G, Holley A. Receptor cell responses to odorants: similarities and differences among odorants. Brain Research 1984;292:283–96.
[4] Leiwereg TJ, Guadagni DG, Harris J, Mon TR, Teranishi R. Chemical and sensory data supporting the difference between the odors of the enantiomeric carvones. Journal of Agricultural and Food Chemistry 1971;19:785–7.
[5] Cain WS. Differential sensitivity for smell: noise at the nose. Science 1977;195:796–8.
[6] Jiang J, Coffey P, Toohey B. Improvement of odor intensity measurement using dynamic olfactometry. Journal of the Air & Waste Management Association 2006;56:675–83.
[7] Eisthen HL. Why are olfactory systems of different animals so similar? Brain, Behavior and Evolution 2002;59:273–93.

[8] Ziegler C. Cell-based biosensors. Fresenius' Journal of Analytical Chemistry 2000;366:552−9.

[9] Baumann W, Lehmann M, Schwinde A, Ehret R, Brischwein M, Wolf B. Microelectronic sensor system for microphysiological application on living cells. Sensors and Actuators B: Chemical 1999;55:77−89.

[10] Owicki JC, Wallace Parce J. Biosensors based on the energy metabolism of living cells: the physical chemistry and cell biology of extracellular acidification. Biosensors and Bioelectronics 1992;7:255−72.

[11] Pancrazio J, Whelan J, Borkholder D, Ma W, Stenger D. Development and application of cell-based biosensors. Annals of Biomedical Engineering 1999;27:697−711.

[12] Ziegler C, Göpel W. Biosensor development. Current Opinion in Chemical Biology 1998;2:585−91.

[13] Ziegler C, Göpel W, Hämmerle H, Hatt H, Jung G, Laxhuber L, et al. Bioelectronic noses: a status report. Part II. Biosensors and Bioelectronics 1998;13:539−71.

[14] Misawa N, Mitsuno H, Kanzaki R, Takeuchi S. Highly sensitive and selective odorant sensor using living cells expressing insect olfactory receptors. Proceedings of the National Academy of Sciences 2010;107:15340−4.

[15] Kuwana Y, Nagasawa S, Shimoyama I, Kanzaki R. Synthesis of the pheromone-oriented behaviour of silkworm moths by a mobile robot with moth antennae as pheromone sensors. Biosensors and Bioelectronics 1999;14(2):195−202.

[16] Davies DR, Chacko S. Antibody structure. Accounts of Chemical Research 1993;26:421−7.

[17] Skottrup PD, Nicolaisen M, Justesen AF. Towards on-site pathogen detection using antibody-based sensors. Biosensors and Bioelectronics 2008;24:339−48.

[18] Velusamy V, Arshak K, Korostynska O, Oliwa K, Adley C. An overview of foodborne pathogen detection: in the perspective of biosensors. Biotechnology Advances 2010;28:232−54.

[19] Goldenberg DM, Kim EE, DeLand FH, Nagell JV, Javadpour N. Clinical radioimmunodetection of cancer with radioactive antibodies to human chorionic gonadotropin. Science 1980;208:1284−6.

[20] Mohammed I, Mullett WM, Lai EP, Yeung JM. Is biosensor a viable method for food allergen detection? Analytica Chimica Acta 2001;444:97−102.

[21] Blake DA, Jones RM, Blake II RC, Pavlov AR, Darwish IA, Yu H. Antibody-based sensors for heavy metal ions. Biosensors and Bioelectronics 2001;16:799−809.

[22] Goldman ER, Medintz IL, Whitley JL, Hayhurst A, Clapp AR, Uyeda HT, et al. A hybrid quantum dot-antibody fragment fluorescence resonance energy transfer-based TNT sensor. Journal of the American Chemical Society 2005;127:6744−51.

[23] Kienberger F, Mueller H, Pastushenko V, Hinterdorfer P. Following single antibody binding to purple membranes in real time. EMBO Reports 2004;5:579−83.

[24] Borisov SM, Wolfbeis OS. Optical biosensors. Chemical Reviews 2008;108:423−61.

[25] Dahm R. Friedrich Miescher and the discovery of DNA. Developmental Biology 2005;278:274−88.

[26] Avery OT, MacLeod CM, McCarty M. Studies on the chemical nature of the substance inducing transformation of pneumococcal types induction of transformation by a desoxyribonucleic acid fraction isolated from pneumococcus type III. The Journal of Experimental Medicine 1944;79:137−58.

[27] E. Chargaff, S. Zamenhof, C. Green, Human Desoxypentose Nucleic Acid: Composition of Human Desoxypentose Nucleic Acid, 1950.

[28] Watson J, Crick F. Molecular structure of nucleic acids: a structure for deoxyribose nucleic acid. Nature 1953;171:3.

[29] Levene P. The structure of yeast nucleic acid. Studies from the Rockefeller Institute for Medical Research 1919;30:221.

[30] Gibson DG, Benders GA, Andrews-Pfannkoch C, Denisova EA, Baden-Tillson H, Zaveri J, et al. Complete chemical synthesis, assembly, and cloning of a *Mycoplasma genitalium* genome. Science Signaling 2008;319:1215.

[31] Dias N, Stein C. Antisense oligonucleotides: basic concepts and mechanisms. Molecular Cancer Therapeutics 2002;1:347−55.

[32] Song K-M, Lee S, Ban C. Aptamers and their biological applications. Sensors 2012;12:612−31.

[33] Song S, Wang L, Li J, Fan C, Zhao J. Aptamer-based biosensors. TrAC Trends in Analytical Chemistry 2008;27:108–17.

[34] Stoltenburg R, Reinemann C, Strehlitz B. SELEX- A (r) evolutionary method to generate high-affinity nucleic acid ligands. Biomolecular Engineering 2007;24:381–403.

[35] Ellington AD, Szostak JW. In vitro selection of RNA molecules that bind specific ligands. Nature 1990;346:818–22.

[36] Keefe AD, Cload ST. SELEX with modified nucleotides. Current Opinion in Chemical Biology 2008;12:448–56.

[37] Tuerk C, Gold L. Systematic evolution of ligands by exponential enrichment: RNA ligands to bacteriophage T4 DNA polymerase. Science 1990;249:505–10.

[38] Ramanathan K, Mangesh A, Yun M, Chen W, Myung NV, Mulchandani A. Bioaffinity sensing using biologically functionalized conducting-polymer nanowire. Journal of the American Chemical Society 2005;127:496–7.

[39] Homola J. Present and future of surface plasmon resonance biosensors. Analytical and Bioanalytical Chemistry 2003;377:528–39.

[40] Fritz J, Baller M, Lang H, Rothuizen H, Vettiger P, Meyer E, et al. Translating biomolecular recognition into nanomechanics. Science 2000;288:316–8.

[41] Harley J, Kenny T. High-sensitivity piezoresistive cantilevers under 1000 Å thick. Applied Physics Letters 1999;75:289–91.

[42] Auroux PA, Koc Y, deMello A, Manz A, Day PJR. Miniaturised nucleic acid analysis. Lab on a Chip 2004;4:534–46.

[43] Sassolas A, Leca-Bouvier BD, Blum LJ. DNA biosensors and microarrays. Chemical Reviews 2007;108:109–39.

[44] Holland CA, Kiechle FL. Point-of-care molecular diagnostic systems – past, present and future. Current Opinion in Microbiology 2005;8:504–9.

[45] Byfield M, Abuknesha R. Biochemical aspects of biosensors. Biosensors and Bioelectronics 1994;9:373–99.

[46] Furton KG, Myers LJ. The scientific foundation and efficacy of the use of canines as chemical detectors for explosives. Talanta 2001;54:487–500.

[47] King TL, Horine FM, Daly KC, Smith BH. Explosives detection with hard-wired moths. Instrumentation and Measurement, IEEE Transactions on 2004;53:1113–8.

[48] Dunn M, Degenhardt L. The use of drug detection dogs in Sydney, Australia. Drug and Alcohol Review 2009;28:658–62.

[49] Gordon RT, Schatz CB, Myers LJ, Kosty M, Gonczy C, Kroener J, et al. The use of canines in the detection of human cancers. The Journal of Alternative and Complementary Medicine 2008;14:61–7.

[50] McCulloch M, Jezierski T, Broffman M, Hubbard A, Turner K, Janecki T. Diagnostic accuracy of canine scent detection in early-and late-stage lung and breast cancers. Integrative Cancer Therapies 2006;5:30–9.

[51] Moser E, McCulloch M. Canine scent detection of human cancers: a review of methods and accuracy. Journal of Veterinary Behavior: Clinical Applications and Research 2010;5:145–52.

[52] Willis CM, Church SM, Guest CM, Cook WA, McCarthy N, Bransbury AJ, et al. Olfactory detection of human bladder cancer by dogs: proof of principle study. BMJ 2004;329:712.

[53] McCulloch M, Turner K, Broffman M. Lung cancer detection by canine scent: will there be a lab in the lab? European Respiratory Journal 2012;39:511–2.

[54] Correa JE, The dog's sense of smell, Alabama Cooperative Extension System UNP-0066, 2005.

[55] Waggoner LP, Johnston JM, Williams M, Jackson J, Jones MH, Boussom T, et al. Canine olfactory sensitivity to cocaine hydrochloride and methyl benzoate, in: Enabling Technologies for Law Enforcement and Security, 1997, pp. 216–226.

[56] Walker DB, Walker JC, Cavnar PJ, Taylor JL, Pickel DH, Hall SB, et al. Naturalistic quantification of canine olfactory sensitivity. Applied Animal Behaviour Science 2006;97:241–54.

[57] Stocker RF. The organization of the chemosensory system in *Drosophila melanogaster*: a review. Cell and Tissue Research 1994;275:3−26.

[58] Sandler BH, Nikonova L, Leal WS, Clardy J. Sexual attraction in the silkworm moth: structure of the pheromone-binding-protein−bombykol complex. Chemistry & Biology 2000;7:143−51.

[59] Getz WM, Smith KB. Olfactory sensitivity and discrimination of mixtures in the honeybee *Apis mellifera*. Journal of Comparative Physiology A 1987;160:239−45.

[60] Habib MK. Controlled biological and biomimetic systems for landmine detection. Biosensors and Bioelectronics 2007;23:1−18.

[61] Poling A, Weetjens B, Cox C, Beyene NW, Bach H, Sully A. Using trained pouched rats to detect land mines: another victory for operant conditioning. Journal of Applied Behavior Analysis 2011;44:351−5.

[62] P.J. Rodacy, S. Bender, J. Bromenshenk, C. Henderson, G. Bender, Training and deployment of honeybees to detect explosives and other agents of harm, in: AeroSense 2002, 2002, pp. 474−481.

[63] Rains GC, Tomberlin JK, Kulasiri D. Using insect sniffing devices for detection. Trends in Biotechnology 2008;26:288−94.

[64] Chamberlain K, Briens M, Jacobs JH, Clark SJ, Pickett JA. Use of honey bees (*Apis mellifera* L.) to detect the presence of Mediterranean fruit fly (*Ceratitis capitata* Wiedemann) larvae in Valencia oranges. Journal of the Science of Food and Agriculture 2012;92:2050−4.

[65] Ulanowska A, Ligor T, Michel M, Buszewski B. Hyphenated and unconventional methods for searching volatile cancer biomarkers. Ecological Chemistry and Engineering S 2010;17:9−23.

[66] Ludwig JA, Weinstein JN. Biomarkers in cancer staging, prognosis and treatment selection. Nature Reviews Cancer 2005;5:845−56.

[67] Horvath G, Järverud GAK, Järverud S, Horváth I. Human ovarian carcinomas detected by specific odor. Integrative Cancer Therapies 2008;7:76−80.

[68] Weetjens B, Mgode G, Machangu R, Kazwala R, Mfinanga G, Lwilla F, et al. African pouched rats for the detection of pulmonary tuberculosis in sputum samples. The International Journal of Tuberculosis and Lung Disease 2009;13:737−43.

[69] Kravets D. Inside the motion picture industry's cuddliest anti-piracy operation: the DVD-sniffing dogs. Wired 2008 July 11.

[70] Pfiester M, Koehler PG, Pereira RM. Ability of bed bug-detecting canines to locate live bed bugs and viable bed bug eggs. Journal of Economic Entomology 2008;101:1389−96.

Further Reading

Human olfactory system
Silverthorn DU. Human Physiology: An Integrated Approach. San Francisco: Pearson/Benjamin Cummings; 2007 Print

Antibodies
Wild D. The Immunoassay Handbook: Theory and Applications of Ligand Binding, ELISA and Related Techniques. fourth ed. Oxford: Elsevier Science; 2013.
Harlow E, Lane DP. Antibodies: A Laboratory Manual. Cold Spring Harbor: Cold Spring Harbor Laboratory Press; 1988.
Alberts B, Johnson A, Lewis J, et al. Molecular Biology of the Cell. fourth ed. New York: Garland Science; 2002.

DNA
Tobin A, Dusheck J. Asking About Life. third ed. Belmont CA: Thomson Learning; 2005.
Campbell N, Reece J, et al. Campbell Biology. nineth ed. Boston: Benjamin Cummings; 2010.

CHAPTER 2

Fundamentals of Nano/Microfabrication and Effect of Scaling

Chapter Outline

2.1 Introduction

Micro-nanotechnology, or the science and technology associated with "scaling-down", is an emerging engineering frontier with revolutionary impact in information technology, manufacturing, and most recently, biomedicine. Microchip-scale fabrication of integrated semiconductor devices has triggered the information technology revolution since the mid-20th century. In 1965, Gordon Moore described that the number of transistors that can be fabricated on a given chip doubles every 18 months (Moore's Law). As social and economic demands of the field changed rapidly over the past decades, the scaling of transistors and other electronics became amazingly smaller and faster than ever predicted. Small-scale engineering can be applied to systems beyond electrical components. The advent of NEMS and MEMS (nano- and micro-electro-mechanical systems) has expanded the trend of "scaling-down" to a broader class of nano/micro functional devices, including: mechanical cantilevers, microfluidic channels, and optical components, among many others. In this chapter, we describe various technologies to "scale down" functional components that are essential for the new generation of molecular sensors.

For the new class of molecular sensors, bionanotechnology, the integration of nanotechnology in biological research and applications, is the focus of growing interest. The combination of nanomaterials and microfabrication with applications in biomedical sensing has opened a new area of research called BioMEMS, μTAS (micro total analysis systems) or LOC (Lab on a Chip) devices (see Fig. 2.1). For example, in biomedical research and clinical applications, knowledge of disease mechanisms on a molecular level is providing incentives for introducing micro/nanotechnology to medical sciences. The trend of "scaling down" is apparent in medical diagnostic technologies, especially for detection of biological molecules. The ability to detect single molecules within a volume of micro- and sub-microliters has made a large impact on the market of biomedical sensing.

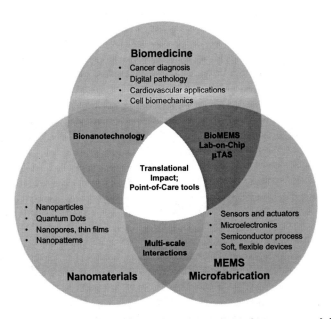

Figure 2.1: BioMEMS and Bionanotechnology: Integration of Nanomaterials and MEMS in Biomedicine.

In the first half of this chapter, we describe principles and practice of nano/microscale fabrication technologies. We start the chapter from discussions on the scale and shape effect in molecular sensors. We consider how scaling down of sensor dimensions affect the overall performance of molecular sensors. We then introduce the basic semiconductor microfabrication processes including photolithography, deposition, etching, and doping, as well as silicon material properties. We will also introduce the MEMS process, including bulk and surface micromachining. In the latter half of this chapter, we describe techniques categorized as "soft lithography," which allow integration of a large variety of non-silicon materials including polymers, nanomaterials and biomaterials onto versatile micro devices. We will discuss soft lithography, including microcontact printing, and other key patterning techniques.

2.2 Scaling in Molecular Sensors

It is well-known that physical characteristics of mechanical systems, electrical systems, animals (or any biological systems) are always dominated by their sizes. Those effects are called **scale effects** or **scaling effects**. Here we focus on the principles and the designs of molecular sensors and discuss how downsizing of sensing elements by microfabrication affects the sensitivity of molecular sensors. Scale effects found specifically for fluidic microsystems will be described in Chapter 3, Sections 3.2.3 and 3.3.1.

A simple, but very important scaling factor is the **surface-to-volume ratio**, also called **surface-area-to-volume ratio**. Most capture and recognition processes of molecular sensors are surface processes where analyte molecules are captured on the functionalized surface of the sensing element, while many transduction techniques utilize changes in physical parameters based on the volume of the sensing element. Such physical parameters include mass, heat capacity, and magnetic force.

Let us consider a model molecular sensor that transduces mass changes due to analytes captured by antibodies on the surface (Fig. 2.2). A good example of such molecular sensors is the quartz crystal microbalance (QCM) based sensor discussed in Chapter 6.

We define the characteristic length of this sensor to be L. The amount of molecules that can be captured is given by the surface area, which is proportional to L^2. The total volume V is proportional to L^3. The sensitivity is estimated to be the volume change ΔV with respect to the original volume V expressed as:

$$\Delta V / V \propto L^{-1}$$

In this idealized model, the sensitivity is proportional to L^{-1}. If the sensor is downsized to 1/10, the sensitivity becomes 10 times higher. Smaller sensors are more efficient for this

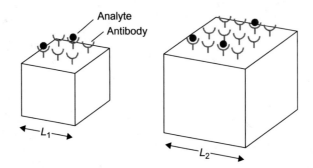

Figure 2.2: Surface Volume Ratio in Molecular Sensors.

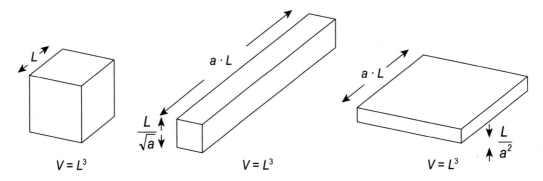

Figure 2.3: Surface Volume Ratios of Different Shapes with the Same Volume.

type of sensor, although, of course, one has to remember there will be always other factors to limit the downsizing of devices.

The example shown in Fig. 2.2 is a case of uniform scaling. The shape of the device is also an important factor to be considered. Advent of nanomaterials has allowed fabrication of sensors with unique shapes. Carbon nanotubes, graphene, and other nanowires are examples of materials with extreme dimensions. Molecular sensors based on those nanomaterials are described in Chapter 6.

In order to simplify the problem, let us consider three materials with the same volume: a cube, a rod and a sheet. When we define the characteristic lengths of the rod and the sheet to be $a \cdot L$ as shown in Fig. 2.3, the surface areas for the three materials become $6L^2$, $4\sqrt{a} \cdot L^2$ and $2a^2L^2$. Since the volumes are the same for all the materials, the sheet is more advantageous in terms of surface volume rate. Graphene, a one-atom thick film of carbon, is thus an excellent candidate as a building block material for very sensitive molecular sensors.

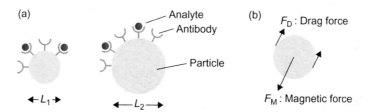

Figure 2.4: Immunomagnetic Separation.

In practical systems, we cannot always simply conclude "the smaller the better" or "the larger the better." The next interesting example is a separation technique based on magnetic particles. Magnetic particles are suspended in a liquid and the surface of the particles is functionalized with antibodies specific to the target analytes (proteins, cells, etc.). A magnetic field is applied to collect nanoparticles that capture analytes. This technique is called **immunomagnetic separation** (Fig. 2.4). Detailed descriptions of the principle and applications of the method are described in Section 3.3.2.

Based on the same discussion we made for Fig. 2.2, smaller particles are more efficient in terms of the amount of analytes that can be captured per the same total volume of the particles, because the surface volume ratio is larger. However, there is another important factor in this model. The magnetic force that acts on a particle is proportional to the volume ($\propto L^3$), while the drag force from the liquid is proportional to the radius ($\propto L$). For the mangetic force, scale effect works in the opposite way, and the smaller particles are more difficult to collect. The optimal particle size largely depends on the system. In fact, the size of particles that have been tested for immunomagnetic separation ranges from a few nanometers to micrometers.

2.3 Microfabrication Basics

Microfabrication is the term used to describe processes of fabricating nano/micrometer scale structures. Traditionally, microfabrication specifically meant fabrication of semiconductor integrated circuits (ICs). It may also be regarded as fabrication processes that involve lithography as the key technology. Now that the world of nano/micro devices has expanded from electrical circuits to include mechanical, optical, chemical, and even biological materials, microfabrication is more widely used for fabrication of any type of nano/microscale functional elements. In this chapter, we first describe basics of conventional silicon microfabrication, followed by fundamentals of silicon-based MEMS. The discussion is then extended to nano/microfabrication of "soft" materials such as polymers, biomolecules, and functional nanomaterials.

2.3.1 Silicon as a Material for Microfabrication

Silicon Crystal Structure

Single crystal silicon has a covalent bonded diamond cubic (zinc blende) structure as shown in Fig. 2.5. A silicon atom has four covalent bonds that create a tetrahedral organization in the crystal. The unit cell can be imagined as two face-centered cubic (FCC) cells with one cell offset by [a/4, a/4, a/4]. The offset is more clearly visualized in the top view.

Silicon has several important direction-dependent characteristics that come from the crystal structure. In order to specify lattice planes, Miller indices are used. A family of lattice planes is determined in the following way:

1. Find the *x*, *y*, and *z* intercepts of the plane: (*a*, *b*, *c*).
2. Take the reciprocals of each intercept: (1/*a*, 1/*b*, 1/*c*).
3. Multiply each reciprocal by the denominator of the smallest fraction until integer values are obtained: (*h*, *k*, *l*).

Miller indices can be used to denote planes and directions in the following way:

Plane:	(*h k l*)
Family of planes:	{*h k l*}
Direction:	[*h k l*]
Family of directions:	<*h k l*>

For example, {*h k l*} denotes the set of all planes equivalent to (*h k l*) by the symmetry of the lattice. Examples of planes in a cubic crystal lattice and corresponding Miller indices are shown in Fig. 2.6.

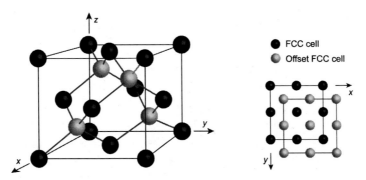

Figure 2.5: Crystal Structure of Single Silicon Crystal.

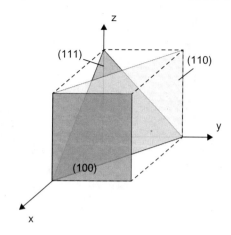

Figure 2.6: Examples of Miller Indices for Planes in a Cubic Crystal Lattice.

2.3.2 Photolithography

Photolithography is a process that uses light, a photomask, and a photosensitive material to facilitate the transfer of a specific pattern onto another layer on the substrate. Photolithography is commonly used because it allows for exact control over the shape of the structure, and the process is suitable for mass production. In principle, the process time does not depend on the number or complexity of the patterns.

Process of Photolithography

A typical process flow of photolithography is shown in Fig. 2.7. It includes the following steps:

1. **Film deposition:** The film to be patterned (film of metal, silicon oxide, silicon nitride, polymers, etc.) is prepared on a wafer by film deposition techniques including **spin coating**, **sputtering**, **evaporation**, and **oxidation**. The wafer surface itself may be patterned.
2. **Spin coating:** The film surface is coated with a very thin photo sensitive layer of polymer called **photoresist** by spin coating (see later). Spin coating is followed by soft baking, which drives away the solvent from the deposited photoresist. Soft baking is usually performed between 90−100 °C.
3. **Exposure:** The photoresist layer is exposed to a pattern of UV light defined by a **photomask**. The photomask is a glass or silica plate with a metal (often chromium) pattern that screens the exposure light. The photomask is usually patterned by **electron-beam lithography**. If the pattern needs to be aligned to already existing

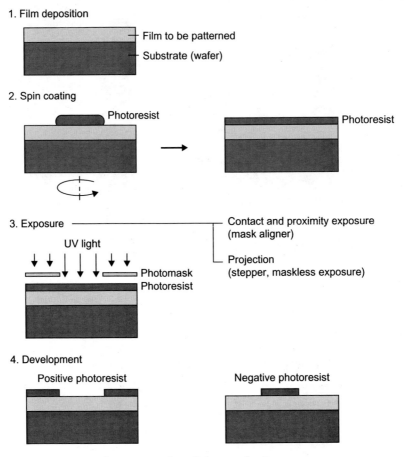

1. Film deposition
    Film to be patterned
    Substrate (wafer)

2. Spin coating
    Photoresist

    Photoresist

3. Exposure —————————————— Contact and proximity exposure
                                  (mask aligner)
    UV light

    Photomask   Projection
    Photoresist   (stepper, maskless exposure)

4. Development
    Positive photoresist   Negative photoresist

Figure 2.7: Photolithography Process.

patterns on the substrate, a **mask aligner** is used to align the mask and the substrate pattern. A mask aligner includes a microscope and mechanical stages that facilitate precise alignment between the photomask and the substrate.

4. **Development:** Part of the photoresist is then removed by a developing solution or a spray developer. A post-exposure bake (PEB) is often performed before development to amplify chemical reactions within the photoresist [1]. For a **positive photoresist**, the pattern left on the wafer is a positive image of the photomask (the exposed area will be removed). For a **negative photoresist**, the pattern is negative (the unexposed area will be removed). After development, **hard baking** is usually performed to complete the process. Hard baking strengthens the photoresist structure and improves adhesion of the resist to the wafer surface. It is typically performed at 120~180°C for 10~20 minutes.

Resolution of Photolithography

The resolution of photolithography is defined by the fineness of the light pattern imaged through the photomask. When the pattern is as fine as the wavelength of the exposure light, the fineness is subject to diffraction of light. Based on the way the diffraction pattern is formed on the photoresist film, exposure can be categorized into two types: (1) contact and proximity exposure and (2) projection exposure.

Contact and Proximity Exposure

Contact and proximity exposure is a method where light patterns are formed in the near field of the photomask and the patterns are described as **Fresnel diffraction** (see Appendix 5 A for details). The minimum resolvable feature size of the Fresnel diffraction pattern can be approximated as $\sqrt{\lambda d}$, where λ is the wavelength, $d = g + t$ is the summation of the photomask−photoresist separation, g, and the photoresist thickness, t. In **contact exposure**, the photomask mechanically contacts the photoresist, i.e. $d = t$. In **proximity exposure**, a small gap is left between the photomask and the wafer to avoid damaging the surface of the photoresist film. For a typical proximity exposure, the thickness of a photoresist film is $> \sim 1\,\mu m$, and the gap is $\sim 10\text{--}20\,\mu m$. Assuming the wavelength of UV light to be 400 nm, the resolution limit of typical proximity photolithography is $\sqrt{0.4 \cdot 20} \approx 3\,\mu m$. Figure 2.8 shows Fresnel diffraction patterns of a 5 μm-wide aperture at the wavelengths of 400 nm calculated for $d = 0.5\,\mu m$ (contact exposure) and $d = 2.5\,\mu m$ (proximity exposure).

Projection Exposure

For high volume production, projection exposure is more commonly utilized because it eliminates mechanical contact between the mask and wafers, reducing the risk of damages.

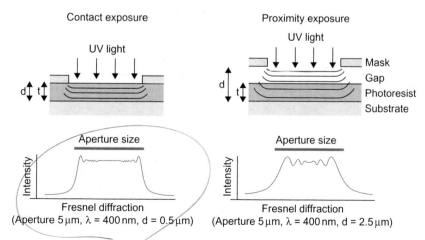

Figure 2.8: Resolution of Photolithography: Contact, and Proximity Exposure.

In projection exposure, an image of a light pattern is projected onto the wafer via projection optics. The image is usually formed centimeters away from the lens, namely the **far field**, and described as **Fraunhofer diffraction** (see Appendix 5 A for details). Fraunhofer diffraction patterns can be calculated by finding the **point spread function (PSF)** of the projecting optics. The PSF is the image of an ideal point light source. Any aperture on the mask can be considered as a combination (or an integration) of multiple point light sources, thus the image of the aperture can be found as the combination (or the integration) of PSFs along the aperture. The PSF of an optical system is defined by the numerical aperture (NA), the wavelength λ, and the refractive index n. The NA is defined with the refractive index and the angle θ made by the radius of the lens and the focal length f in the following way (see Fig. 2.9):

$$NA = \sin \theta \tag{2.1}$$

$$\sin \theta = \frac{a}{\sqrt{f^2 + a^2}} \approx \frac{a}{f} \tag{2.2}$$

The resolution of an ideal optical system is limited only by the diffraction of light, or **diffraction limited**. The PSF of a diffraction limited optical system is called the **Airy disc** or **Airy pattern**. The resolution is estimated as the distance δ between the center and the first zero of the airy disc, which is approximated as:

$$\delta = 0.61 \times \frac{\lambda}{n \cdot NA} \tag{2.3}$$

An example of the PSF calculated for wavelength $\lambda = 400$ nm, the refractive index $n = 1$, and NA $= 0.7$ are shown in Fig. 2.9.

Stepper is most commonly used for very large scale integration (VLSI) mass production. It projects optically reduced images of a photomask, or also called reticle, onto a wafer. A small portion of the wafer is exposed for each exposure, and the wafer is scanned stepwise

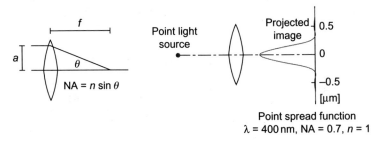

Figure 2.9: Numerical Aperture (Left) and Point Spread Function (Right).

for the entire surface to be exposed (Fig. 2.10(a)). Figure 2.10(b) shows another example of projection exposure using the digital mirror device (DMD) from Texas Instruments instead of a conventional photomask. It can be considered as a microscopic version of the projection system commonly used in movie theaters. An array of switchable micro mirrors directs the patterns of light onto a wafer. The PSF and the size of a single mirror define the resolution. An application of this method is Affymetrix gene chip described in Chapter 2, Section 2.5.3 and Chapter 3, Section 3.3.3. Other types of maskless exposure systems include laser spot scanned exposure systems.

As one can see in Eq. 2.3, shorter wavelengths are desirable for finer patterns. However, in general, when the wavelengths are shorter, the lights are more difficult to generate or handle. The light source most commonly available for photolithography is a mercury lamp, which has the emission spectrum with material specific peaks. The strongest peaks in the UV region are called g-line (436 nm), h-line (405 nm) and i-line (365 nm). For the much shorter wavelength region called deep UV, KrF (248 nm) or ArF (130 nm) excimer lasers are used. Another method to improve the resolution is increase the refractive index n. The index of water for UV light at wavelengths of ~ 200 nm is measured to be ~ 1.45 [2]. Advanced steppers allows exposure of wafers immersed in water, which is called **immersion lithography.** The achieved resolution, as of 2013, is in the range of ~ 30 nm.

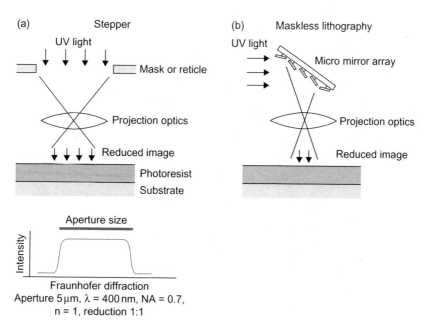

Figure 2.10: Projection Exposure. (a) Stepper. (b) Maskless Exposure.

Electron-beam (e-beam) lithography is a technique that provides better resolution than photolithography. E-beam lithography is the maskless, direct writing of patterns with an electromagnetically focused scanning e-beam. An e-beam sensitive resist is used for transfer of patterns. The wavelengths of an e-beam is as small as 12×10^{-12} m with an acceleration voltage of 10 kV (note that the wavelengths of a matter wave depends on its momentum, i.e. the acceleration voltage), and is not a practical limiting factor of resolution.

The primary limitation of e-beam lithography is the higher cost associated with the expensive apparatus and the low throughput. Precise control of e-beam must be made under high vacuum. Unlike photolithography, the process time of e-beam lithography is proportional to the area of exposure.

2.3.3 Deposition

Spin Coating

Spin coating is a technique used to spread uniform thin films on flat substrates by centrifugal force. The apparatus used for spin coating is called a spin coater, or a spinner. A solution of material is dispensed onto the center of a wafer, which is then rotated at a high speed.

Rotation is continued until the excess solution spins off the substrate and the desired thickness of the film is left on the substrate. The applied solvent is usually volatile and evaporates during deposition. The two main factors that define film thickness are the spin speed and the viscosity of the solution. Other factors that may need to be considered include spin time, solution density, solvent evaporation rate, and surface wettability. For a typical case of photoresist deposition (example: AZ3312, viscosity 18 cP), rotation of 2000−6000 rpm is used for 30−60 seconds to obtain a 1.5 μm ~ 1 μm thick film. Softbaking (a drying process at a relatively low baking temperature of ~100 °C) is usually performed after deposition to evaporate the residual solvent and anneal the mechanical stress. After the soft baking process, the photoresist coating becomes photosensitive. Under-soft baking may leave considerable amount of solvent which prevents exposure light from reacting the photoresist. On the other hand, over-soft baking may reduce the resolution of photoresists by either changing the developer solubility or damaging the resist.

Spin coating is one of the easiest ways to deposit various types of materials that are in solution form. Deposition of polymers, nanoparticles, and biomaterials is commonly performed via spin coating. Regardless of the materials that are deposited, spin coating usually requires a drying or baking process.

Thermal Oxidation

Thermal oxidation is the process of growing a thin layer of oxide on the surface of a wafer. Here we discuss thermal oxidation of silicon. Silicon oxide, or more specifically silicon dioxide (SiO_2), has excellent properties that makes silicon-based microfabrication very efficient. Such properties include:

- **High resistivity.** Silicon dioxide is a very good insulator with a resistivity more than 1^{20} Ωcm.
- **Etching selectivity.** Silicon dioxide is resistant to most etching solutions used for metals or silicon. In addition, it can be etched with HF, which does not affect Si.
- **Material.** Silicon dioxide is produced from silicon atoms of the wafer. The process can be clean, as it does not require any chemicals other than oxygen and water.

The process is usually performed at a temperature between 900 and 1100°C. It uses either dry oxygen (**dry oxidation**) or oxygen with water vapor (**wet oxidation**). For dry oxidation, the reaction is simply:

$$Si + O_2 \rightarrow SiO_2$$

For wet oxidation, the following reaction also occurs:

$$Si + 2\,H_2O \rightarrow SiO_2 + 2\,H_2$$

The most commonly used mathematical model to estimate oxide thickness is the Deal−Grove model [3], which is given by:

$$t = \frac{X^2 + AX}{B}$$

where t is the process time, X is the oxide thickness and A and B are the parameters determined for each oxidation condition (wet or dry oxidation, temperature, wafer types, etc). Very roughly speaking, the film thickness X is proportional to $\sqrt{t}$ (or $X \approx \sqrt{Bt}$) because it is related to the depth of oxide diffusion into silicon (see diffusion in Chapter 3, Section 3.2.1.). The oxidation speed is faster for a higher temperature, i.e. the value of B becomes smaller for higher temperatures.

Figure 2.11 shows oxide thickness for dry and wet thermal oxidation reported by Deal and Grove [3]. Wet oxidation is much faster than dry oxidation, but the obtained oxide films tend to be more porous. Dry oxidation should be used for key elements (such as gate oxide, see Chapter 3) that define device functionality.

Another important characteristic of a thermally grown oxide film is that it becomes thicker than the portion of the silicon wafer that is used to produce the oxide. The atomic and molecular densities of Si and SiO_2 are 5×10^{22}/cm^2 and 2.3×10^{22}/cm^2, respectively. This means a SiO_2 film is twice as thick as a Si film for the same amount of silicon atoms (The number of Si atoms per volume in Si is twice the number of SiO_2 molecules per volume in SiO_2).

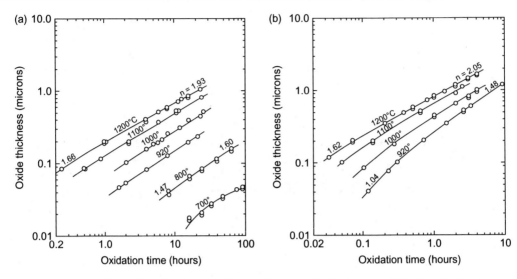

Figure 2.11: Thermal Oxidation of Silicon. (a) Wet Oxidation and (b) Dry Oxidation.
From [3].

Evaporation

Another method of deposition is **evaporation**, which is commonly used to deposit thin metal films. When a material (or source) is heated in a vacuum, it evaporates and releases vapor particles, which travel straight to the target substrate, on which they condense back into a solid state.

If traveling particles collide with background gas or other unknown particles, they may react with them to form undesirable films. Higher vacuum results in better quality films. Vacuum pressure of about 10^{-4} Pa (or 10^{-6} Torr) or lower is usually achieved before heating the source material. Evaporation is a type of deposition called physical vapor deposition (PVD), since the process ideally does not involve any chemical reactions.

There are two main types of evaporation apparatuses, **e-beam evaporation** and **resistive heat evaporation**, which are illustrated in Fig. 2.12. A quartz crystal microbalance (**QCM**)-based deposition monitor is commonly used for both types. The QCM sensor can measure the mass of the film deposited, which is correlated to the thickness (see Chapter 6, Section 6.3.3 for details on QCM sensors).

E-beam Evaporation

E-beam evaporation uses thermal emission of electrons (an electron beam, or e-beam) from a filament to heat samples. A magnetic field is used to bend the electron beam by an angle of 270° to eliminate direct contamination from the filament (Fig. 2.12(a)). Electron beams

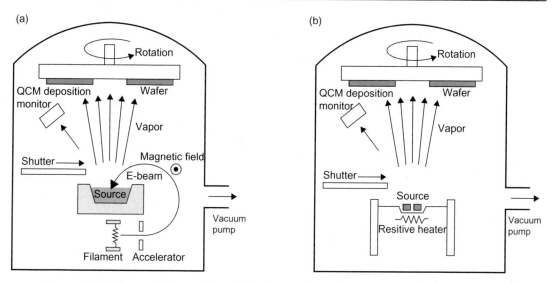

Figure 2.12: Thermal Evaporation Techniques. (a) E-beam Evaporation; (b) Resistive Evaporation.

Table 2.1: Materials and Deposition Rates Used in the Deposition Tool at the Microelectronics Research Center (MRC) at the University of Texas at Austin

Material	Ag	Al	Au	Cr	Ni
Rate (Å/s)	0.5−0.8	0.2−0.5	0.8−1.0	0.2−0.6	0.7−1.0

can be used for materials that require high temperatures for evaporation. They are also suitable for deposition of larger sample volumes.

Table 2.1 shows examples of materials and deposition rates used in the deposition tool at Microelectronics Research Center (MRC) at the University of Texas at Austin. Note that the values are just examples of typical values, and different values may be used depending on applications or apparatus.

Other materials that can be deposited with the method include: Co, Ge, Fe, Hf, In, Ir, Mo, Pd, Pt, Pb, Sn, Si, Ti, Ta, W, Zn, Ni/Cr, Al_2O_3, In_2O_3, SnO, SiO_2.

Resistive Heat (Joule Heat) Evaporation

The other type of evaporation utilizes Joule heat induced by a current passing through a filament or a metal plate (called an evaporation boat) which holds the material to be evaporated (Fig. 2.12(b)). Because the composition of the tool is simple, this method is often used for low cost or small sample volume deposition systems. Metals with relatively lower melting points, such as Al, Au, Ag, Cu, and Ti, are usually deposited with this method. It is also suitable for deposition of organic materials [4,5] which require a "mild" heat process rather than more intense e-beam heating.

Problems Associated with Evaporation

Since the evaporation deposition is performed at a high temperature, when the wafer is brought back to a room temperature, internal stress is induced due to the difference between thermal expansion coefficients of the wafer and the deposited material. This stress, or the mismatch at the interface, could bend the wafer (see Chapter 6 Section 6.2.2 for Stoney's formula and problem 6.13 for thermally introduced stress), or let the deposited film peels off the substrate by itself. For this reason, evaporation is not suitable for depositing a very thick layer. For deposition of metal layers thicker than a few micrometers, electroplating may be used.

Sputtering

Sputtering is also categorized as PVD. In sputtering, atoms are ejected from a solid target material due to plasma bombardment. Materials that are deposited by sputtering include metals, metal alloys, metal oxide, metal nitride, silicon, silicon oxide, and silicon nitride.

The ejected material ballistically travels from the target onto the substrate. In sputtering deposition, the "target" is the target of bombardment and is the source of the material to be deposited (see Fig. 2.13).

In a typical process, the chamber is first evacuated and the sputter gas is then introduced. A DC or RF voltage is applied between the target and the substrate to establish a plasma. An inert gas such as argon is usually used as the sputter gas, but reactive gases such as oxygen can be also mixed with the sputter gas. When reactive gases are introduced, the deposition involves chemical reactions. For example, if a metal is sputtered with oxygen, deposition of

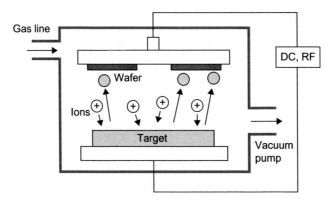

Figure 2.13: Sputtering.

a metal oxide film may occur. Co-sputtering is a method where multiple targets are used simultaneously to allow the deposition of a variety of mixed materials.

Chemical Vapor Deposition

Chemical vapor deposition (CVD) is a process for thin film deposition that includes chemical reaction of volatile precursors. The term CVD covers a large variety of methods for different types of materials. Materials that are deposited by CVD include polysilicon, silicon oxide, silicon nitride, carbon nanotubes, graphene, diamond, and polymers. Most of the CVD processes for microfabrication are performed in a vacuum chamber where the rate and purity of chemical reaction can be precisely regulated via temperature and pressure control. In contrast with PVD processes, which tend to involve vapor particles with directional larger momentum, CVD processes usually provide conformal, multidirectional deposition.

Plasma-enhanced CVD (PECVD) is a process that utilizes plasma to enhance chemical reaction. PECVD processes allow film deposition at lower temperatures, which is crucial in processing materials with lower temperature resistance. The electrodes and an RF power source similar to the ones used for sputtering or plasma etching are employed in a PECVD chamber. In fact, many commercial apparatuses for PECVD allows plasma-based dry etching (see also Section 2.3.4) in the same chamber without breaking vacuum.

Here are some important examples of chemical reactions used for CVD.

Polysilicon

Polysilicon, poly-Si, or polycrystalline silicon, is a material consisting of a number of smaller crystals or crystallites. Deposition of polysilicon is one of the most important CVD processes used in microfabrication. Polysilicon is composed from trichlorosilane ($SiHCl_3$) or silane (SiH_4) through the following reactions:

$$SiHCl_3 \rightarrow Si + H_2 + HCl$$
$$SiH_4 \rightarrow Si + 2 \cdot H_2$$

Polysilicon is usually deposited by LPCVD. It can be also deposited by PECVD, or crystallization (SPC) of amorphous silicon, either of which requires relatively high temperatures of at least 300°C.

Amorphous silicon

Amorphous silicon (a-Si) is a non-crystalline form of silicon which can be deposited in thin films by PECVD at a very low temperature, such as 75°C. The low process temperature makes it possible to deposit a-Si films on large areas of glass or plastic substrates. A-Si is often used for the active layer in thin-film transistors (TFTs) for liquid-crystal displays

(LCDs). The largest issue is that due to the disordered nature of the material, the field effect mobility of a-Si is orders of magnitude smaller than that of poly-Si.

Silicon Dioxide

Silicon dioxide can be deposited by different processes. Commonly used sources are silane (SiH_4) and oxygen or dichlorosilane ($SiCl_2H_2$) and nitrous oxide (N_2O) in the following reactions:

$$SiH_4 + O_2 \rightarrow SiO_2 + 2 \cdot H_2$$
$$SiCl_2H_2 + 2 \cdot N_2O \rightarrow SiO_2 + 2N_2 + 2HCl$$

Silicon Nitride

Silicon nitride is often used as a passivation layer for integrated circuits or micro mechanical structures. Compared to silicon dioxide, silicon nitride has a significantly better diffusion resistant against water and sodium. Silicon nitride is also used for a dielectric material for microelectronic, or photonic applications. The most common form of silicon nitride for microfabrication is Si_3N_4, which is typically formed by either LPCVD or PECVD based on the following reactions.

$$3\ SiH_4(g) + 4\ NH_3(g) \rightarrow Si_3N_4(s) + 12\ H_2(g)$$
$$3\ SiCl_4(g) + 4\ NH_3(g) \rightarrow Si_3N_4(s) + 12\ HCl(g)$$
$$3\ SiCl_2H_2(g) + 4\ NH_3(g) \rightarrow Si_3N_4(s) + 6\ HCl(g) + 6\ H_2(g)$$

Electroplating

Electroplating is an electrodeposition process to produce a layer of metal, and has been used for various purposes [6]. For nano/microfabrication, it is often used to deposit thicker (typically $>5\ \mu m$) metal films, since It is difficult to obtain films thicker than a few micrometers with evaporation or sputtering, because of the internal thermal stress induced during deposition (see Chapter 6, Section 6.2.2 for Stoney's formula and problem 6.13 for thermally introduced stress). Electroplating can be performed at room temperature, and the effect of thermal stress is much smaller than other methods.

One application of electroplating is to create elements that conduct larger current. An example for such an element is a microelectrical coils [7,8]. Another is to form a thicker structure that will demonstrate larger magnetic force. They may be used for magnetic force based actuators [9]. It is also used to construct a metallic mold for plastic injection molding [10].

Several materials have been used for electroplating. Table 2.2 lists some examples showing different materials used for specific purposes. For microfabrication, Cu, Ni, and Ni-Fe are among the most commonly utilized materials.

Table 2.2: Principles of Electroplating

Electroplating	Material	Application
Microfabrication	Cu	Deposition of thick conductive metal layers
Microfabrication	Ni, Ni-Fe	Deposition of magnetic structural layers, metallic mold
Engineering/functional coatings	Au, Ag, Pb, Ru, Rh, Pd, Os, Ir, and Pt	Enhancing specific properties of the surface (e.g. solderability)
Decorative/protective coatings	Au, Ag, Cu, Ni, Cr, Zn, and Sn	Attractive appearance
Sacrificial coatings	Zn and Cd	Protection of the base metal
Alloying coating	Ni-Fe, Au-Cu-Cd, and Cu-Sn	Multiple purposes

Chemical Reaction

The process is the deposition of a metallic film onto a wafer which act as a cathode (negative electrode) in an aqueous solution of a salt of metal to be deposited. The electrochemical reaction between the metal M and its ions in the aqueous solution can be represented in simple formulae as below.

$$M^{z+}(aq) + ze^- \rightarrow M(s)$$
$$M(s) \rightarrow M^{z+}(aq) + ze^-$$

where M^{z+} is the metal ion and z is the number of electrons involved in the reaction. The amount of metal deposited is defined by the current applied to the electrode. Figure 2.14 shows an electrolytic cell used for electroplating.

The chemical reaction of copper, for example, is as follows.

$$\text{Cathode:} \quad Cu^{2+}(Aq) + 2e^- \rightarrow Cu(s)$$
$$\text{Anode:} \quad Cu(s) \rightarrow Cu^{2+}(Aq) + 2e^-$$

Factors need to be considered to obtain desired film deposition include current density, electroplating time, and temperatures. Current efficiency is ratio of the chemical change to the total chemical change that can be expressed as follows:

$$\text{Current efficiency} = 100 \times W_{\text{Act}}/W_{\text{Theo}}$$

where W_{Act} is the weight of metal actually deposited, and W_{Theo} is the corresponding weight expected from Faraday's laws. In addition, current density is defined as current in amperes per unit area of the electrode.

An example of conditions for the copper electroplating with an electroplating solution containing 200g/L of $CuSO_4 \cdot 5H_2O$ and 43mL/L of 98% H_2SO_4 is shown in Table 2.3.

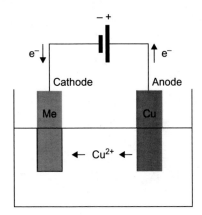

Figure 2.14: Electroplating Cell.

Table 2.3: Conditions for Copper Electroplating

Temperature	$25\,°C$
Current density	$20\ mA/cm^2$
Electroplating rate	$0.34\ \mu m/min$

Fabrication of Electroplated Microstructures

Utilizing electroplating and photolithography technologies, metallic microstructures can be fabricated for several applications, such as fabrication of an electromagnet (microcoil). Figure 2.15 shows the fabrication process for a copper micro coil. The process is performed in the following steps:

a. Etching plating molds into silicon wafers
b. Wafer oxidation and seed layer deposition
c. Patterning seed later, leaving Cu/Cr on mold bottom only
d. Electroplating Cu from mold bottom up, forming Cu coils in silicon wafer
e. Electroplating Cu above the wafer surface
f. Polishing to level Cu lines with wafer surface
g. Removing silicon sidewalls between Cu lines.

2.3.4 Etching

Etching is a process that transfers a two dimensional pattern onto a structural layer underneath the top masking layer. In a typical process, the masking layer is a photoresist pattern. Any type of film that is resistant to the etching process could also be used as the

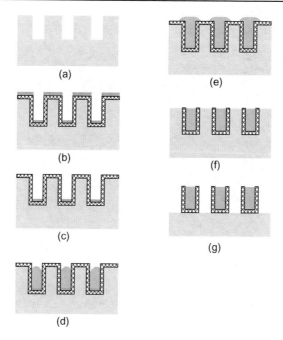

Figure 2.15: Fabrication Process of Electroplated Copper Coil [11]. (a) Etching of Plating Molds. (b) Seed Layer Deposition. (c) Patterning of Seed Later. (d) Electroplating of Cu Forming Cu Coils in Silicon Wafer. (e) Electroplating Cu Above the Wafer Surface. (f) Polishing to Level Cu. (g) Removing Silicon Sidewalls.

masking material. Silicon dioxide, silicon nitride, metal films are often used. Most etching processes in microfabrication are categorized as either wet etching or dry etching.

Wet Etching

Wet etching is a patterning process that utilizes a chemical solution, or an **etchant**, to cut or "etch" metals. It is based on the same idea as that used for old master prints. The film to be patterned is masked by another layer of patterns which is resistant to the etchant (Fig. 2.16(a)). The masking layer often is a patterned photoresist film, and the layer to be patterned is a film of a metal such as aluminum or copper. Fig. 2.16(a) shows a case where a substrate of a thick material is being etched, while Fig. 2.16(b) is a case where a film deposited on a substrate is being patterned.

Photoresist may not provide sufficient protection if etching time is very long, or if the etchant is a strong base solution which can easily damage the photoresist. An example of such a scenario is wet etching of silicon, where silicon dioxide is used as the masking layer, and potassium hydroxide (KOH) or tetramethylammonium hydroxide (TMAH) is used as

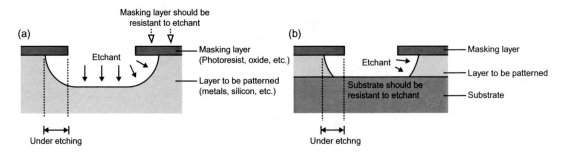

Figure 2.16: Wet Etching (Side View). (a) Etching of a Substrate or a Thick Material. (b) Etching of a Film Deposited on a Substrate.

the etchant. Photoresist layers are easily removed by these solutions, and do not function as a masking layer. In this case, the process becomes a multi-step etching: the silicon dioxide layer is first patterned using a photoresist as the masking layer and HF as the etchant. The silicon layer is then etched using silicon dioxide as the mask, which is highly resistant to KOH or TMAH.

Selectivity

The term **selectivity** is used to express the ratio of etch rates in different materials. We have to consider the etch rates of three materials: namely, the masking layer, the layer to be etched, and the bottom material. The masking layer must be resistant to the etchant, while the layer to be patterned should be etched as efficiently as possible to avoid damages to other layers. The higher etch rate for the layer to be patterned and the slower for the masking layer are desirable. Photoresists are resistant to acids, but are slowly damaged. The main structural layer may be damaged, once a pinhole is created on the photoresist masking layer. The bottom surface is often silicon or silicon dioxide, which is not affected by most of acids.

Isotropic Wet Etching

Isotropic wet etching is a process based on chemical reactions which do not have directional dependence. Most metal wet etching processes are considered isotropic as shown in Fig. 2.16(a).

The minimum feature size, (i.e. the fineness of the final outcome), largely depends on the thickness of the layer to be patterned. In general, thinner films provide finer patterning of features. One reason is that etching of thicker films requires thicker masking layers that can endure the longer processing time. As we described earlier, a thicker photoresist results in poorer resolution in terms of feature size. More importantly, isotropic or non-directional chemical reaction causes underetching or undercutting of materials (see Figs 2.16(a) and (b)). Patterning of thicker layers tends to result in longer process times and more underetching.

Silicon Anisotropic Wet Etching

Some crystalline materials are etched at very different rates depending on crystal faces that are exposed. Silicon wafer is a single-crystal material and allows very high anisotropic etching. This anisotropy results in a cavity with a trapezoidal cross-section.

Potassium hydroxide (KOH) or tetramethylammonium hydroxide (TMAH) is usually used as the etchant. Silicon oxide or nitride is used for the masking layer.

KOH shows a faster etch rate and a higher anisotropy. The etch rate of a 30% KOH solution at 70 °C in $<1\,0\,0>$ directions is 0.8 µm/min, which is more than 150 times higher than in $<1\,1\,1>$ directions [12], while the $<1\,0\,0>/<1\,1\,1>$ etch ratio of TMAH is $\sim 10-50$ [13]. One issue is that KOH etches silicon dioxide slowly (a few nanometers per minute), which may not be negligible for some applications. For deeper etching (>100 µm), silicon nitride should be used as the masking material with KOH etching. TMAH has good etching selectivity between silicon and silicon dioxide. Thermally grown oxide can be used as a mask when etching as deep as the thickness of a silicon wafer (~ 500 µm).

Figure 2.17(a) shows a typical anisotropic etching process with a (1 0 0) wafer. Etching proceeds in the $<1\,0\,0>$ direction and is guided by the (1 1 1) surface on all four sides. During the etching process, the (1 0 0) surface will shrink as the four (1 1 1) surfaces meet at the apex of the inverse pyramid structure. Etching in $<1\,1\,1>$ direction is much slower, but still appears and is observed as undercut. Considering the angle between a (1 0 0) surface and a (1 1 1) surface, which is 54.7° (see Problem 2.8), the shape of the etched cavity can be roughly estimated (see Fig. 2.17(b)). Etching of a (1 1 0) wafer is a little more complicated than that of a (1 0 0) wafer (see Fig. 2.17(c)). Several studies have been made to analyze general cases of silicon anisotropic etching [14–17]. Numerical analysis provides good prediction for the shapes of etched wafers [18–20].

Dry Etching

Dry etching, or **plasma etching**, is an etching process that utilizes free radicals produced by plasma. Dry etching is preferred in modern VLSI processes because it can be more precisely controlled by adjusting parameters such as gas pressure, temperature, and electric field distribution. It is performed in a vacuum chamber equipped with gas lines, pressure and temperature controllers, electrodes and power sources. The system often resembles the ones designed for sputtering or PECVD.

Reactive ion etching (RIE) (Fig. 2.18) is a type of plasma etching that utilizes combinations of isotropic chemical reactions and anisotropic or directional physical

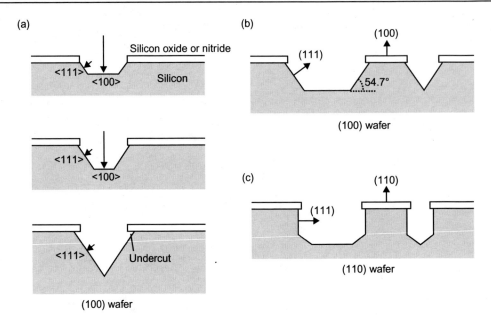

Figure 2.17: Anisotropic Wet Etching of Silicon (Side View). (a) Etching Processds of a (100) Wafer. (b) Because of the Anisotropy of the Etch Rate, (111) Surface Appears with an Angle of 54.7°. (c) Etching of a (110) Wafer.

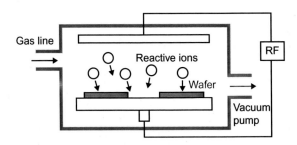

Figure 2.18: Reactive Ion Etching (RIE).

bombardment of ions. Most plasma etching processes used in microfabrication are categorized as RIE, since they usually utilize chemical reactions which enhance the etching rate and material selectivity. Removing atoms purely by physical bombardment of an inert gas such as Ar is specifically called **ion milling** or **sputter etching**.

Unless specifically modified, plasma etching involves unidirectional etching. High aspect ratio plasma etching is crucial for realization of thick and fine structures. Etching conditions, including types of reactive gases, pressures, temperatures, and RF power and DC bias, have been thoroughly studied to fabricate such deeper structures.

Deep Reactive-Ion Etching (DRIE)

Deep reactive-ion etching (DRIE) is a type of plasma etching which allows creation of deep and steep-sided holes and trenches. The term DRIE is often specifically used to describe the silicon etching process invented by Robert Bosch GmbH (the Bosch process) [21].

The Bosch process is a time-multiplexed etching that repeatedly alternates between two modes to achieve holes with nearly vertical sidewalls. The process typically consists of the following two steps:

1. **Etching**. The process is basically a standard RIE process that usually uses SF_6 as the reactive gas.
2. **Deposition**. This process deposits a thin passivation layer of fluoride-based substance similar to Teflon.

Optimized combination of these two steps allows protection of sidewalls as well as deeper etching into the bottom of the holes. As a result of the two steps, the Bosch process usually creates undulating sidewalls. The aspect ratios of the trenches or beams patterned with DRIE can be greater than $50 \sim 100$. Figure 2.19 shows SEM photographs of a micromachined ring gyroscope [22].

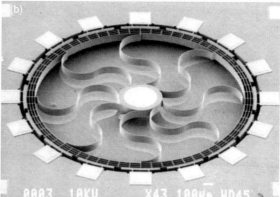

Figure 2.19: High Aspect Ratio Structure Patterned by DRIE [22].

2.3.5 Lift Off

Lift-off is a process of creating patterns by using an inversely patterned sacrificial layer first created on the substrate. A typical process flows in the following steps:

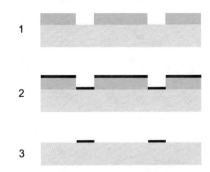

Figure 2.20: Lift off Process. 1. Patterning of a Sacrificial Layer. 2. Deposition of a Metal Film. 3. Remove Unwanted Metal Parts

1. The inverse sacrificial layer is deposited and patterned.
2. A thin film of the target material is deposited.
3. The sacrificial layer is removed, leaving a desired pattern of the target material.

A simplified schematic of the process is shown in Fig. 2.20.

Sacrificial Layer

There are several requirements for the material of a sacrificial layer:

- It has to be easily patterned
- The film has to be thick enough to "cut" the layer of target material into the desired pattern
- It has to be easily removed without damaging the desired pattern.

A layer of a patterned photoresist or an e-beam resist is usually used as the sacrificial layer. After deposition of the target material, the sacrificial layer is removed by sonication in an organic solvent such as acetone. A strong oxidizing agent such as piranha (mixture of H_2SO_4 and H_2O_2) may be used when the target material is gold or chromium.

Final Layer

In order to efficiently remove the sacrificial layer and the unwanted part of the target material, the final pattern layer has to be thinner than the sacrificial layer. For that reason, the lift off process is usually used to pattern layers thinner than ~ 200 nm.

2.3.6 Doping

Ion doping is an important process that defines the electrical characteristics of semiconductor materials. **Doping** is a method that intentionally introduces impurities called

dopants into an intrinsic semiconductor. Details of the electrical properties of doped semiconductors are described in Chapter 4. Here we describe the techniques used to dope semiconductors with dopants.

Thermal Diffusion

Thermal diffusion is a process that diffuses impurity atoms into intrinsic materials at high temperatures of 800–1000 °C. A patterned layer of silicon oxide (SiO_2) is used as a mask against thermal diffusion. Impurity atoms are selectively introduced in the area of open windows on the masking layer.

Atoms used for n-type doping of silicon are phosphorus, arsenic, antimony, and bismuth (group V elements). Atoms used for p-type doping are boron, aluminum, and gallium (group III elements). Boron and phosphorus are the most commonly used dopants. The dopant atoms can be introduced in solid, gaseous or liquid form. The following are materials used for boron and phosphorus doping:

	Boron	Phosphorus
Solid	B_2O_3, BN	P_2O_5
Gas	BCl_3	PH_3
Liquid	BBr_3	$POCl_3$

The diffusion process is mathematically described by Fick's first law. Details are described in Section 3.2.1.

Ion Implantation

In **ion implantation**, dopant ions are accelerated in an electrical field and implanted into a solid. Ions of boron, phosphorus or arsenic created from a gas source are usually used as a dopant. Figure 2.21 shows a schematic of a typical ion implanter.

Ions of the desired element are initially formed from the source atoms. The ions are then electrostatically accelerated through an ion beam analyzer, which allows only desired ions to pass through the slit. The analyzer comprises a magnet and utilizes the fact that a charged particle in a magnetic field undergoes a rotational motion with a radius defined by the mass, the charge, the velocity of the particle, and the magnetic field intensity.

The ion beam is then shaped by a series of electrostatic and magnetic lens to scan the surface of the substrate.

Advantages of using ion implantation include the precise control over implantation depth and amount of doped ions used. The implantation depth is related to the acceleration voltage, which is typically 10–100 keV, while the amount of ions (called the **dose**) can be monitored by measuring the ion current. Patterned silicon dioxide layer is used as the

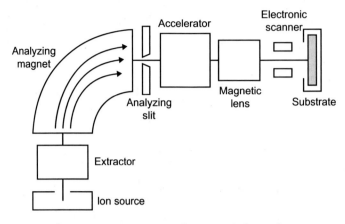

Figure 2.21: Apparatus for Ion Implantation.

masking layer. Ion implantation must be followed by a high temperature process called **annealing**, which reconstructs a crystal lattice of silicon and the implanted atoms.

Summary of Microfabrication Techniques

- Lithography
  - UV-photolithography (contact and proximity exposure, projection expoposure)
  - E-beam lithography
- Film deposition
  - Spin coating
  - Thermal oxidation
  - PVD (evaporation, sputter)
  - CVD (LPCVD, PECVD)
  - Electroplating
- Etching
  - Wet etching (isotropic etching, anisotropic etching)
  - Dry etching (plasma etching, RIE, DRIE)
- Doping
  - Ion implantation
  - Thermal diffusion

2.4 Micro-electro-mechanical Systems (MEMS)

In this section, we describe the basics of MEMS in the classical meaning; i.e. microscale mechanical and electrical components designed and fabricated based on the silicon microfabrication technology. Although the MEMS processes are becoming more and more complicated, they are roughly divided into two main methods: **bulk micromachining** and **surface micromachining**. We explain these two methods by introducing pressure sensors and electrostatic actuators, which are two basic but most important examples of micromachined devices.

Fabrication techniques for wider varieties of materials including polymers and biomaterials will be discussed in Section 2.5.

2.4.1 Bulk Micromachining

Piezoresistivity

The **piezoresistive effect** in silicon and germanium was first reported in 1954 [23], which led to the development of high-sensitivity silicon-based strain gauges.

The piezoresitive effect is the change in the electrical resistivity of a semiconductor on application of mechanical stress. It occurs because the mobility of majority carrier (either electrons or holes. See Chapter 4, Section 4.3.1 for semiconductor basics.) is affected by mechanical stress (Note that if the surface is stretched, the stress is positive. If compressed, it is negative.).

- In p-type Si, hole mobility decreases: The resistivity R increases.
- In n-type Si, electron mobility increases: The resistivity R decreases.

Sensors that measure strains based on the change in resistance are called **strain gauges**. Other types of strain gauges include metal wire-based gauges, which are also very commonly used. Metal wire-based gauges change resistivity simply due to physical changes in dimensions.

One problem associated with semiconductor strain gauges are the fact that they are sensitive to temperature change. As we will discuss in Chapter 4, Carrier mobilities of semiconductors are highly dependent on temperature, which means resistivity changes are dependent on temperature. One solution is to utilize the fact that temperature dependence of piezoresistive coefficients becomes small at high doping with sacrificed sensitivity. Another is to integrate temperature compensation circuitry, which is used in most commercially available pressure sensors.

The level of thermal noise defines the theoretical sensitivity limit of a semiconductor based strain gauges, or most types of semiconductor based sensors. At high temperatures, the noise levels become higher (i.e. the sensitivity becomes lower). Details of thermal noise and sensor sensitivities are described in Chapter 1, Section 1.4.2.

Pressure Sensor

The discovery of piezoresistivity in silicon made a large impact on the development of microsensors, because it showed the possibility of monolithic integration of mechanical structures (membranes or cantilevers), sensing elements (piezoresistors) and processing circuitries (electrical wiring, amplifiers, and other processing circuitry) into a single silicon chip. Figure 2.22 shows the basic structure of a silicon-based pressure sensor, which is the most typical application of MEMS-based sensors. A membrane is created on the silicon substrate by anisotropic wet-etching or DRIE. The sensor should be integrated in a way that the membrane deforms due to the difference between two pressures, P_1 of chamber 1 and P_2 of chamber 2. In many cases, either chamber 1 or 2 holds a vacuum or atmospheric pressure that serves as a reference. In this example, the main structure, including the membrane and the sensors, are made from a part of the silicon substrate. The microfabrication process that integrates main structures into the etched substrate is called **bulk micromachining**. Fabrication of a silicon membrane-based pressure sensor is the earliest and most important example of bulk micromachining.

Sensing areas are created on top of the substrate by doping of impurities. Upon application of pressure, the membrane surface is either stretched or compressed. The resistivity of sensing elements changes due to the piezoresistive effect. The doped areas are designed to be longer in the direction of deformation. Serpentine designs may be also used to obtain a large effective length of the sensing element. Serpentine shapes are very common for metal-based strain gauges as well. For the details of piezoresitivity-based force/pressure sensing, see Chapter 6.

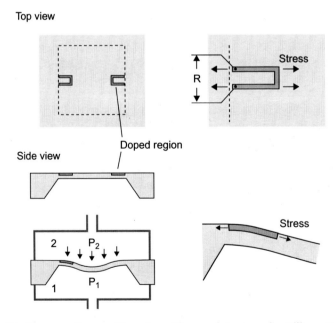

Figure 2.22: Pressure Sensor Based on Piezoresistors and a Silicon Membrane.

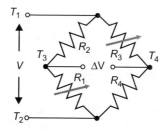

Figure 2.23: Wheatstone Bridge.

Most piezoresistive sensors use **Wheatstone bridges** as shown in Fig. 2.23. In this example, a fixed, known voltage is applied between terminals T_1 and T_2. The changes in resistances R_3 and R_4 can be measured as the change in voltage between terminals T_3 and T_4.

The change in the voltage is given by following:

$$\Delta V/V = (R_1 R_3 - R_2 R_4)/[(R_1 + R_2)(R_3 + R_4)]$$

Usually ΔV is very small and amplified by a differential amplifier such as an instrumentation amplifier. For the details of mechanical strain and the change in the resistance, see Chapter 6, Section 6.2.5.

2.4.2 Surface Micromachining

Surface Micromachining

Surface micromachining is a method that creates microstructures by the etching of thin layers that have been deposited on the surface of a wafer. The most important technique in surface micromachining is the introduction of the sacrificial layer, which allows fabrication of suspended, or free-standing, movable structures. Polysilicon and silicon oxide are usually used for the structural layer and the sacrificial layer, respectively. The key to choosing appropriate materials is the etching selectivity. The sacrificial layer has to be easily removed by a process that does not affect the structural layer.

A simple fabrication process for a surface micromachined cantilever is shown in Fig. 2.24. The process flows in the following way:

1. A sacrificial layer of oxide and a structural layer of polysilicon are deposited on the surface of the wafer by CVD. Thermal oxidation may be used to deposit the oxide layer, and **wafer bonding** may be used to obtain a thicker structural layer. The

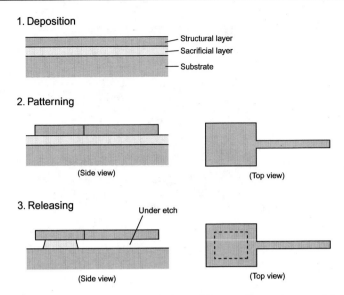

1. Deposition
— Structural layer
— Sacrificial layer
— Substrate

2. Patterning
(Side view)
(Top view)

3. Releasing
Under etch
(Side view)
(Top view)

Figure 2.24: Microcantilever Fabricated by Surface Micromachining.

thickness of the structural layer can range from ~ 100 nm to $\sim 100\ \mu$m, but typically $5 \sim 10\ \mu$m. A high-temperature treatment called **annealing** is usually performed to reduce internal stress which causes unwanted deformation of the structure after releasing.

2. The polysilicon layer is then patterned using RIE to form a beam with an anchor pad.
3. The wafer is then wet etched to remove the oxide layer under the beam. The etching has to be stopped before it removes all the oxide from under the larger anchor pad, while the smaller parts are already released from the substrate.

In this scenario, finding a proper etching time is very difficult, because the etching process usually does not proceed uniformly on a wafer. It is possible that in some areas, structures that have to be released are still fixed on the substrate, while some of the parts that should be anchored are completely released and separated from the substrate.

An improvement to the above process is to create small through holes on the sacrificial oxide layer in the areas of anchor pads before the deposition of the structural polysilicon layer (see Fig. 2.25). In this way, the top polysilicon layer is directly connected to the bottom substrate and will not be detached by the releasing process. Additional photomask for the anchoring holes and mask alignment for multiple masks will be needed. Fabrication of multilayered, complicated structures is possible based on the method [24]. Figure 2.26 shows a multilayered anchored microstructure fabricated at Sandia National Laboratories.

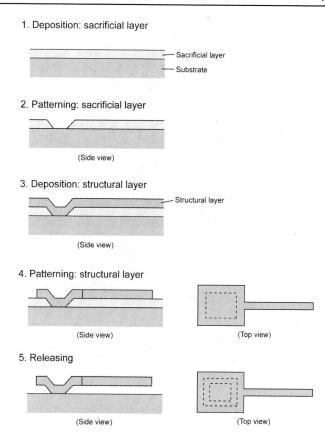

1. Deposition: sacrificial layer

— Sacrificial layer
— Substrate

2. Patterning: sacrificial layer

(Side view)

3. Deposition: structural layer

— Structural layer

(Side view)

4. Patterning: structural layer

(Side view)   (Top view)

5. Releasing

(Side view)   (Top view)

Figure 2.25: Fabrication of an Anchored Microstructure.

Electrostatic Actuators

The simplest and most important form of an electrostatic actuator (Fig. 2.27) is a parallel plate capacitor. It can be easily constructed by surface micromachining, and is a fundamental building block for many MEMS electrostatic devices.

Let us consider a parallel plate capacitor with the dimensions (area $S = a \times b$, gap x) shown in Figure 2.27(a). The capacitance is:

$$C = \varepsilon_0 \frac{S}{x},$$

(2.4)

where ε_0 is vacuum permittivity.

The charge on the capacitor at an applied voltage v is:

$$q(v) = Cv$$

(2.5)

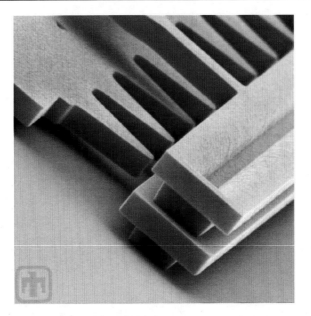

Figure 2.26: Multilayered Anchored Microstructures.
Image from (http://mems.sandia.gov/gallery/images_linear_racks.html). Courtesy of Sandia National Laboratories, SUMMiT™ Technologies, www.mems.sandia.gov

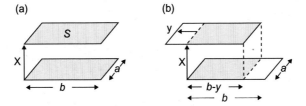

Figure 2.27: Electrostatic Actuator Based on a Parallel Plate Capacitor. (a) Force in the x Direction and (b) Force in the y Direction.

The energy stored in the capacitor at an applied voltage $v = V$ is

$$U = \int_0^{q(V)} v \cdot \mathrm{d}q = \int_0^{q(V)} \frac{q}{C} \cdot \mathrm{d}q = \frac{1}{2C} q(V)^2 = \frac{1}{2} CV^2 \qquad (2.6)$$

From Eqs 2.4 and 2.6,

$$U = \frac{1}{2} \varepsilon_0 \frac{S}{x} V^2 \qquad (2.7)$$

When one of the two plates is allowed to move, the capacitor works as an actuator. The force vector needed to counter the electrostatic force is given as

$$\mathbf{F} = -\Delta U \tag{2.8}$$

or each component of the vector is given by:

$$F_u = -\frac{\partial U}{\partial u} = -\frac{\partial C}{\partial u} V^2 \tag{2.9}$$

The direction of the force to be considered depends on the design of the actuator. Force in the x-direction is

$$F_x = -\frac{\partial U}{\partial x} = \frac{\varepsilon_0 S}{2x^2} V^2 \tag{2.10}$$

If we consider the motion in the y-direction (see Fig. 2.27(b)), the overlapping area is given as $S = a(b - y)$. The force in the y-direction is:

$$F_y = \frac{\partial U}{\partial y} = -\frac{\partial}{\partial y} \frac{\varepsilon_0 a(b-y)}{2x} V^2 = \frac{\varepsilon_0 a}{2x} V^2 \tag{2.11}$$

Note that Eqs 2.10 and 2.11 are given as the forces needed to counter the electrostatic force. *The electrostatic forces for this model are always attractive.* Most MEMS electrostatic actuators are designed based on attractive forces described as either Eqs 2.10 or 2.11. Repulsive electrostatic forces do exist, of course, but utilization of repulsive forces [25,26] is not as straightforward as this simple model.

Pull-in Limit

In one practical case, one plate is suspended by a spring with a spring constant k, and moves only in the x-direction. The other plate is fixed (see Fig. 2.28(a)). The initial gap between the two plates is x_0. The force F_E needed to counter the electrostatic force is

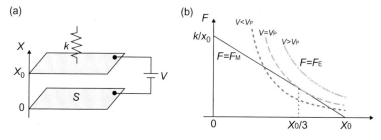

Figure 2.28: (a) Simple Model of an Electrostatic Actuator. (b) Diagram Showing the Pull-in Limit.

proportional to the square inverse of the gap x, while the restoring mechanical force F_M from the spring increases linearly.

$$F_E = \frac{\varepsilon_0 S}{2x^2} V^2 \tag{2.12}$$

$$F_M = k(x_0 - x) \tag{2.13}$$

Above a critical "pull-in" limit, the electrostatic force dominates and moving plate will be forced to hit the bottom plate. At the stability limit ($x = x_P$, $V = V_P$), the forces and the gradients are equal:

$$F_M|_{x=xP} = F_E|_{x=xP} \tag{2.14}$$

$$\frac{\partial}{\partial x} F_M \bigg|_{x=x_p} = \frac{\partial}{\partial x} F_E \bigg|_{x=x_p} \tag{2.15}$$

Plug Eqs 2.12 and 2.13 into Eqs 2.14 and 2.15,

$$\frac{\varepsilon_0 S}{2x_P^2} V_P^2 = k(x_0 - x_P) \tag{2.16}$$

$$-\frac{\varepsilon_0 S}{x_P^3} V_P^2 = -k \tag{2.17}$$

Solve Eqs 2.16 and 2.17 for x_P and V_P, to obtain

$$V_P = \frac{2}{3} \sqrt{\frac{2kx_0^3}{3\varepsilon_0 S}} \tag{2.18}$$

$$x_P = \frac{2}{3} x_0 \tag{2.19}$$

When voltage V is applied, there are three cases depending on the relationship between V and V_p. The three cases are summarized in Fig. 2.28(b).

1. $V < V_p$
   There are two solutions for $F_M = F_E$. One of them gives a stable solution to define the position of the actuator (see Problem 2.10).
2. $V = V_p$
   There is one solution for $F_M = F_E$, which is not stable. The actuator will be pulled in.
3. $V > V_p$
   There is no solutions for $F_M = F_E$. The actuator will be pulled in.

Nonlinear Effect

As given in Eq. 2.9, the electrostatic force is not linearly proportional to the applied voltage V, but to V^2. We need to take the nonlinear effect into account when we design an electrostatic actuator. Typical driving voltage applied to an electrostatic actuator is an AC signal with a DC offset given by:

$$V(t) = V_{DC} + V_{AC} \cos \omega t \qquad (2.20)$$

Plug Eq. 2.20 into Eq. 2.9:

$$F_E(t) = -\frac{\partial C}{\partial u}(V_{DC} + V_{AC} \cos \omega t)^2$$

$$= -\frac{\partial C}{\partial u}\left\{ \left(V_{DC}^2 + \frac{1}{2}V_{AC}^2\right) + 2V_{DC} \cdot V_{AC} \cos\omega t + \frac{1}{2}V_{AC}^2 \cos^2(2\omega t)\right\} \qquad (2.21)$$

The electrostatic force is accompanied by the underlined 2ω term. It may be minimized by applying a larger DC offset to make $2V_{DC} \cdot V_{AC} \gg \frac{1}{2}V_{AC}^2$. However, it may not be neglected especially when 2ω is close to one of the resonant frequencies of the mechanical system.

Examples of Micro Electrostatic Actuators

"Vertical" actuation: A simple example of an actuator which works in the direction perpendicular to the capacitor plates (as modeled by Eq. 2.10 is a cantilever and an electrode as shown in Fig. 2.29(a). It can be also used to model the electrostatic force between a moving plate and the bottom substrate as shown in Fig. 2.29(b). There are two main issues associated with this mode of actuation.

1. The electrostatic force $-(\partial C/\partial x)V^2$ is dependent on the gap x.
2. The moving range before pull-in is limited only to 1/3 of the initial gap x_0.

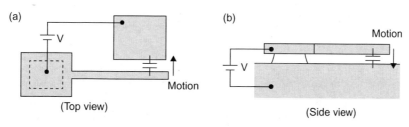

Figure 2.29: Examples of "Vertical" Actuation. The Actuator Motion Is Perpendicular to the Capacitive Plates.

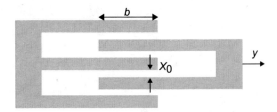

Figure 2.30: Comb Drive Actuator. Lengths *b*, *x0*, and *y* Are Overlapping Length, Gap Distance, and *y* Displacement, Respectively.

These characteristics make the dynamic control of the system very difficult. The advantage of vertical actuation is the simple structure and low operation voltages. It is often used for on-off type devices such as RF switches [27,28], optical switching mirrors [29], and optical gratings [30].

"Parallel" actuation: The electrostatic force parallel to the capacitor plates (as modeled by Eq. 2.11) is important in describing the function of a **comb actuator**, also called as a **comb-drive actuator**, which is a very common form of MEMS electrostatic actuators. Figure 2.30 shows a simplified model of a comb actuator.

The capacitance of a comb-drive actuator is given as

$$C = \varepsilon_0 \frac{Na(b-y)}{x_0} \tag{2.22}$$

where N is the number of gaps and a is the thickness of the device. The electrostatic force is

$$F_y = \frac{\partial C}{\partial y} V^2 = \frac{\varepsilon_0 a N}{2x_0} V^2 \tag{2.23}$$

It is easy to obtain larger electrostatic force by increasing the number of gaps. Another very important characteristic is that $-\partial C/\partial y$ is independent of the displacement y. The electrostatic force is determined only by the voltage. Figure 2.31 shows actuation of a very early comb drive actuator reported by Tang et al. in 1990 [31].

Electrostatic Actuators

Consider an electrostatic actuator with the overlapping surface area of $S = W$ (in x direction) $\times L$ (in y direction) and initial gap of g (in z direction). It is suspended by a spring with a constant K, and voltage V is applied.

(Continued)

Electrostatic Actuators (Continued)

Electrostatic Force

$$F_x = \frac{\varepsilon_0 S}{2g^2} V^2 \quad \text{(in } x \text{ direction)}$$

$$F_y = \frac{\varepsilon_0 W}{2g^2} V^2 \quad \text{(in } x \text{ direction)}$$

Pull in Limit

$$V_P = \frac{2}{3} \sqrt{\frac{2K \cdot g^3}{3\varepsilon_0 S}}$$

Maximum displacement before pull in: 1/3g (or minimum gap: 2/3g)

50 μm

Figure 2.31: Actuation of a Comb Drive Actuator.

2.5 Soft Lithography

Soft lithography is a collective name for several techniques such as **replica molding**, **micro-contact printing (μCP)**, or micro-transfer molding [32,33]. The main idea behind all these patterning techniques is to use a patterned elastomer, which is "soft" and compatible with delicate nano/biomaterials, as a stamp, mold, or mask to generate micropatterns and microstructures. Advantages of soft lithography are that it is relatively inexpensive, easy to use, and straightforward to apply. Some conventional techniques such as cast molding, embossing, and injection molding, which are commonly used for mass production of everyday plastic parts, may also be considered as soft lithography.

The most important element in soft lithography is the elastomeric mold, which is usually composed of polydimethylsiloxane (PDMS). PDMS elastomers are advantageous because of their elasticity, ease of use and biocompatibility. PDMS has many other properties that make it desirable for soft lithography. PDMS has low interfacial free energy and is largely inert. PDMS is also hygroscopic and is easily gas permeable. Other attributes of PDMS include good thermal stability, optical transparency, isotropic, durability, and malleability.

2.5.1 Microcontact Printing

μCP is a microscale stamping technique for the transfer of molecules or polymers onto a surface [34−37]. An elastomeric stamp is "inked" with the material to be transferred. The stamp is then pressed against the substrate to transfer the pattern onto the substrate.

Stamp Preparation

The procedure for creating the elastomeric stamp is outlined in Fig. 2.32. It is formed by a silicon wafer template called the master mold. Any physical patterns of solid materials such as silicon, silicon oxide, silicon nitride, metals, and polymers can be used as the master, unless they create permanent bonding with the cured elastomer. An epoxy-based, high aspect ratio photoresist called SU-8 [38] is most commonly used to create the master patterns. Once the master pattern is obtained, the surface is passivated to avoid the adhesion of the elastomer. Hydrogen fluoride [39], sodium dodecylsulfate [40], or

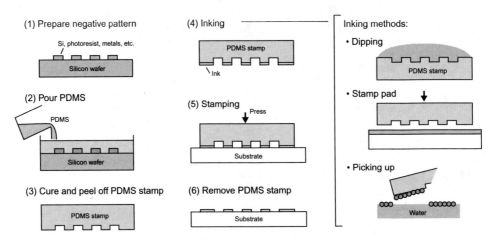

Figure 2.32: Microcontact Printing.

hydrofluorocarbon-based plasma treatment [41] have been used for surface passivation. A mix of the elastomer and a curing agent is then poured over the master mold and heated so the liquid elastomer mix solidifies. This mix can also be called the prepolymer. The elastomer is peeled off once it is solidified.

Inking

A film or a solution of the materials to be patterned is inked on the PDMS stamp. The stamp can be simply dipped in a solution and dried [42], or one can use a "stamp pad" where the stamp is placed and contacts the ink [43]. Other techniques utilize the Langmuir−Blodgett technique to form a molecular film which is picked up by the stamp [44,45].

Stamping

The inked stamp is then inverted onto a substrate. A gentle pressure is applied typically for ∼ 30 seconds so that the material transfers onto the substrate. The stamping process transfers the raised part of the stamp onto the substrate.

Materials

Materials that have been patterned by microcontact printing include self-assembled monolayers (SAMs) [46−48], Au nanoparticles [44,45], and colloidal quantum dots [49−51]. Patterning of biomolecules may be the most important application of microcontact printing, since most conventional lithography processes are not suitable for delicate biomolecules because chemical processes used for wet or dry etching may damage such sensitive materials. Microcontact printing has been utilized to provide controlled biocompatible surfaces [52,53], or patterned immunoassays [54,55]. Microcontact printing techniques have also been used to deposit DNA molecules. Bernard et al demonstrated the microcontact printing of DNA on APTES modified substrates [56]. The PDMS stamp was chemically modified to allow DNA molecules to stick. It was demonstrated that microcontact printing has an advantage of reduction in process time over microspotting.

2.5.2 Soft Lithography-Based Microfabrication

The vacuum processes used for silicon-based microfabrication are not cost effective for production of micro devices for biosample testing, which often need to be disposable. The main components of such analytical tools are microfluidic systems.

Fabrication of microfluidic devices is a good demonstration of soft lithography techniques such as replica molding, embossing, in situ construction, and injection molding.

Advantages of soft lithography based micromachining include high precision, low cost, and chemical-stability of the materials.

Replica Molding

Replica molding is a type of replication method. A part of the procedure is similar to fabrication of elastomeric stamps used for microcontact printing (compare Fig. 2.32 and Fig. 2.33). A master mold is replicated in PDMS by casting and curing the PDMS pre-polymer. The master mold can be a rigid surface or an elastomeric surface. This process results in a negative replica. In many microfluidic devices, channels are fabricated as negative replicas of microfabricated "hills".

To create positive replica, another curable material such as epoxy-based polymers, polymethylmethacrylate (PMMA), polyurethane is cast onto the PDMS negative mold. It is then cured, and peeled away to become the positive replica of the original master mold. The success of this procedure is largely based on physical factors such as Van der Waals interactions, wetting, and kinetic factors. The replica molding technique allows for more accurate replication of small feature sizes compared to photolithography [33]. Replication of feature sizes as small as 10 nm has been reported with using an elastomeric PDMS mold [57,58]. The value of using replica molding as a microfabrication process is its ease of use, precise replication of three-dimensional topographies, accuracy when used to provide multiple exact copies of the master mold, and relative low cost [33].

Microtransfer Molding

Microtransfer molding is a type of replica molding that relies on the transfer of a pattern onto a substrate [59] (Fig. 2.33(b)). It can be used on nonplanar surfaces relatively easily. In this method, liquid prepolymers are filled in the PDMS mold and the excess liquid is removed. The mold is then placed on a substrate. The prepolymer inside the mold is either heated or irradiated by UV light so that it is cured and solidified. When the mold is

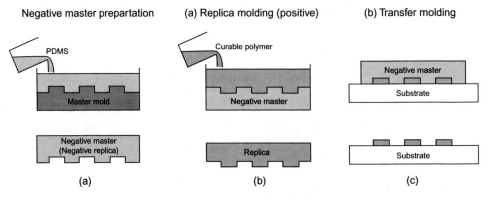

Figure 2.33: Diagram of the Procedures of Replica Molding and Microtransfer Molding.

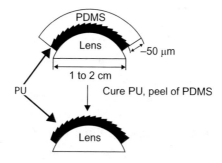

Figure 2.34: Microtransfer Molding onto a Curved Surface [62].

carefully peeled off, the pattern will remain on the surface of the substrate. The technique can be used as a substitute for photolithography. Using this approach, it is possible to generate patterns over relatively large surface areas within a short period of time. It can produce features as small as 40 nm [60]. This technique has been used to fabricate optical waveguides, couplers, and interferometers from organic polymers [61]. Disadvantages of this approach are that microstructures created by this technique usually have a 100 nm film between the raised features. This residual layer can be removed by O_2 reaction ion etching. One of interesting characteristics of microtransfer molding is its ability to generate three-dimensional patterns on non-planar substrates layer by layer. Figure 2.34 shows an example of microtransfer molding where optical gratings made of UV-curable polyurethane (PU) are replicated onto the curved surface of a lens [62].

Embossing Techniques

Embossing or nanoimprinting is the process of indenting a pattern onto a plastic flat surface by pressing a stamp into it. Nanoimprinting has been observed to create patterns with the aspect ratio of 10:1 and features as small as 10 nm [63,64]. The most common technique is **hot embossing**, which is used to manufacture CDs and DVDs. The method involves the transfer of a pattern from a micromachined quartz or metal master to a pliable plastic sheet. The material is heated and pressed by the patterned mold. After cooling, the substrate is solidified and the mold is removed. Polymethylmethacrylate (PMMA) is often used as the plastic material. Hot embossing is low cost and provides more material options, but it is not a timely method for changing designs in small scale laboratory tests.

Another method reported by Xia and Whiteside [33] utilizes a strong solvent that degrades or softens the substrate without affecting the mold. First, a solvent is used to wet the surface of the PDMS mold. The solvent dissolves or softens parts of the substrate and the resulting fluid or gel is molded by the PDMS mold. As the solvent evaporates, the substrate solidifies to form a pattern. The parallel lines created by the method measured as small as 60 nm wide and 50 nm high [65].

Injection Molding

Injection molding is the most widely used technique for manufacturing plastic parts. Thermoplastic polymers are heated past their glass transition temperature to be soft and pliable. The molten material is then injected into a mold cavity where it cools and hardens to form the configured shape. An example of injection molding is described for polymer based cantilevers in Chapter 6, Section 6.2.3. The cooling of the plastic occurs in only a few minutes. Injection molding is compatible with a wide choice of materials [65].

2.5.3 Immobilization and Patterning Techniques for Biomolecules

The recent development of biomaterials based molecular sensors has been significant. These devices allow for gene analysis, detection of biological warfare agents, detection of genetic mutations, and rapid disease diagnosis.

Immobilization of molecular probes on the substrate is a critical step in designing molecular sensors. Microcontact printing or soft lithography, as discussed in the previous section, is an efficient way of patterning or immobilizing biomolecules. However, handling of hundreds or thousands of molecular probes on a substrate is still one of the most challenging and diverse fabrication design considerations in molecular sensors such as DNA microarrays [66,67]. In this section, we describe techniques that are used to immobilize biomaterials for molecular sensors.

Immobilization Techniques

Adsorption

The adsorption technique is believed to be the simplest method [68]. For example, DNA probes contain negatively charged groups, and if the surface of the biosensor or microarray is positively charged, an ionic interaction will occur. A positively charged surface or material, like chitosan, will allow for this simple interaction to occur [69]. Another adsorption technique is physical adsorption. Physical absorption of DNA has been demonstrated on screen-printed graphite electrodes [70]. This technique works by immersing the screen printed electrode in a DNA containing solution overnight, and rinsing out the unadsorbed DNA.

Thiol—Gold Bonding

One of the most important techniques in biomolecule immobilization is utilization of thiol—gold interactions. A thiol is a compound that contains a carbon-bonded sulfhydryl (R—SH) group. Thiol groups show a strong attraction for noble metal surfaces, which allows covalent bonds to form between sulfur and gold atoms. The technique is important because gold films patterned with conventional photolithography lead to patterning of

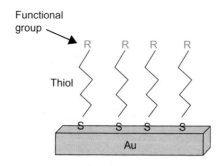

Figure 2.35: Self-assembled Monolayer (SAM) of Thiol.

biomolecules on integrated devices. The technique is most commonly utilized to create stable self-assembled monolayers (SAMs) as shown in Fig. 2.35 [71,72]. It is also used to conjugate gold nanoparticles with functional biomolecules [73,74]. This idea is applied to DNA probes, by modifying them with thiol groups, which are then immobilized on surfaces covered with a gold layer [75].

Avidin–Biotin Interactions

Avidin–biotin interactions are also very commonly utilized for surface functionalization. Biotin is a small molecule and avidin is a tetrameric protein that has four binding sites, which biotin has a high affinity for. Biotin is able to bind to any of the four sites.

There are various ways in which avidin-biotin interactions aid in binding [67,76–78]. DNA probes are bound to self-assembled monolayers of 2-mercaptoethanol and 11-mercaptoundecanoic acid through this avidin–biotin interaction [76]. In another method, biotinylated polypyrrole film is first deposited on a metal surface by electropolymerization of monomers bearing biotin entities, which anchor to avidin added in the next step. The avidin initially has four sites for biotin–avidin interaction. After its immobilization on the biotinylated polypyrrole, avidin has three remaining binding sites to bind to DNA probes that have been biotinylated [77] (see Fig. 2.36). Electrode functionalization for quartz crystal microbalance (QCM) sensors (see Chapter 6, Section 6.3.3) are one of the most commonly studied forms of DNA-based molecular sensors utilizing avidin–biotin interactions [78].

Other Materials for Surface Treatment

A substrate, such as gold or silicon, can be functionalized with several types of active molecules. Those molecules that form SAMs are often utilized for this purpose. Common materials used for surface functionalization is summarized in Box 2.1.

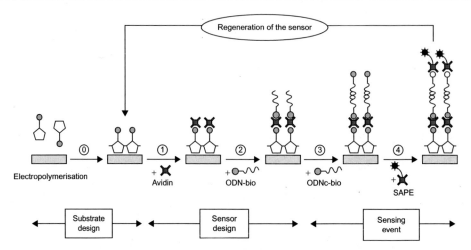

Figure 2.36: Avidin−Biotin Interaction for DNA Immobilization.
From [77].

Inkjet Printing

Inkjet printing is a non-contact method which is ideal for immobilization in biological applications because it prevents cross-contamination from the surface being printed on. It delivers small droplets in reproducible volumes. Ink jet printing is also very cheap as the devices are already mass produced commercially. The wide availability of this technology makes it a very low cost solution.

Some types of inkjet printers utilize thermal bubble formation as the propelling force to eject droplets. Due to the nature of ink jet printing, the materials being ejected from the nozzle experience high shear stresses that could cause denaturation [83]. The denaturation of biological materials due to heat transfer may also be a concern, although no noticeable degradation was observed according to Okamoto et al [84]. Other issues that still remain with ink jet printing technologies include possible cross-contamination of different materials [85]. The system may require cleaning each time before loading another substrate. Even with washing, it is difficult to completely remove the entire sample from the nozzle. Some new designs employ multiple reservoirs and nozzles for different materials.

Photolithography and Direct Photochemistry

In a conventional photolithographic method, a photoresist is first added over the substrate and patterned to create chemically active sites on the substrate where adhesion-promoting silane, usually amino terminated, is bound. The photoresist is then etched away chemically, and a hydrophobic or adhesion resistant silane is bound to the rest of the substrate [86]. In this way, the biomolecules of interest can be fixed on the "adhesion promoted" areas.

Box 2.1 Materials Used for Surface Modification

Poly(ethylene glycol) (PEG)

HO–CH₂CH₂–OH
Ethylene glycol
× n

HO–(CH₂CH₂O)ₙ–H
Polyethylene glycol (PEGₙ)

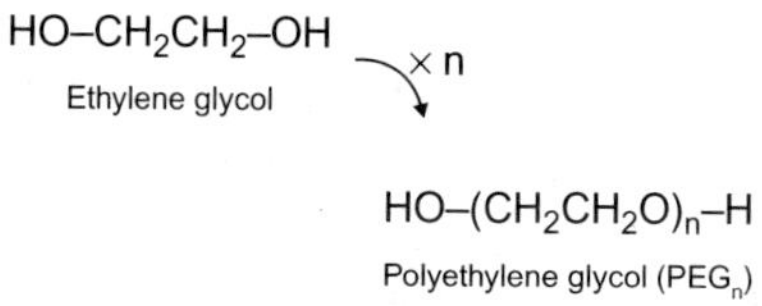

Example of thiol-PEG-acid

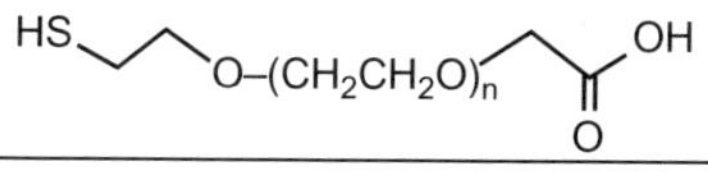

HS ~~~ O–(CH₂CH₂O)ₙ ~~~ OH
O

An amphiphilic polymer, the PEG backbone forms a hydration layer, useful for creating a hydrophilic surface passivation. Either end of a PEG chain can be modified with functional groups such as amine (−NH₂), biotin, acid (−COOH), and thiol (−SH). It can be prepared in various lengths. After modification with oxysilane, PEG SAMs can be formed on Si. The PEG layer prevents nonspecific adsorption of cells and proteins [79].

(3-Aminopropyl)triethoxysilane (APTES)

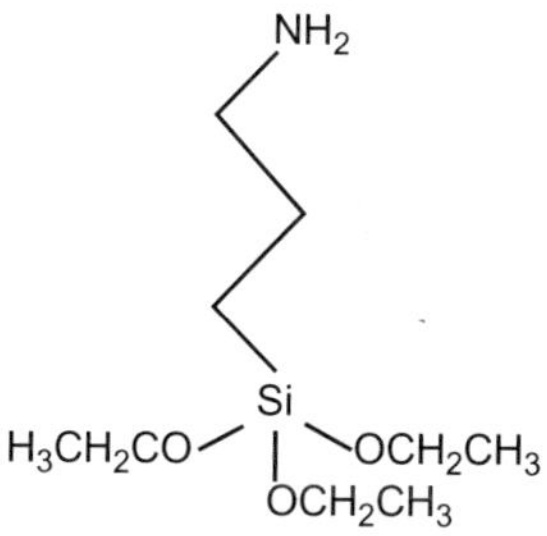

NH₂

Si
H₃CH₂CO — | — OCH₂CH₃
OCH₂CH₃

APTES is an aminosilane and is commonly used for affinity-based biosensors. The amine group can form covalent bonds with carboxyl groups, which are found in many biomolecules. APTES has been used to promote polymer film and protein adhesion to glass and to create films responsive to light and temperature. Metal nanoparticles have also been attached via APTES [80]. HDMS is an organosilicon terminated with methyl groups.

Hexamethyldisilazane (HDMS)

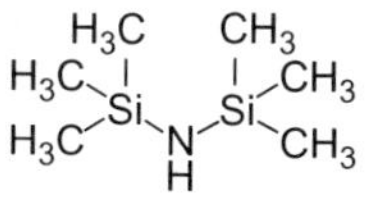

H₃C   CH₃
H₃C\ | | /CH₃
   Si   Si
H₃C/ \N/ \CH₃
      H

A hydrophobic compound, HDMS protects the surface from water vapor adsorption, reduces surface conductivity, and increases stored charge stability [81]. HMDS treatment is usually applied to bare silicon wafers to promote adhesion of photoresist.

Poly-L-lysine

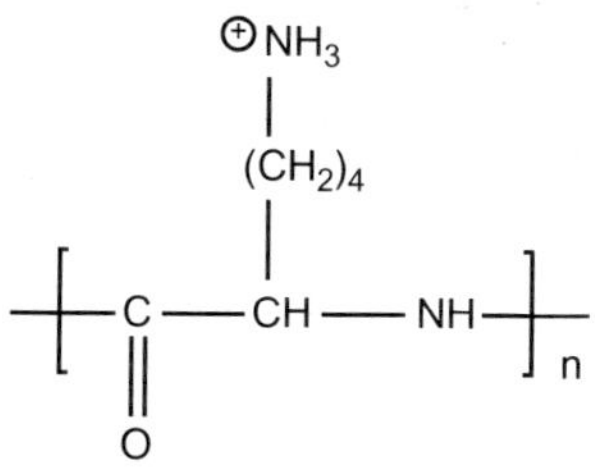

⊕NH₃
|
(CH₂)₄
|
—[C — CH — NH]ₙ—
   ‖
   O

Poly-lysine is a polycationic that promotes cell adhesion and improves protein coating. Positively charged surface ions electrostatically attach to the cell membrane, which is negatively charged [82]. It is often used to treat glass slides for optical microscopy.

In direct photochemistry, photoresist layers are not needed. Instead, a surface is coated with photosensitive molecules and then exposed to UV using a photomask. The molecules exposed to the UV light become chemically active and can bind biological molecules such as DNA or proteins. One popular commercial product synthesized in this way is the Affymetrix gene chip. The thousands of 25-sequence oligonucleotide probes are synthesized in situ as follows [87]. A silicon surface is coated with molecules that have photo-removable protecting groups. Exposure to a UV light pattern removes groups, activating selected sites which bind a nucleotide. The nucleotide itself also contains a photo-removable group, which will be exposed in subsequent iterations to add more nucleotides. One of the major breakthroughs in this method was the development of "virtual" photomasks. Traditionally photomasks were made using chromium or glass masks, which is problematic for DNA microarrays since a different mask is needed for adding each different type of nucleotide. In 1999, Singh-Gasson et al first demonstrated the use of a computer generated mask to synthesize an oligonucleotide array [87]. The mask was created on a computer then relayed to a digital micro-mirror array device (DMD) made by Texas Instruments (See Fig 2.10(b) for schematic). Since this breakthrough, all commercial microarrays that use photolithography are made this way, saving considerable time and money without sacrificing feature size and density.

Scanning Probe Microscopy (SPM) for Biomolecule Patterning

Scanning probe microscopy (SPM) is a type of microscopy that images a surface using a physical probe that scans the specimen. Atomic force microscopy (AFM, see Chapter 6, Section 6.2.1) is especially important for analysis of biomolecules because it can operate in ambient air or even a liquid environment.

There are several microscopy techniques built upon the AFM technology. Some allow manipulation of biomolecules by taking advantage of the force sensing capability of AFM probes [88]. Dip pen nanolithography [89] is among such techniques. It can be used to pattern DNA and peptides, creating DNA and protein microarrays [90].

Near-field scanning optical microscopy (NSOM, see Chapter 5, Section 5.8), or scanning near-field optical microscopy (SNOM), is also based on the AFM technology. The probe in NSOM is a nanometer-sized optical aperture created at a scanning tip, which enables optical imaging beyond diffraction limit [91]. This configuration was used to build scanning near-field photolithography (SNP) [92]. SNP utilizes a NSOM probe coupled with a UV laser to overcome the diffraction-limit that is present when using photomasks. This system is able to create a pattern of SAMs with a resolution of 20 nm, which is comparable to electron-beam lithography.

Methods for Patterning of Soft Materials

- Soft lithography
  - Microcontact printing (μCP)
  - Molding techniques
- Immobilization techniques
  - Adsorption
  - Thiol—gold binding
  - Avidin—biotin interactions
  - Other surface modification techniques (PEG, HMDS, APTES, Poly-L-lysine)
- Other techniques
  - Inkjet printing
  - Direct photochemistry
  - Scanning probing microscopy

2.6 "Top Down "and "Bottom Up" Approaches

There are two approaches to fabricate nanomaterial-based devices. Most lithography based processes we described in this chapter are regarded as a part of "top down" approach, where the devices are formed in the way as designed, usually with computer-aided design tools. They can be readily fabricated in large numbers and are suitable for mass production.

Many of nanomaterial based devices introduced in this book are based on what is called "bottom up" approach, where nanomaterials are first chemically (or biologically) synthesized and transferred to a substrate, or directly synthesized on a substrate. The electrodes, or other supporting structures, have to be aligned to randomly positioned materials, or materials have to be somehow aligned to already-patterned structures for testing. Many even simply take chances and find useful, or acceptable, positioning of nanomaterials and electrodes from random arrays of scattered materials. Figure 2.37 shows a simplified example of a bottom up process, where synthesized nanowires are simply dropcast on a substrate.

The advantage of the bottom up approach is that the abundant array of emerging new materials, such as nanoparticles, nanowires, polymers, or proteins can be tested in an already well-studied configuration of an analytical method. Silicon nanowire based FET sensors (discussed in Chapter 4, Section 4.3.4) are a very good example to see how a bottom up approach is successfully used in testing of new materials. The readers will find that many different materials have been tested in the almost identical configuration of the FET.

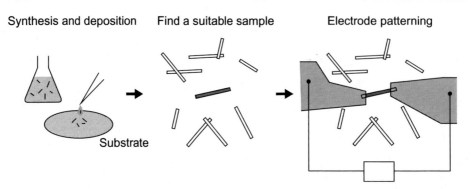

Synthesis and deposition   Find a suitable sample   Electrode patterning

Substrate

Figure 2.37: Concept of the Bottom Up Approach.

The drawback of the bottom up approach is their incompatibility with conventional mass production technologies. Study of nanomaterial positioning techniques that are compatible with the contemporary mass production constitute an important part of nanotechnology research. Chemical binding techniques along with surface modification (see Section 2.5.3), microcontact printing, inkjet printing, and other soft lithography techniques are among those studies that try to bring nanomaterials into the context of contemporary mass production technologies.

On the other hand, the minimum feature size of top-down fabricated structures are approaching or even overcoming the size of nanomaterials. Intel processors based on 22 nm technology [93] went on sale worldwide in 2012. Of course it does not mean any nanostructure can be built based on the extension of the conventional silicon microfabrication process. Several nanomaterials introduced in this book show attractive characteristics that will not be easily replaced by conventional silicon devices. The key issue in nanotechnology is that how we can combine those two trends of bottom up and top down technologies.

Top Down or Bottom Up?

- Top down approach
  1. Conventional lithography based fabrication suitable for mass production.
  2. Minimum feature sizes are approaching the scales of nanomaterials.
  3. Choice of materials are limited
- Bottom up approach
  1. Materials (nanoparticles, nanowires, graphene, polymers, proteins, DNAs/RNAs) are synthesized/obtained through a non-standard fabrication process.
  2. Materials need to be aligned to the electrodes, or electrodes need to be aligned to the randomly positioned nanomaterials.

2.7 Conclusion

In this chapter, we described the nano/microscale fabrication processes, originating from the mainstream silicon photolithography to manufacture microelectronics, and the MEMS process, including bulk and surface micromachining. We also discussed various soft lithography techniques, including microcontact printing, and other key patterning techniques with many emerging applications in BioMEMS or micro TAS. Finally, we discussed the issues related to downsizing of molecular sensors and related effect.

Problems

P2.1 Effect of Scaling

Figure P2.1 shows an example of cantilever based molecular sensor, where the top surface of the cantilever is functionalized with antibodies. According to Stoney's formula (see Chapter 6, section 6.2.2 for details), deflection δ at the end of a cantilever is expressed as a function of the surface stress $\sigma_{surface}$

$$\delta = \frac{3L^2}{Et^2} \cdot \sigma_{surface},$$

where L, t and E are the length, thickness and Young's modulus of the cantilever, respectively. Discuss the scale and shape effect of the sensor, assuming that $\sigma_{surface}$ is simply a function of analyte concentration and independent of the size of the cantilever.

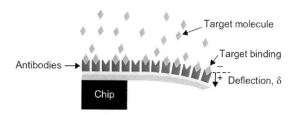

Figure P2.1

P2.2 Crystal Planes and Miller Index

1. What are the Miller Indices of the plane (a) and (b) shown in Fig. P2.2?
2. Draw a (1 0 0) plane and a (1 1 1) plane in the graph in Fig. P2.3 and find the angle between them.

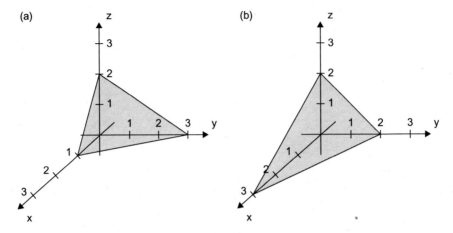

Figure P2.2

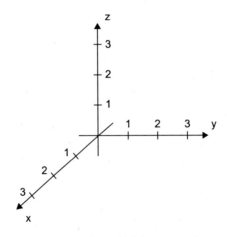

Figure P2.3

P2.3 Microfabrication

Answer the following questions:

1. Name four methods of thin film deposition commonly used for microfabrication.
2. What is a photomask?
3. What is wet etching?
4. What is dry etching?
5. What is a sacrificial layer?

P2.4 Photolithography

proximity → contact

1. The resolution limit of proximity photolithography is typically 3 μm. Explain where this limitation comes from.
2. Name at least three methods that can overcome the above resolution limit and explain how the better resolutions are obtained.

P. 53 -54

increase NA, decrease λ

change numerical aperture

(E-beam)

P2.5 Silicon Micromachining

1. Explain why silicon is an excellent material for integrated circuits (ICs).
2. Explain why silicon is also often used for micromechanical systems. Are there any disadvantages of using silicon?

P2.6 Patterning of Metal Layers

1. Come up with a fabrication process to obtain patterns of 50-nm-thick chromium and 150-nm-thick nickel on the same glass substrate.
2. What if the nickel pattern has to be as thick as 10 μm with the chromium pattern remains to be 50 nm thick.

P2.7 Electroplating

How much current is needed to electroplate a copper film at a deposition rate of 0.5 μm/ min on the entire top surface of a two inch wafer?

P2.8 Wet Anisotropic Etching of Silicon

You want to make a through hole on a (1 0 0) wafer with the thickness of 500 μm (Fig. P2.4). If the aperture at the bottom surface should be a 100 μm × 100 μm square, what is the size of the aperture at the top surface? Assuming (1 0 0)/(1 1 1) etch rate to be 50, what is the optimal size of the opening on the top masking oxide layer?

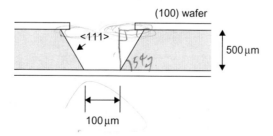

Figure P2.4

P2.9 Microelectromechanical Systems

1. Explain the difference between surface micromachining and bulk micromachining.
2. Name three practical example devices for each of surface micromachining and bulk micromachining.

P2.10 Micro Electrostatic Actuators

If the applied voltage v is smaller than the pull in limit v_p, there are two mathematical solutions for the gap x of a parallel plate actuator suspended by a spring (Fig. P2.5). Only one of them gives the stable solution. Discuss why.

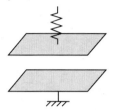

Figure P2.5

P2.11 Micro Electrostatic Actuators

The electrostatic parallel plate actuator shown in Fig. P2.6 has the width of $w = 0.1$ mm and the length of $L = 1$ mm. The original gap between the plates $g = 15$ μm, the equivalent spring constant $k = 5$ N/m, the dielectric constant of air is $\varepsilon = 8.85 \times 10^{-12}$ F/m.

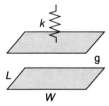

Figure P2.6

1. Find the displacement of the actuator when a voltage of $V = 30$ V is applied.
2. Find the pull-in limit x_{PI} and V_{PI}.

P2.12 Scaling and Micro Actuators

1. Prove that electrostatic force of a parallel plate actuator is a surface force ($\sim L^2$). Consider the breakdown electric field.

2.   Is the stored energy under constant voltage also proportional to L^2?
3.   Build a mathematical model of an electrostatic rotary motor, and discuss scaling in force, torque and power. Discuss if the friction scale along with power generated by the electrostatic micromotor.

P2.13 Soft Lithography

Describe advantages and disadvantages of soft lithography compared to conventional photolithography or e-beam lithography.

P2.14 Soft Lithography

Name at least four methods to fabricate micropatterns of multiple antibodies on a single silicon wafer. Discuss advantages and disadvantages of each method.

P2.15 Top Down and Bottom Up Fabrication

Give the typical sizes of the following objects:

virus, antibody, diameter of a human hair, white blood cell, red blood cell

For each of above, find an artificial device or material with a comparable size. Explain whether it is fabricated/synthesized in a top down approach or bottom up approach.

References

[1]   Postnikov SV, Stewart MD, Tran HV, Nierode MA, Medeiros DR, Cao T, et al. Study of resolution limits due to intrinsic bias in chemically amplified photoresists. Journal of Vacuum Science & Technology B: Microelectronics and Nanometer Structures 1999;17:3335–8.
[2]   Daimon M, Masumura A. Measurement of the refractive index of distilled water from the near-infrared region to the ultraviolet region. Applied Optics 2007;46:3811–20.
[3]   Deal BE, Grove A. General relationship for the thermal oxidation of silicon. Journal of Applied Physics 1965;36:3770–8.
[4]   Gundlach DJ, Lin Y-Y, Jackson TN, Nelson S, Schlom D. Pentacene organic thin-film transistors-molecular ordering and mobility. Electron Device Letters, IEEE 1997;18:87–9.
[5]   Tang C, VanSlyke S. Organic electroluminescent diodes. Applied Physics Letters 1987;51:913.
[6]   Schlesinger M, Paunovic M. Modern Electroplating, 55. New York: Wiley; 2011.
[7]   Wu J, Quinn V, Bernstein GH. Powering efficiency of inductive links with inlaid electroplated microcoils. Journal of Micromechanics and Microengineering 2004;14:576.
[8]   Chiou C-H, Huang Y-Y, Chiang M-H, Lee H-H, Lee G-B. New magnetic tweezers for investigation of the mechanical properties of single DNA molecules. Nanotechnology 2006;17:1217.
[9]   Judy JW, Muller RS, Zappe HH. Magnetic microactuation of polysilicon flexure structures. Journal of Microelectromechanical Systems 1995;4:162–9.
[10]   Romankiw L. A path: from electroplating through lithographic masks in electronics to LIGA in MEMS. Electrochimica Acta 1997;42:2985–3005.

[11]   Chiou C-H, Lee G-B. A micromachined DNA manipulation platform for the stretching and rotation of a single DNA molecule. Journal of Micromechanics and Microengineering 2005;15:109.

[12]   Sato K, Shikida M, Matsushima Y, Yamashiro T, Asaumi K, Iriye Y, et al. Characterization of orientation-dependent etching properties of single-crystal silicon: effects of KOH concentration. Sensors and Actuators A: Physical 1998;64:87−93.

[13]   Tabata O, Asahi R, Funabashi H, Shimaoka K, Sugiyama S. Anisotropic etching of silicon in TMAH solutions. Sensors and Actuators A: Physical 1992;34:51−7.

[14]   Pal P, Sato K, Shikida M, Gosálvez MA. Study of corner compensating structures and fabrication of various shapes of MEMS structures in pure and surfactant added TMAH. Sensors and Actuators A: Physical 2009;154:192−203.

[15]   Sato K, Shikida M, Yamashiro T, Asaumi K, Iriye Y, Yamamoto M. Anisotropic etching rates of single-crystal silicon for TMAH water solution as a function of crystallographic orientation. Sensors and Actuators A: Physical 1999;73:131−7.

[16]   Bean KE. Anisotropic etching of silicon. IEEE Transactions on Electron Devices 1978;25:1185−93.

[17]   Seidel H, Csepregi L, Heuberger A, Baumgärtel H. Anisotropic etching of crystalline silicon in alkaline solutions I. Orientation dependence and behavior of passivation layers. Journal of the Electrochemical Society 1990;137:3612−26.

[18]   Than O, Büttgenbach S. Simulation of anisotropic chemical etching of crystalline silicon using a cellular automata model. Sensors and Actuators A: Physical 1994;45:85−9.

[19]   Gosálvez M, Nieminen R, Kilpinen P, Haimi E, Lindroos V. Anisotropic wet chemical etching of crystalline silicon: atomistic Monte-Carlo simulations and experiments. Applied Surface Science 2001;178:7−26.

[20]   Zhu Z, Liu C. Micromachining process simulation using a continuous cellular automata method. Journal of Microelectromechanical Systems 2000;9:252−61.

[21]   Laerme F, Schilp A, Funk K, Offenberg M. Bosch deep silicon etching: improving uniformity and etch rate for advanced MEMS applications. In: Twelfth IEEE international conference on micro electro mechanical systems, 1999. MEMS'99., 1999, pp. 211−216.

[22]   Ayazi F, Najafi K. High aspect-ratio combined poly and single-crystal silicon (HARPSS) MEMS technology. Journal of Microelectromechanical Systems 2000;9:288−94.

[23]   Smith CS. Piezoresistive effect in Germanium and Silicon. Physical Review 1954;94:42.

[24]   Sniegowski J, Rodgers M, Multi-layer enhancement to polysilicon surface-micromachining technology. in: Electron Devices Meeting, 1997. IEDM'97. Technical Digest., International, 1997. pp. 903−906.

[25]   Lee KB, Cho Y-H. Laterally driven electrostatic repulsive-force microactuators using asymmetric field distribution. Journal of Microelectromechanical Systems 2001;10:128−36.

[26]   Suzuki Y, Miki D, Edamoto M, Honzumi M. A MEMS electret generator with electrostatic levitation for vibration-driven energy-harvesting applications. Journal of Micromechanics and Microengineering 2010;20:104002.

[27]   Pacheco SP, Katehi LP, Nguyen C-C. Design of low actuation voltage RF MEMS switch. In: Microwave symposium digest. 2000 IEEE MTT-S International, 2000, pp. 165−168.

[28]   Goldsmith CL, Yao Z, Eshelman S, Denniston D. Performance of low-loss RF MEMS capacitive switches. Microwave and Guided Wave Letters, IEEE 1998;8:269−71.

[29]   Van Kessel PF, Hornbeck LJ, Meier RE, Douglass MR. A MEMS-based projection display. Proceedings of the IEEE 1998;86:1687−704.

[30]   Bloom DM. Grating light valve: revolutionizing display technology. Electronic Imaging'97 1997;165−71.

[31]   Tang WC, Nguyen T-CH, Judy MW, Howe RT. Electrostatic-comb drive of lateral polysilicon resonators. Sensors and Actuators A: Physical 1990;21:328−31.

[32]   Madou M. Fundamentals of Microfabrication: The Science of Miniaturation. 2nd ed. Boca Raton, Fl: CRC Press; 2000.

[33] Xia Y, Whiteside GM. Soft lithography. *Angewandte Chemie, International Edition in English* 1998;551−72.

[34] Quist AP, Pavlovic E, Oscarsson S. Recent advances in microcontact printing. Analytical and Bioanalytical Chemistry 2005;381:591−600.

[35] Singhvi R, Kumar A, Lopez GP, Stephanopoulos GN, Wang DI, Whitesides GM, et al. Engineering cell shape and function. Science 1994;264:696.

[36] Branch DW, Corey JM, Weyhenmeyer JA, Brewer GJ, Wheeler BC. Microstamp patterns of biomolecules for high-resolution neuronal networks. Medical and Biological Engineering and Computing 1998;36:135−41.

[37] Visconti D. Sub-micron lithography on proteins by room temperature transfer molding. Synthetic Metals 2003;137:.

[38] Lorenz H, Despont M, Fahrni N, LaBianca N, Renaud P, Vettiger P. SU-8: a low-cost negative resist for MEMS. Journal of Micromechanics and Microengineering 1997;7:121.

[39] Pavlovic E, Quist AP, Gelius U, Nyholm L, Oscarsson S. Generation of thiolsulfinates/thiolsulfonates by electrooxidation of thiols on silicon surfaces for reversible immobilization of molecules. Langmuir 2003;19:4217−21.

[40] Lauer L, Klein C, Offenhäusser A. Spot compliant neuronal networks by structure optimized micro-contact printing. Biomaterials 2001;22:1925−32.

[41] Hosokawa K, Maeda R. A pneumatically-actuated three-way microvalve fabricated with polydimethylsiloxane using the membrane transfer technique. Journal of micromechanics and microengineering 2000;10:415.

[42] Pompe T, Fery A, Herminghaus S, Kriele A, Lorenz H, Kotthaus JP. Submicron contact printing on silicon using stamp pads. Langmuir 1999;15:2398−401.

[43] Libioulle L, Bietsch A, Schmid H, Michel B, Delamarche E. Contact-inking stamps for microcontact printing of alkanethiols on gold. Langmuir 1999;15:300−4.

[44] Santhanam V, Andres RP. Microcontact printing of uniform nanoparticle arrays. Nano Letters 2004;4:41−4.

[45] Santhanam V, Liu J, Agarwal R, Andres RP. Self-assembly of uniform monolayer arrays of nanoparticles. Langmuir 2003;19:7881−7.

[46] Biebuyck HA, Larsen NB, Delamarche E, Michel B. Lithography beyond light: microcontact printing with monolayer resists. IBM Journal of Research and Development 1997;159−70.

[47] Rogers JA, Nuzzo RG. Recent progress in soft lithography. Materials Today 2005;8:50−6.

[48] Jeon NL, Finnie K, Bradshaw K, Nuzzo RG. Structure and stability of patterned self-assembled films of octadecyltricholorosilane formed by contact printing. Langmuir 1997;13:3382−91.

[49] Kim LA, Anikeeva PO, Coe-Sullivan SA, Steckel JS, Bawendi MG, Bulovic V. Contact printing of quantum dot light-emitting devices. Nano Letters 2008;8:4513−7.

[50] Gopal A, Hoshino K, Kim S, Zhang X. Multi-color colloidal quantum dot based light emitting diodes micropatterned on silicon hole transporting layers. Nanotechnology 2009;20:235201.

[51] Gopal A, Hoshino K, Zhang X. Photolithographic patterning of subwavelength top emitting colloidal quantum dot based inorganic light emitting diodes on silicon. Applied Physics Letters 2010;96:131109.

[52] Chen CS, Mrksich M, Huang S, Whitesides GM, Ingber DE. Geometric control of cell life and death. Science 1997;276:1425.

[53] Théry M, Pépin A, Dressaire E, Chen Y, Bornens M. Cell distribution of stress fibres in response to the geometry of the adhesive environment. Cell Motility and the Cytoskeleton 2006;63:341−55.

[54] Blinka E, Loeffler K, Hu Y, Gopal A, Hoshino K, Lin K, et al. Enahanced microcontact printing of protiens on nanoporous silica surface. Nanotechnology. 2010;21(41):415302.

[55] Ng E, Gopal A, Hoshino K, Zhang X. Multicolor microcontact printing of proteins on nanoporous surface for patterned immunoassay. Applied Nanoscience 2011;1:79−85.

[56] Lange SA, Benes V, Kern DP, Hörber JKH, Bernard A. Microcontact printing of DNA molecules. Analytical Chemistry 2004;76:1641−7.

[57] Loo Y, Willett RL, Baldwin KW, Rogers JA. Interfacial chemistries for nanoscale transfer printing. Journal of the American Chemical Society 2002;124:7654.

[58] Xia Y, McClelland JJ, Gupta R, Qin D, Zhao XM, et al. Replica molding using polymeric materials: a practical step towards nanomanufacturing. Advances in Materials 1997;9:147−9.

[59] Zhao XM, Xia Y, Whiteside GM. Fabrication of three-dimensional micro-structures: microtransfer molding. Advances in Materials 1996;8:837−40.

[60] Thibault C, Severac C, Trevisiol E, Vieu C. Microtransfer molding of hydrophobic dendrimer. Micro Engineering 2006;83:1513−6.

[61] Zhao XM, Smith SP, Waldman SJ, Whitesides GM, Prentiss M. Demonstration of waveguide couplers fabricated using microtransfer molding. Applied Physics Letters 1997;71:1017−9.

[62] Xia Y, Kim E, Zhao X-M, Rogers JA, Prentiss M, Whitesides GM. Complex optical surfaces formed by replica molding against elastomeric masters. Science 1996;273:347−9.

[63] Gates B, Xu Q, Love C, Whitesides GM. Unconventional nanofabrication. Annual Reviews in Materials Research 2004;34:339−72.

[64] Chou SY, Krauss PR, Renstrom PJ. Imprint of sub-25 nm Vias and trenches in polymers. Applied Physics Letters 1995;67:3114−6.

[65] Xia Y, Whiteside GM. Soft lithography. Annual Reviews in Materials Science 1998;:153−84.

[66] Heller MJ. DNA microarray technology: devices, systems, and applications. Annual Review of Biomedical Engineering 2002;4:129−53.

[67] Sassolas A, Leca-Bouvier BD, Blum LJ. DNA biosensors and microarrays. Chemical Reviews 2008;108:109−39.

[68] Rasmussen SR, Larsen MR, Rasmussen SE. Covalent immobilization of DNA onto polystyrene microwells−the molecules are only bound at the 5' end. Analytical Biochemistry 1991;198:138−42.

[69] Yi HM, Wu LQ, Sumner JJ, Gillespie JB, Payne GF, Bentley WE. Chitosan scaffolds for biomolecular assembly: coupling nucleic acid probes for detecting hybridization. Biotechnology and Bioengineering 2003;83:646−52.

[70] Azek F, Grossiord C, Joannes M, Limoges B, Brossier P. Hybridization assay at a disposable electrochemical biosensor for the attomole detection of amplified human cytomegalovirus DNA. Analytical Biochemistry 2000;284:107−13.

[71] Tour JM, Jones L, Pearson DL, Lamba JJ, Burgin TP, Whitesides GM, et al. Self-assembled monolayers and multilayers of conjugated thiols, alpha, omega, dithiols, and thioacetyl-containing adsorbates. Understanding attachments between potential molecular wires and gold surfaces. Journal of the American Chemical Society 1995;117:9529−34.

[72] Bain CD, Biebuyck HA, Whitesides GM. Comparison of self-assembled monolayers on gold: coadsorption of thiols and disulfides. Langmuir 1989;5:723−7.

[73] Brust M, Walker M, Bethell D, Schiffrin DJ, Whyman R. Synthesis of thiol-derivatised gold nanoparticles in a two-phase liquid−liquid system. Journal of the Chemistry Society, Chemical Communications 1994;801−2.

[74] Daniel M-C, Astruc D. Gold nanoparticles: assembly, supramolecular chemistry, quantum-size-related properties, and applications toward biology, catalysis, and nanotechnology. Chemical Reviews-Columbus 2004;104:293.

[75] Charles PT, Vora GJ, Andreadis JD, Fortney AJ, Meador CE, Dulcey CS, et al. Fabrication and surface characterization of DNA microarrays using amine- and thiol-terminated oligonucleotide probes. Langmuir 2003;19:1586−91.

[76] Pan S, Rothberg L. Chemical control of electrode functionalization for detection of DNA hybridization by electrochemical impedance spectroscopy. Langmuir 2005;21:1022−7.

[77] Dupont-Filliard A, Roget A, Livache T, Billon M. Reversible oligonucleotide immobilisation based on biotinylated polypyrrole film. Analytica Chimica Acta 2001;449:45−50.

[78] Caruso F, Rodda E, Furlong DN, Niikura K, Okahata Y. Quartz crystal microbalance study of DNA immobilization and hybridization for nucleic acid sensor development. Analytical Chemistry 1997;69:2043−9.

[79] Yang Z, Galloway JA, Yu H. Protein interactions with poly (ethylene glycol) self-assembled monolayers on glass substrates: diffusion and adsorption. Langmuir 1999;15:8405−11.

[80] Howarter JA, Youngblood JP. Optimization of silica silanization by 3-aminopropyltriethoxysilane. Langmuir 2006;22:11142−7.

[81] Chen X, Zhang J, Wang Z, Yan Q, Hui S. Humidity sensing behavior of silicon nanowires with hexamethyldisilazane modification. Sensors and Actuators B: Chemical 2011;156:631−6.

[82] Mazia D, Schatten G, Sale W. Adhesion of cells to surfaces coated with polylysine. Applications to electron microscopy. The Journal of Cell Biology 1975;66:198−200.

[83] Barbulovic-Nad I, Lucente M, Sun Y, Zhang MJ, Wheeler AR, Bussmann M. Bio-microarray fabrication techniques−A review. Critical Reviews in Biotechnology 2006;26:237−59.

[84] Okamoto T, Suzuki T, Yamamoto N. Microarray fabrication with covalent attachment of DNA using bubble jet technology. Nature Biotechnology 2000;18:438−41.

[85] Allain LR, Askari M, Stokes DL, Vo-Dinh T. Microarray sampling-platform fabrication using bubble-jet technology for a biochip system. Fresenius Journal of Analytical Chemistry 2001;371:146−50.

[86] Blawas A, Reichert W. Protein patterning. Biomaterials 1998;19:595−609.

[87] Singh-Gasson S, Green RD, Yue YJ, Nelson C, Blattner F, Sussman MR, et al. Maskless fabrication of light-directed oligonucleotide microarrays using a digital micromirror array. Nature Biotechnology 1999;17:974−8.

[88] Fotiadis D, Scheuring S, Müller SA, Engel A, Müller DJ. Imaging and manipulation of biological structures with the AFM. Micron 2002;33:385−97.

[89] Piner RD, Zhu J, Xu F, Hong S, Mirkin CA. Dip-pen nanolithography. Science 1999;283:661−3.

[90] Lee K-B, Park S-J, Mirkin CA, Smith JC, Mrksich M. Protein nanoarrays generated by dip-pen nanolithography. Science 2002;295:1702−5.

[91] Betzig E, Lewis A, Harootunian A, Isaacson M, Kratschmer E. Near field scanning optical microscopy (NSOM): development and biophysical applications. Biophysical Journal 1986;49:269−79.

[92] Sun SQ, Chong KSL, Leggett GJ. Nanoscale molecular patterns fabricated by using scanning near-field optical lithography. Journal of the American Chemical Society 2002;124:2414−5.

[93] Damaraju S, George V, Jahagirdar S, Khondker T, Milstrey R, Sarkar S, et al. A 22 nm IA multi-CPU and GPU system-on-chip. In: Solid-state circuits conference digest of technical papers (ISSCC), 2012 IEEE International, 2012, pp. 56−57.

Further Reading

Madou MJ. Fundamentals of microfabrication: the science of miniaturization. Boca Raton: CRC Press; 2002.
Kovacs GT. Micromachined transducers sourcebook. NY: WCB/McGraw-Hill New York; 1998.

CHAPTER 3

Microfluidics and Micro Total Analytical Systems

Chapter Outline

3.1 Introduction

Microfluidics is an enabling technology to perform biological and chemical experiments at greatly reduced spatial scales, with minimal material consumption and high-throughput, through controlling flows in microchannels with characteristic dimensions ranging from millimeters to micrometers. It allows for handling of fluid with volumes in the range of nano- to microliters ($10^{-9}-10^{-6}$ L) or smaller. Microfluidics is key to advancing molecular sensors based on bioassays including immunoassay, cell separation, DNA amplification and analysis, among many other examples. Microfluidic systems process a large number of parallel experiments rapidly with a small amount of reagent and automate chemical, biological, and medical applications on a large-scale with low cost. For example, reducing the reaction chamber size by a factor of 10 increases the reaction rate by a factor of 100 because of the smaller characteristic length of the system decreases diffusion time. In addition to faster reaction times, the amount of analyte and reagents required is also reduced proportional to the reduction of the reaction chamber volume. Not only does this reduce the cost of the test by reducing the required amount of chemicals, it also allows more types of tests to be conducted in parallel with the same amount of sample.

In this chapter, we first discuss basic theories of fluid dynamics that are essential to characterize microfluidic systems, and describe the solutions to simple cases of incompressible laminar flows. We then introduce examples of microfluidic devices that are related to molecular sensing. The examples include microfluidic circuits for protein analysis, where molecular interactions are measured by manipulating reagent solutions through multiple microchannels and chambers. Other examples include cell sorters and particle separators for microfluidic cellular analysis.

3.2 Microfluidics Fundamentals

3.2.1 Diffusion

Fick's first law of diffusion states that the rate of transfer of molecules from regions of high concentration to regions of low concentration is proportional to the concentration gradient (see Fig. 3.1). In a one-dimensional model of diffusion, the law is expressed as:

$$J = -D\frac{\partial \phi}{\partial x},\tag{3.1}$$

Where

   J is the flux [(number of molecules) per (unit area) per (unit time). Example: mol/m^2/s],
   D is the diffusion coefficient [length^2/time. Example: m^2/s], and
   ϕ is the concentration [(number of molecules) per (unit volume). Example: mol/m^3]

Note the negative sign on the right side of Eq. 3.1. The flux J is positive when the gradient $\frac{\partial \phi}{\partial x}$ is negative, because diffusion occurs to level the gradient.

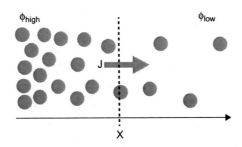

Figure 3.1: One-dimensional Model of Diffusion. The Rate of Transfer Is Proportional to the Concentration Gradient.

Fick's second law can be derived from Fick's first law. Consider a small one dimensional control volume from x to $x + \Delta x$ with transport only by diffusion (Fig. 3.2). The change in the concentration of molecules with time $\left(= \frac{\partial \phi}{\partial t}\right)$ is the difference between the flux going in $(= J_{in})$ and the flux going out $(= J_{out})$.

$$\frac{\partial \phi}{\partial t} = J_{in} - J_{out} \tag{3.2}$$

By using the first law (Eq. 3.1)

$$\frac{\partial \phi}{\partial t} = -D\left(\frac{\partial \phi}{\partial x}\bigg|_{x=x} - \frac{\partial \phi}{\partial x}\bigg|_{x=x+\Delta x}\right)\bigg/\Delta x \tag{3.3}$$

When Δx approaches zero, Eq. 3.3 becomes

$$\frac{\partial \phi}{\partial t} = D\frac{\partial^2 \phi}{\partial x^2}, \tag{3.4}$$

Eq. 3.4 is called **Fick's second law**, or also called as a **diffusion equation**.

A simple, but useful case of diffusion is where we consider a semi-infinite region in $x > 0$ with a fixed concentration at the boundary $(\phi(0, t) = \phi_0)$, as shown in Fig. 3.3. The solution to this problem is given as:

$$\phi(x, t) = \phi_0 \cdot \text{erfc}\left(\frac{x}{2\sqrt{Dt}}\right), \tag{3.5}$$

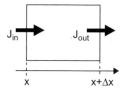

Figure 3.2: Fick's Second Law. The Concentration of a Small One-dimensional Volume Is Considered.

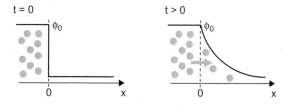

Figure 3.3: Diffusion from a Constant Source. See Text for Details.

where erfc is the complementary error function defined as

$$\mathrm{erfc}(x) = \frac{2}{\sqrt{\pi}} \int_x^\infty e^{-t^2}\,dt \quad (\,= 1 - \mathrm{erf}(x)) \tag{3.6}$$

The graph of $Y = \mathrm{erfc}(X)$ is plotted in Fig. 3.4. As one can expect, $\phi(x, t)$ at a constant position $x = c$ is a monotonically increasing function of time t. It is useful to discuss the characteristic diffusion length $\lambda = 2\sqrt{Dt}$, where the concentration satisfies $\phi(\lambda, t) = \phi_0 \cdot \mathrm{erfc}(1)$. As the time t increases, the length λ becomes larger, which means molecules diffuse further.

Summary of Diffusion

Let J, D, ϕ, t be flux, diffusion coefficient, concentration, and time, respectively.

Fick's First Law

$$J = -D\frac{\partial \phi}{\partial x}$$

Fick's Second Law

$$\frac{\partial \phi}{\partial t} = D\frac{\partial^2 \phi}{\partial x^2}$$

Diffusion Length

Diffusion length is proportional to $\sqrt{Dt}$

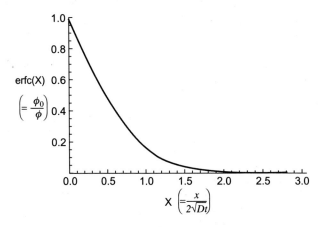

Figure 3.4: Complementary Error Function as a Solution to the Diffusion Equation.

3.2.2 Laminar Flow and the Hagen–Poiseuille Equation

Viscosity

Viscosity quantifies the resistance of a fluid which is being deformed by stress. A simple measurement of viscosity is illustrated with a flow of a viscous fluid in the space between two parallel plates, one of which is moving with a constant velocity U and the other stationary.

In this case, flow velocity u_x in the x direction changes from $u_x = 0$ to $u_x = U$ linearly from $y = 0$ to y $= H$, where H is the distance between the two plates. Such a parallel flow with a constant velocity gradient is called a Couette Flow (Fig. 3.5).

Let us define shear stress τ as $\tau = F/A$, where F is the force needed to move the top plate at a velocity U, and A is the area of the plate. Isaac Newton found that the shear stress τ is proportional to the velocity gradient U/H for many fluids. The viscosity μ of a fluid is defined as the coefficient of this linear relationship, namely:

$$\tau = \mu \frac{U}{H}. \tag{3.7}$$

The term μ is also called the dynamic viscosity. A fluid that satisfies Eq. 3.7 is called a Newtonian fluid.

Next, we discuss a more general case of a flow in a channel, where a fluid flows in parallel layers or laminae with no disruption between the layers. Such a flow is called a laminar flow. In most cases of microdevices, a flow can be considered laminar (we discuss this issue again with Reynolds number in the next section). The laminar flow can be considered to be layers of thin films (see Fig. 3.6). The walls are stationary and the fluid is driven by the pressure gradient along the channel. The important assumption is that the flow velocity is continuous at the boundary between the fluid and the wall — i.e., the fluid has zero velocity relative to the boundary. This assumption is called the no-slip condition. This condition defines the velocity u_x at the walls.

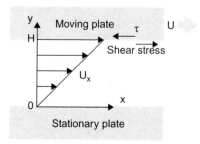

Figure 3.5: Viscosity Is a Couette Flow and the Definition of Viscosity.

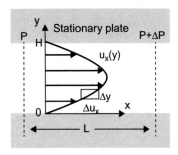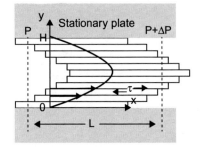

Figure 3.6: Laminar Flow Can Be Considered as Layers of Thin Films.

Unlike the case with a Couette flow, the velocity gradient is not constant. However, if we look at a layer thin enough ($\Delta y \to 0$), the same idea as the Couette flow can be used to define the viscosity. The local velocity gradient can be defined as (see the small square in Fig. 3.6 left).

$$\frac{\partial u_x}{\partial y} = \lim_{\Delta y \to 0} \frac{\Delta u_x}{\Delta y}. \tag{3.8}$$

The shear stress acting between two adjacent layers in Fig. 3.6 right is

$$\tau(y) = \mu \frac{\partial u_x}{\partial y}. \tag{3.9}$$

Eq. 3.9 can be used to form a differential equation to find the velocity profile $u_x(y)$. Eq. 3.7 is a special case of Eq. 3.9.

Laminar Flow in a Pipe

Let us solve the velocity profile of a stationary laminar flow in a pipe. The viscosity of the fluid is μ, the flow rate is Q, the pipe length is L, the pipe radius is R, the pressures at $x = 0$ and $x = L$ are $P = P$ and $P = P + \Delta P$, respectively. Since the flow is laminar, we can assume that there are cylindrical layers of liquid as shown in Fig. 3.7.

Let us consider the forces acting on a thin cylindrical volume from $r = r$ to $r = r + \Delta r$ (Fig. 3.8). The force at the inside interface is

$$F_r = -A_r \tau_r = -2\pi r L \cdot \tau_r = -2\pi r L \cdot \mu \frac{du_x}{dr} \bigg|_r , \tag{3.10}$$

where A_r is the area of the interface. The sign is negative because the force acts in the opposite direction of $\frac{du_x}{dr}$. In this case, the flow is faster inside, i.e. $\frac{du_x}{dr} < 0$, and the force F_r from the inside acts in the direction to drive the volume. The force at the outside interface $r = r + \Delta r$ is:

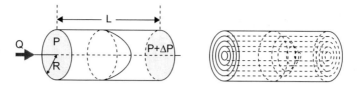

Figure 3.7: Laminar Flow in a Pipe.

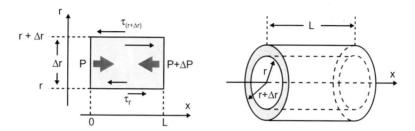

Figure 3.8: Forces Acting on a Cylindrical Layer.

$$F_{r+\Delta r} = 2\pi(r + \Delta r)L \cdot \tau_{r+\Delta r} = 2\pi(r + \Delta r)L \cdot \mu \frac{du_x}{dr}\bigg|_{r=r+\Delta r} \tag{3.11}$$

The force $F_{r+\Delta r}$ from the outside acts in the direction to slow down the flow. Note that the area of the interface are different from Eq. 3.10.

The actual force that drives the volume is the pressure difference between $x = x$ and $x = x + \Delta x$.

$$F_P = -\{\pi(r+\Delta r)^2 - \pi r^2\}\Delta P = -(2\pi r \Delta r + \Delta r^2)\Delta P \tag{3.12}$$

The sign is negative because when the fluid flows in the positive direction of the x axis, the pressure at $x = L$ is lower than at $x = 0$, i.e. $\Delta P < 0$. The term Δr^2 will be neglected.

Since there is no acceleration in the stationary flow, the net force is zero.

$$0 = F_P + F_r + F_{r+\Delta r} \tag{3.13}$$

Plugging Eqs. 3.10, 3.11, and 3.12 into Eq. 3.13 gives

$$-(2\pi r \Delta r + \Delta r^2)\Delta P - 2\pi rL \cdot \mu \frac{du_x}{dr}\bigg|_r + 2\pi(r + \Delta r)L \cdot \mu \frac{du_x}{dr}\bigg|_{r+\Delta r} = 0$$

$$\text{or} \quad -(r + \Delta r)\frac{\Delta P}{\mu L} + r\left(\frac{du_x}{dr}\bigg|_{r+\Delta r} - \frac{du_x}{dr}\bigg|_r\right)\bigg/ \Delta r + \frac{du_x}{dr}\bigg|_{r+\Delta r} = 0 \tag{3.14}$$

When $\Delta r \to 0$, this becomes

$$r\frac{\Delta P}{\mu L} = r\frac{d^2 u_x}{dr^2} + \frac{du_x}{dr},$$

which can be rearranged to

$$r\frac{\Delta P}{\mu L} = \frac{d}{dr}r\frac{du_x}{dr} \tag{3.15}$$

Integrating Eq. 3.15 twice gives

$$\frac{d}{dr}u_x(r) = \frac{1}{2\mu}r\frac{\Delta P}{L} + A\frac{1}{r} \tag{3.16}$$

and

$$u_x(r) = \frac{1}{4\mu}r^2\frac{\Delta P}{L} + A\ln(r) + B \tag{3.17}$$

Two important assumptions give the boundary conditions for Eq. 3.17.

1. Axial symmetry: the velocity profile $u_x(r)$ is symmetric about the x axis, which requires

$$\frac{d}{dr}u_x(0) = 0 \tag{3.18}$$

2. No slip condition: the velocity is zero at the wall,

$$u_x(R) = 0 \tag{3.19}$$

From Eqs. 3.16–3.19, we obtain

$$u_x(r) = -\frac{1}{4\mu}r^2\frac{\Delta P}{L}(R^2 - r^2) \tag{3.20}$$

The fluid is moving fastest at the center (r = 0),

$$u_{max}(r) = -\frac{1}{4\mu}r^2\frac{\Delta P}{L}R^2 \tag{3.21}$$

We can obtain the total flow rate by adding up the contribution from each cylinder.

$$Q = \int_{r=0}^{r=R} 2\pi r \cdot u_x(r) \cdot dr = -\frac{\Delta P \pi R^4}{8\mu L} \tag{3.22}$$

In many practical cases, the flow velocity U refers to the average velocity, namely

$$U = Q/\pi R^2 = -\frac{\Delta P R^2}{8\mu L} \tag{3.23}$$

Note that U is different from the velocity at the center. The average velocity U is actually ½ of the center velocity.

Hagen–Poiseuille Equation

From Eqs. 3.22 and 3.23, we obtain the **Hagen–Poiseuille equation**

$$\Delta P = -\frac{8\mu L}{R^2} U \tag{3.24}$$

$$\text{or} \quad \Delta P = -\frac{8\mu L}{\pi R^4} Q \tag{3.25}$$

The Hagen–Poiseuille equation describes the relationship between pressure, fluidic resistance and flow rate, analogous to voltage, resistance, and current, respectively, in Ohm's law for electrical circuits ($V = RI$). Both electrical resistance and fluidic resistance are proportional to the length of the device. One important difference between Hagen–Poiseuille equation and Ohm's law is that, unlike the electrical resistance, the fluidic resistance is not inversely proportional to the cross-sectional area πR^2, but to πR^4 (for Ohm's law, see Chapter 4, Section 4.3.1). The difference arises from the non-slip condition of the laminar flow. It should be noted that the Hagen–Poiseuille equation applies only to laminar flows in a pipe. As in many cases of microfluidic devices, a flow between two parallel plates is also practically important. The velocity profile of the flow between two parallel plates is found to be parabolic, following the same steps as in the discussion earlier (see problem 3.4).

Hagen–Poiseuille Equation

Consider stationary laminar flow in a pipe with viscosity μ, flow rate Q, average velocity U, pipe length L, pipe radius R, and pressure difference ΔP.

The **Hagen–Poiseuille equation** is given as:

$$\Delta P = -\frac{8\mu L}{R^2} U \quad \text{or} \quad \Delta P = -\frac{8\mu L}{\pi R^4} Q$$

3.2.3 Reynolds Number and Scaling Law

Navier–Stokes Equations

The Navier–Stokes equations are the basic governing equations for the motion of fluid substances. They relate the three-dimensional components (u, v, w) of the velocity vector **v**, pressure p and density ρ as functions of the position (x, y, z) and the time t. Because there are five unknown parameters (u, v, w, p and ρ), five independent equations are necessary to describe a flow field. The general form of the Navier–Stokes equations consist of one continuity equation, three equations of motion (for x, y, z), and one energy equation.

Here we consider a simplified form of the Navier–Stokes equations for an unsteady incompressible flow of a Newtonian fluid, where ρ is constant.

The continuity equation is:

$$\frac{\partial u}{\partial x} + \frac{\partial v}{\partial y} + \frac{\partial w}{\partial z} = 0 \tag{3.26}$$

The equations of motion are:

$$\rho\left(\frac{\partial u}{\partial t} + u\frac{\partial u}{\partial x} + v\frac{\partial u}{\partial y} + w\frac{\partial u}{\partial z}\right) = -\frac{\partial p}{\partial x} + \mu\left(\frac{\partial \tau_{xx}}{\partial x} + \frac{\partial \tau_{xy}}{\partial y} + \frac{\partial \tau_{xz}}{\partial z}\right) + f_x \tag{3.27}$$

$$\rho\left(\frac{\partial v}{\partial t} + u\frac{\partial v}{\partial x} + v\frac{\partial v}{\partial y} + w\frac{\partial v}{\partial z}\right) = -\frac{\partial p}{\partial y} + \mu\left(\frac{\partial \tau_{xy}}{\partial x} + \frac{\partial \tau_{yy}}{\partial y} + \frac{\partial \tau_{yz}}{\partial z}\right) + f_y \tag{3.28}$$

$$\rho\left(\frac{\partial w}{\partial t} + u\frac{\partial w}{\partial x} + v\frac{\partial w}{\partial y} + w\frac{\partial w}{\partial z}\right) = -\frac{\partial p}{\partial z} + \mu\left(\frac{\partial \tau_{xz}}{\partial x} + \frac{\partial \tau_{yz}}{\partial y} + \frac{\partial \tau_{zz}}{\partial z}\right) + f_z \tag{3.29}$$

Each term in the equations is explained in the following way:

1. Equation 3.26 states that the total of the incompressible flows coming in and flowing out of a control volume dxdydz should be zero, namely

$$du \cdot dydz + dv \cdot dxdz + dw \cdot dxdy = 0$$

2. The left sides of Eqs. 3.27–3.29 represent acceleration, where $\frac{\partial()}{\partial t}$ is the local acceleration, and $u\frac{\partial()}{\partial x} + v\frac{\partial()}{\partial y} + w\frac{\partial()}{\partial z}$ is the convective acceleration. The term $\frac{\partial()}{\partial t}$ is similar to one in a typical form of the equation of motion for a single particle, namely $m\frac{d\mathbf{v}}{dt} = F$. In the case of fluid dynamics, substances around $P(x,y,z)$ at time $t = t$ moves to $P(x + dx, y + dy, z + dz)$ at time $t = t + dt$. In order to consider the acceleration of the substances, one has to take this translation into account.

$$d\mathbf{v} = \mathbf{v}(x + dx, y + dy, z + dz; t + dt) - \mathbf{v}(x, y, z; t)$$

$$= \frac{\partial \mathbf{v}}{\partial x}dx + \frac{\partial \mathbf{v}}{\partial y}dy + \frac{\partial \mathbf{v}}{\partial z}dz + \frac{\partial \mathbf{v}}{\partial t}dt$$

Now, the acceleration can be expressed in the following way with the convective acceleration term:

$$\frac{d\mathbf{v}}{dt} = \frac{\partial \mathbf{v}}{\partial t} + \frac{\partial \mathbf{v}}{\partial x}\frac{dx}{dt} + \frac{\partial \mathbf{v}}{\partial y}\frac{dy}{dt} + \frac{\partial \mathbf{v}}{\partial z}\frac{dz}{dt}$$

$$= \frac{\partial \mathbf{v}}{\partial t} + \frac{\partial \mathbf{v}}{\partial x}u + \frac{\partial \mathbf{v}}{\partial y}v + \frac{\partial \mathbf{v}}{\partial z}w$$

$-\frac{\partial p}{\partial (\ )}$ in the right sides of Eqs (3.27)–(3.29) represent the pressure gradient, which works in the same way as the discussion for a laminar flow in pipe.

$\mu\left(\frac{\partial \tau_{x(\ )}}{\partial x} + \frac{\partial \tau_{y(\ )}}{\partial y} + \frac{\partial \tau_{z(\ )}}{\partial z}\right)$ is the viscous force (Fig. 3.9). Note $\tau_{ij} = \tau_{ji}$.

$f_{(\ )}$ is the volume force (/unit volume) acting on the fluid. Volume forces include gravitational force, electrostatic force and magnetic force.

There are many cases where the Navier–Stokes equations can be expressed in much simpler forms. In fact, the Hagen–Poiseuille equation is the solution to a very simple case of one-dimensional Navier–Stokes equation, where both the acceleration and the volume force are zero (see also problem 3.6).

Reynolds Number

The meaning of each term in the Navier–Stokes equations (3.27)–(3.29) is clearly visible in the following way as explained in the previous section.

(intertial force) = (pressure) + (viscous force) + (volume force)

It is useful to find ratios between two terms to know which part plays a dominant role to define the characteristics of a flow. The Reynolds number Re is a measure showing the ratio of the inertial force to the viscous force. It is defined by the following formula:

$$Re = \frac{Ud}{(\mu/\rho)} = \frac{Ud}{v},\qquad(3.30)$$

Where U is a characteristic velocity of the fluid, and d is a characteristic length. Kinematic viscosity $v = \mu/\rho$, which is the dynamic viscosity divided by the density, is often used when discussing Reynolds numbers. The characteristic length, or characteristic dimension, indicates the scale of the system of concern. It has to be chosen based on the scale of the phenomenon being discussed. There are conventions for types of objects which are often considered in fluidic systems. For example, the radius or the diameter is used for spheres or circles, and the cord length is used for aircraft wings. For flow in a pipe or a microchannel,

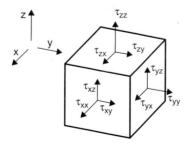

Figure 3.9: Viscous Force on a Control Volume.

the internal diameter or an equivalent diameter is generally used. The average velocity of the object relative to the fluid is usually chosen as the characteristic velocity.

When the Reynolds number is small, the effect of viscous force is larger resulting in heavy damping, and the flow tends to be steadier. The Reynolds number Re is used to determine if flow is laminar, transient or turbulent. When the Reynolds number is small enough, the flow is laminar and we can neglect the inertial force. When the Reynolds number is large enough, the flow is turbulent and we can neglect the viscous force. In the transient region, we have to consider all the terms. For example, flow in a pipe is considered to be laminar when $Re < 2300$, transient when $2300 < Re < 4000$, and turbulent when $Re > 4000$. These critical Reynolds numbers change [1] depending on the criteria such as shape of the flow, parameters to be focused, and choice of the characteristic velocity and length. As Eq. 3.30 implies, smaller systems tend to have smaller Reynolds numbers. Reynolds numbers are smaller than 10^3 for most of microdevices or molecular sensors and the flow is considered laminar.

Reynolds numbers are also related with behavior of creatures with different scales. Bacteria have an extremely small Reynolds number ($\sim 10^{-5}$). As the size of an animal becomes larger, the Reynolds number becomes larger (for whales, $Re = 10^{7-9}$), and the way the animal swims or flies changes (see also problem 3.5). There are several interesting articles about the scale of animals [2] and Reynolds numbers [3].

3.3 Microfluidics for Molecular Sensors

Microfluidics, as a miniaturized and high throughput technology, has been receiving prominent attention and has been experiencing rapid growth. Growth in microfluidics has expanded in research towards its potential applications as well as optimization of microfluidic channel designs for specific applications. Microfluidic devices have been utilized for lab-on-a-chip (LOC) or micro total analysis systems (μTAS), where one or more laboratory functions are integrated on a single chip of centimeter to millimeter scale [4–7]. The beauty of using microfluidic devices is that there are various attractive and advantageous characteristics, including down-scaling and miniaturization. Microfluidics utilizes and consumes less fluid volumes, and requires fewer materials to make the actual device, making the device and procedures more cost effective and capable of being mass produced [8–10]. In many situations they can be made disposable, hence creating and generating a market with constant need. Miniaturization brings molecules or samples closer together for more effective, efficient, and rapid interactions. Downsizing of microfluidic channels results in faster analysis and response times, owing to higher surface to volume ratios, shorter diffusion distances, and smaller heating capacities. Miniaturization also increases the portability of microfluidic devices because the devices are compact yet fully functional by itself.

Microfluidics involves fluid characteristics on a small scale. When scaling down, the fundamental physics changes rapidly. Reynolds numbers are small and fluid flow is laminar, which implies no turbulence, and the geometry of the channel defines the flow field. Viscous forces tend to dominate over inertial forces, and mixing of fluid by diffusion becomes dominant over that by advection.

In this section, we introduce basics of microfluidics and practical examples of technologies that are or potentially will be utilized for molecular sensing applications.

3.3.1 Microfluidic Device Basics

Fabrication of Microfluidic Devices

Most microfluidic devices are fabricated by replica molding techniques [11,12] 2.5.2 (see Chapter 2, Section 2.5.2) because it allows for simple, low cost prototyping of micro channels. Figure 3.10 shows a typical fabrication procedure of microfluidic chip. The polydimethylsiloxane (PDMS) microchannel is replicated from an SU-8 [13] photoresist negative pattern. Holes for tube connection are then mechanically punched through the channel. The channel is fixed onto a substrate, which often is a glass slide. There are several techniques for bonding the top micro channel and the bottom substrate. Typically,

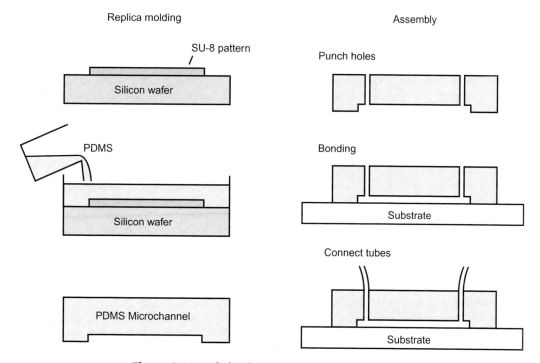

Figure 3.10: Fabrication of a PDMS Microchannel.

the PDMS part is O_2 plasma treated and pressed onto the glass substrate at $\sim 100\,°C$ to create permanent bonding. Simple mechanical clamps are also commonly used.

Other techniques for microchannel fabrication include lithography techniques such as hot embossing [14], injection molding [15], and bulk micromachining of silicon. Anisotropic wet etching [16,17] or DRIE [18,19] creates grooves on silicon substrates. See Chapter 2 for details of each method.

Pumping

Most microfluidic systems are pressure-driven by motor controlled syringe pumps or a peristaltic pump, in which rotating rollers squeeze a flexible tube. In microscale, we need to take fluidic resistances into account. Let us take a look at the Hagan−Poiseuille law again:

$$\Delta P = -\frac{8\mu L}{\pi R^4}Q \tag{3.31}$$

It relates pressure drop and the flow rate for steady, laminar flow in a circular tube. When we scale down the radius of the pipe R to $R/2$, the pressure needed to drive the liquid at the same flow rate Q becomes 2^4 times larger. Surface-driving forces such as capillary forces or electrostatic forces are sometimes employed in accomplishing the microfluidic manipulation.

Mixing in Microchannels

Diffusion and Scale Effect

Mixing of multiple reagents is a critical step for many of micro analytical systems. However, laminar flows in microfluidic systems make efficient mixing difficult. Laminar layers glide past each other, and the only exchange between them is accomplished by diffusion. The **Péclet number** Pe is a dimensionless number defined as the ratio of the rate of advection by the flow to the rate of diffusion:

$$Pe = \frac{LU}{D}, \tag{3.32}$$

Where L is the characteristic length of the system, U is the flow velocity and D is the diffusion coefficient. When Pe is larger, mixing by diffusion is more difficult or less efficient. Diffusion coefficient D is a parameter independent of the size of a system. As the system becomes larger, L and U become larger, and diffusion becomes less significant.

As we discussed in Section 3.2.1 diffusion length is defined as $L_D = 2\sqrt{Dt}$ ($\sqrt{2Dt}$ may be used for other boundary conditions). If we approximate $U \approx L/t$, Pe can be rewritten as:

$$Pe = \frac{LU}{D} \approx \frac{L^2}{Dt} \approx \frac{L^2}{L_D^2} \tag{3.33}$$

In this way, the Péclet number can be considered as a measure to compare the system lengths and the diffusion length. For example, a molecule that diffuse 1 μm in 1 ms will take 100 ms for 10 μm and 100 000 s for 10 mm. For larger systems, the Péclet number becomes larger and using diffusion for transport is unrealistic. Another good example to discuss the scale effect is the cardiovascular system of animals. Larger animals (mammals, birds, reptiles, amphibians, fish) need a circulatory system driven by a heart to transport oxygen and other substances, while smaller animals (e.g., many invertebrates such as insects) have open circulatory system that relies more on simple diffusion. There is an interesting argument that the distance effects on diffusion limits the maximum size of many invertebrates and skin-breathing vertebrates. Some researchers believe that the evolution of extinct giant insects was a result of the higher oxygen partial pressure, which occurred during Carboniferous and early Permian [20].

Mixing in Microchannels

Design of micro mixers has been an important topic of microscale engineering. Even in microscale, simple diffusion may not provide sufficient mixing of liquids. Different methods have been introduced for efficient mixing in microscale [19]. One design solution to sufficient mixing is to create a serpentine channel that will mix the two fluids as they pass through the bends (Fig. 3.11).

The goal of designing an efficient micromixer is to increase the area interfacing two flows. One extension of the serpentine channel is to create a three-dimensional structure as shown in Fig. 3.12(a) [22]. Figure 3.12(b) shows comparison of mixing rates with a straight channel, a two-dimensional square-wave channel, and a three-dimensional serpentine channel. It shows the average intensity of reacted phenolphthalein in the three channels after the streams have been in contact for 18 mm. The three-dimensional channel produces 16 times more reaction than a straight channel and 1.6 times more than the two dimensional square-wave channel. One drawback in this design is the difficulty in fabrication. As we discussed in Chapter 2, photolithography only allows design of two dimensional structures. Multilayers of PDMS films patterned separately were aligned and bonded to create the three-dimensional channel.

Figure 3.11: Mixing in a Serpentine Microchannel.

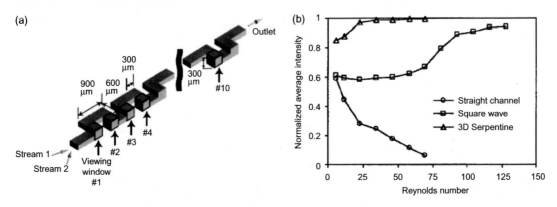

Figure 3.12: Three-dimensional Serpentine Micromixer.
From [21].

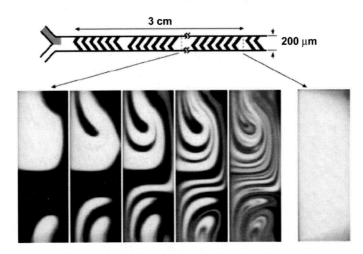

Figure 3.13: Chaotic Mixer.
Staggered herringbone mixer. Each cycle cuts the thickness of the layers in half and double the interfacial area. *From [23].*

Another method is to use staggered grooves called a herringbone or chaotic mixer (Fig. 3.13). The grooves are fabricated on the top wall of the channel and drive alternating helical flows. Each cycle cuts the thickness of the layers in half and double the interfacial area. In principle, the channel length required for mixing increases only logarithmically with the Péclet number [23].

Other methods include placing pillar shaped obstacles in the channel. Optimization of the layout of pillar arrays is discussed in [24]. Because micro pillars or other obstacles can be easily implemented into a microchannel by a simple single mask process, it is one of the most commonly utilized designs to promote micro fluidic reactions [25,26].

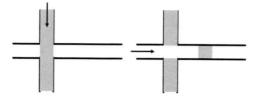

Figure 3.14: Laminar Flow Allows "Cropping" a Part of Flow.

The slow diffusion process can be used positively for localized transport. In a microchannel, as shown in the Fig. 3.14, laminar layers glide an intersection nearly undisturbed. By switching the flow direction a packet of liquid of the other channel can be extracted. This packet can be used to transport a known amount of molecules or cells for analysis. An improved technique to avoid diffusion is the use of two phase flows where oil works as the carrier medium and water work as the sample medium [27].

Another interesting use of diffusion in a microchannel is the H filter. It uses different diffusivities of particles. As we discussed in Section 3.2.1, the length of diffusion can be characterized by $\lambda = 2\sqrt{Dt}$, where the diffusion constant D is dependent on the size and shapes of the particles. According to the Stokes–Einstein equation, the diffusion constant of spherical particles with the radius r in liquid is given as:

$$D = \frac{k_B T}{6\pi\eta r} \tag{3.34}$$

where k_B is Boltzmann constant, T is the absolute temperature, η is viscosity.

Figure 3.15 shows the diagram of the H filter. As the medium flows through the filter, the molecules of interest, which have a larger diffusivity, spread across the channel. Larger waste products have a smaller diffusivities and remain confined in the initial stream [29].

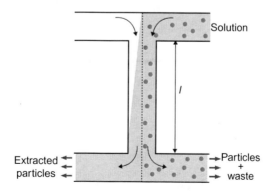

Figure 3.15: Diagram of the H Filter.
Adapted from [28].

Important Dimensionless Numbers in Microfluidics

Reynolds Number
The ratio of the inertial force to the viscous force:

$$\text{Re} = \frac{Ud}{(\mu/\rho)} = \frac{Ud}{v},$$

Where U is a characteristic velocity of the fluid, d is a characteristic length, and $v = \mu/\rho$ is kinematic viscosity. The flow is considered to be laminar when Re is small (typically $Re < {\sim}10^3$).

Péclet Number
The ratio of advection to diffusion, or a measure to compare the system lengths and the diffusion length:

$$Pe = \frac{LU}{D} \approx \frac{L^2}{Dt} \approx \frac{L^2}{L_D{}^2},$$

Where L is the characteristic system length, U is the flow velocity and D is the diffusion coefficient, L_D is the diffusion length.

3.3.2 Microfluidic Cell Separation and Detection

Advantages of miniaturizing bioanalytical tools include improved performance, speed and throughput, reduced costs and reagent consumption, and the possibility of multiplexing (micro-total analysis systems). Here, we describe techniques for microfluidic-based cell separation techniques and their applications. We first describe the principle of conventional flow cytometry or fluorescence-activated cell sorting (FACS). We then will look at affinity-mediated cell separation, magnetic activated cell separation, electrophoresis, and mechanical property-based separation. We examine the working principles behind the various microfluidic cell separation devices and how they are implemented. We will discuss the methods used to fabricate the various microfluidic channel-based cell separation systems, as well as the advantages and disadvantages of the techniques. Most importantly, we will look at the different applications in which the specific technique has been used for, including important research outcomes, commercially available devices, and any novel or recent topics of interest. We also look at the various design parameters and how the improvements or changes to the parameters improve cell separation in microchannels.

Flow Cytometry

Flow cytometry is a technology used to count and sort cells, and detect biomarkers [30–32]. Flow cytometry is routinely used in clinical practice, such as diagnosis of blood cancers, and is used in basic research. In a modern flow cytometer, particles (cells) pass through the observation area at a rate of several thousand particles per second and can be actively separated based on the specified optical properties.

Principles

A system for flow cytometry, or a **flow cytometer**, is a combination of (1) a flow chamber (or tube), (2) a laser-based fluorescence and/or light scattering microscope, (3) and a sorting mechanism.

1. Hydrodynamic focusing, which locates flowing cells in the center of the chamber, is a key technique used in the flow chamber. Chamber diameters are typically hundreds of micrometers. In hydrodynamic focusing, a fluid containing cells is injected into the center of a faster flowing sheath flow. The sheath and sample streams form a two-layer laminar flow. Because the sheath stream flows faster, the core stream velocity increases due to the viscous force acting between the two layers. On the other hand, the volume flow rate of the core stream is the same as the injection rate, reducing the cross sectional area of the core stream. As a result, the cells are "focused" into the center of the flow chamber. The hydrodynamic focusing allows the cells to be in the precise focal point of the laser based microscope.

2. Optical microscopes are very similar to standard fluorescence and light scattering microscopes in composition (see Chapter 5, Section 5.6.2 for fluorescence microscope and 5.7.2 for light scattering microscope). A photomultiplier tube (PMT) or an avalanche photodiode (APD), instead of a CCD in conventional microscopes, is used as the detector. Instead of recording 2D images, the PMT or APD detectors record photo intensities with higher sensitivity and can process thousands of cells in a second. Figure 3.16 shows the optical observation schematic. Lasers with different wavelengths may be used for multiple fluorescent markers. A forward scatter detector is located on the same optical axis as the laser illumination. It measures forward-scattered lights, while the direct illumination is screened by a shutter. Side observation optics with a

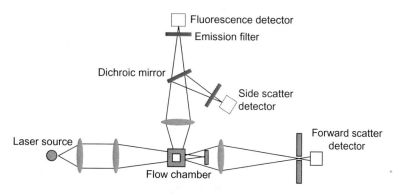

Figure 3.16: Optics of a Flow Cytometer.

side fluorescence detector and a side scatter detector is used to measure fluorescence and side-scattered light. A dichroic mirror (see Chapter 5.6.2) reflects scattered lights to the side scatter detector and transmits fluorescence signals to the fluorescence detector.

3. The sorting system is based on the technique developed for inkjet printers. Fulwyler first demonstrated successful separation of mouse and human erythrocytes in 1965. Figure 3.17 shows the diagram of the flow cytometric cell sorting. As the fluid is ejected from a nozzle as a liquid jet, the flow chamber is vibrated at typically 18 kHz to cut the fluid into a stream of droplets. Flow rates can be adjusted so that there are zero or one cell in a droplet. Each individual droplet can be either positively or negatively charged to be redirected by the electrostatic field created by two deflection plates and collected into tubes.

Commercially Available Systems

Flow cytometry systems from Becton, Dickinson and Company (BD) are among the most commonly used apparatuses. BD commercialized fluorescence-activated cell sorting (FACS), which is now a general name for flow cytometers. Other companies that produce flow cytometers include Beckman Coulter, Sony, and Millipore.

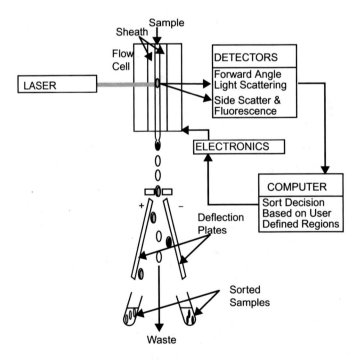

Figure 3.17: Cell Sorting Mechanism.
From [31].

Microfluidic Flow Cytometry

Several attempts have been made to create fluorescence-activated microfluidic systems for cell sorting. Figure 3.18 shows examples of microfluidic flow cytometry systems reported in literature. In an early study by Fu et al., electro-osmotic flow was induced by applying high voltages between Pt electrodes inserted into the microchannel. The sample throughput (10−20 cells/s) was limited by the slow flow rate controlled by the applied voltages [132].

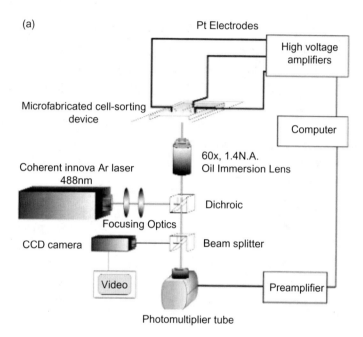

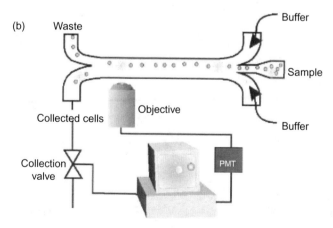

Figure 3.18: Microfluidic Flow Cytometers.
(a) from [31]. (b) from [32].

Wolff et al. used a switchable valve attached on the waste outlet and forced cells of interest to the collecting channel. The system had a high throughput (12 000 cells/s) with an acceptable 100-fold enrichment of fluorescent beads [33]. Wang et al used optical forces, which are commonly used for optical trapping of living cells, for active control of cell routing. They demonstrated sorting of HeLa cells with a throughput of ~ 100 cells/s with a $\sim 60-70$-fold enrichment [34].

Affinity Mediated Separation

Principle

Due to the natural high selectivity and specificity, antibody–antigen interactions have been widely used and integrated with microfluidic channels for cell separation purposes. Figure 3.19 illustrates the fundamental mechanism of the affinity mediated cell separation [35]. Antibodies are attached to the inner surface of the microchannel. When the biological samples are flown through the channel, antibodies selectively bind antigens of target cells. As a result, target cells are captured on the surface while non-target cells flow through. The process may be followed by a washing step where non-specifically bound cells are removed by flowing a buffer solution.

The efficacy of the affinity mediated separation is dependent on the antibody–antigen bonding between the target cells and channel substrate. Therefore, increasing the probability of physical contact between target cells and capture structure is critical in optimizing the separation performance. Meanwhile, the fluid shear stress inside the microchannel needs to be adjusted to favor target cell capture and non-target cell removal. Methods including (1) increasing the contact area [26,36], and (2) introducing in-channel mixing [39] have been demonstrated to improve the capture efficiency.

Applications

Selective cell separation from biological body fluids has broadly impacted medicine, enabling a variety of analyses with high accuracy. Several devices have been proposed for cell separation based on the cell affinity.

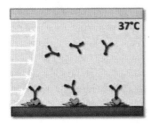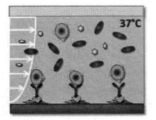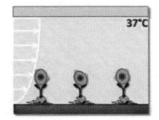

Figure 3.19: Schematic Showing an Affinity Based Cell Separation System.
From [35].

Moon et al. developed a portable microchip counting platform for HIV monitoring based on the CD4$^+$ T-lymphocyte counting [37]. The target CD4$^+$ cells were captured from unprocessed fingerprick volume of HIV-infected whole blood using anti-CD4 antibody, which was previously immobilized on the surface of a microfluidic channel. After capture, the cells were imaged using a lensless CCD platform with large field of view. The shadows of cells were automatically detected and counted by an automatic recognition program. Chin et al. proposed a microfluidic system [38] that can detect human immunodeficiency virus (HIV) using only 1 µL of unprocessed whole blood. Reduction of silver ions onto gold nanoparticles in an immunosandwich procedure was adopted to amplify the signals. These types of detecting methods with extremely low cost are promising in areas where resources are limited [133].

Microchannels with different designs have been introduced to isolate rare circulating tumor cells (CTCs) from human blood samples [26,39,40]. Viable CTCs were separated from peripheral whole blood samples, based on the interactions between target CTCs and anti-EpCAM antibodies coated on the surface of the microchannel, without pre-labeling processes. To enhance the contact between the CTCs and microchannel wall, micro-structures, such as the micropillars [26], herringbone grooves [39], and 3D nanostructures [40], were implemented into the microchannel. CTCs were successfully revealed from cancer patient blood samples. To further improve the separation selectivity, aptamer was integrated to detect multiple types of CTCs simultaneously [41]. Aptamers that bind to different type of cells selectively were immobilized to different regions in the microchannel. Hence, different cell lines can be sorted into independent fractions.

Magnetic Cell Separation

Principle

Magnetic separation has been utilized in microfluidic flow cytometry for increased separation. In magnetic sorting, sample cells are first incubated with magnetic particles or beads with the recognition molecule, typically antibodies, for magnetic labeling. Then, a magnetic field gradient is used to isolate the magnetic beads, which in turn picks out the cells. Cells labeled with magnetic particles are separated by magnetic field gradients that trap or alter the flow of cells loaded with selective particles [42]. Magnetic sorting can be operated in either a serial or parallel manner, which results in a higher throughput.

Magnetic particles are typically functionalized with antibodies for target proteins. The size of the particles, or beads, tested in those studies ranged from 100 nm [134] to 4.5 µm [43] (see discussion in Chapter 2, Section 2.2.). Dynabeads from Life Technologies are among the most commonly used commercially available magnetic particles.

Cell separation can be categorized into positive selection and negative selection. One type of separation is positive selection in which desired cells are magnetized. In negative selection, undesired cells are targeted and removed via magnetization. Immunomagnetic cell separation is the most common method and utilizes antigen presence on the surface of the cell as the cell marker [44]. Magnetic separation is a powerful medical diagnostic tool, but is also used in laboratory settings to isolate pure sources of cells for experimental purposes. Conventionally, separation has been performed with a conical tube based setup where strong magnets (~ 1 T) attract magnetized cells toward the inside wall of the tube [45]. Commercially available detection kits including MACS (magnetic-activated cell sorting) and CellSearch use the conical tube based separation setup. MACS [45] is a general-purpose cell sorting kit that has been utilized for various types of cells [46−50]. As of 2013, the CellSearch system [135] is the only US Food and Drug Administration (FDA) cleared methodology for enumeration of circulating tumor cells (CTCs) in blood. Detection of CTCs in blood [51−55] is becoming a very important research field. We will describe other techniques for CTC separation in later sections.

For sensor applications, cell separation with microfabricated analytical devices is the main concern. Here we mainly describe magnetic cell separation performed in microfluidic systems.

Magnetic separation is affected by the level of magnetic force as well as the fluid drag force. In order for separation to occur, the magnetic force must balance the viscous drag. The magnetic force is determined by the magnetic dipole (m) and the magnetic field (B) as given in the equation below [56,57]:

$$\mathbf{F}_m = (\mathbf{m} \cdot \nabla)\mathbf{B} \tag{3.35}$$

The total moment of a cell magnetized with magnetic nanoparticles can be expressed as:

$$\mathbf{m} = \frac{V\Delta\chi}{\mu_0}\mathbf{B} \tag{3.36}$$

where $\Delta\chi$ is effective magnetic volumetric susceptibility of the cell, $\mu_0 = 4\pi \times 10^{-7}$ $T \cdot m \cdot A^{-1}$ is the magnetic permeability of vacuum. $\Delta\chi$ is defined by the amount of magnetic particles attached and the volume of the cell in the following way:

$$\Delta\chi_C = N\frac{R_F^3}{R_C^3}\Delta\chi_F, \tag{3.37}$$

where N is the number of magnetic particles attached to the cell, R_F and R_C are radii of the Ferrofluid particle and the cancer cell, respectively, and $\Delta\chi_F$ is the volumetric susceptibility of the Ferrofluid particle.

Assuming $\nabla \times \mathbf{B} = 0$,

$$(\mathbf{B} \cdot \nabla)\mathbf{B} = \frac{1}{2}\nabla(\mathbf{B} \cdot \mathbf{B}) \tag{3.38}$$

From Eqs. 3.35, 3.36 and 3.38, the force acting on the cell given in the following way [57]

$$\mathbf{F}_m = \frac{V \Delta \chi_C}{2\mu_0} \nabla B^2 \tag{3.39}$$

where, B is the magnetic field intensity, or the magnitude of the magnetic field **B**.

Drag force $\mathbf{F}_d$ from the medium is given by

$$\mathbf{F}_d = 6\pi \cdot \eta \cdot R_C \cdot \Delta \mathbf{v} \tag{3.40}$$

where η the medium viscosity, $\Delta \mathbf{v}$ is the cell velocity relative to the medium. When we assume a quasi-static motion, the two forces equal each other, namely

$$\mathbf{F}_d = \mathbf{F}_m \tag{3.41}$$

Substituting Eqs. 3.39 and 3.40 into Eq. 3.41 and using $V = (4/3)\pi R^3$ gives the instant relative velocity of cells:

$$\Delta \mathbf{v} = \frac{R^2 \Delta \chi_C}{9\mu_0 \eta} \nabla B^2 \tag{3.42}$$

Eq. 3.42 defines the motion of the cell in a microchannel.

Application

Ingber *et al.* developed a microfluidic device which removes magnetized *Escherichia coli* bacteria from flowing solutions containing red blood cells [58]. The multiport channel shown in Fig. 3.20 is based on a typical design used for several types of microfluidics-based separation. The same group also demonstrated a blood cleansing device that removes *Candida albicans* fungi from flowing human whole blood with over 80% clearance at a flow rate of 20 mL/h [136].

Furdui et al reported an integrated silicon microchip for separation of Jurkat cells from reconstituted horse blood samples as well as human blood (about 1:10 000 ratio of Jurkat cells to blood cells) [59]. Zborowski et al developed a device for blood screening to test for human malaria. The device exploits the fact that *Plasmodium* species parasites produce hemozoin that gives magnetic susceptibility to red blood cells [60].

Immunomagnetic separation of cancer cells has also been studied by several groups. Zborowski et al demonstrated separation of MCF7 cells (breast cancer cell line) from mixtures of human leukocytes [61]. Figure 3.21 is the microfluidic device developed by the authors' group [62,64].

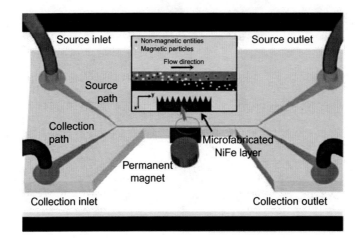

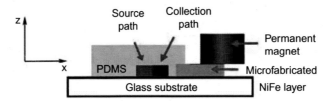

Figure 3.20: Magnetic Separation in Microchannels.
From [58].

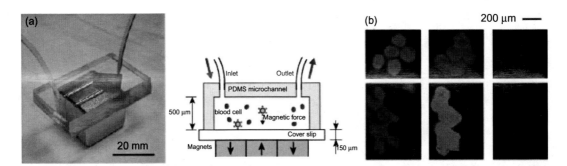

Figure 3.21: Microchip-Based Immunomagnetic Assay for Detection of Circulating Tumor Cells. (a) Schematic of the Experimental Setup and (b) Cancer Cells Captured from Patient Blood Samples.
(a) From [62]; (b) from [63].

Labeled cells are effectively captured because the microchannel is thin and flat and arrayed magnets with alternate polarities provide a sharp magnetic field gradient. Cell capture rates of 90% and 86% from whole blood flowing at 10 mL/h were demonstrated for COLO205 (colon cancer) and SKBR3 (breast cancer) cells. The screening system has been tested for clinical screening and successfully isolated CTCs from the blood samples of cancer patients with breast, lung and prostate cancers [63].

Ahn et al developed a total analytical system that integrates components such as microvalves, flow sensors, and filters for magnetic bead-based biochemical detection [65]. Other examples of the use of magnetic force with microfluidic devices have been reviewed in the literature [66].

Electrophoresis

Principle

Electrophoresis is a technique where an applied electric field separates target cells from the medium. A particle, or a cell, suspended in a medium of different dielectric characteristics is electrically polarized in an alternating electrical field [67]. The particle polarization and the external electrical field induce a lateral dielectrophoretic (DEP) force, which directs the particle toward the minimum of dielectric potential, and electrorotation (ROT) that rotates the particle. The magnitude and direction of these forces for living cells are dependent on cell characteristics such as composition, morphology, and phenotype, and can be used for cell separation. Specific capacitance of a cell membrane is given by the morphology including microvilli, membrane folds, and blebbing.

With regards to the separation mechanics, forces acting on a cell include gravitational force, DEP force, fluid drag, and hydrodynamic lift effects. The DEP force acting on a particle with a dipole moment $\mathbf{P}$ in the electric filed $\mathbf{E}$ is given as

$$\mathbf{F}_{DEP} = (\mathbf{P} \cdot \nabla)\mathbf{E} \tag{3.43}$$

Note $\mathbf{P}$ is a function of $\mathbf{E}$. Under a uniform electric field $\mathbf{E}$ between two parallel plate electrodes, the attraction forces acting on a neutral particle from the two electrodes equal each other (see Fig. 3.22(a)), and no net force is observed. When the two electrodes are asymmetric (see Fig. 3.22(b)), the two forces are different and the particle is attracted to one of them.

For the case with a cell suspended in a medium, the average DEP force can be written as

$$\mathbf{F}_{DEP} = 4\pi r^3 \varepsilon_m \alpha \cdot \nabla E^2_{RMS} \quad \text{or} \quad \mathbf{F}_{DEP} = 4\pi r^3 \varepsilon_m \alpha V^2 (k_x \mathbf{a}_x + k_x \mathbf{a}_Y) \tag{3.44}$$

where ε_m is the dielectric permittivity of the eluate, V is the peak electrode voltage, and α is the real part of the Clausius—Mossotti factor expressing the effective polarizability of the cell in the buffer. k_x and k_y are the maximum vertical and horizontal components, respectively, of the field non-uniformity factor of ∇E^2_{RMS} for an applied 1 V peak—peak voltage. They depend on the geometry of the electrodes and the relative position of the cell. They become larger when the cell is closer to the electrodes.

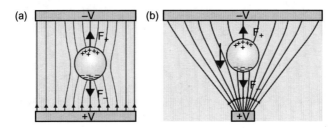

Figure 3.22: The Principle of Electrophoresis. (a) No net force acts on a particle in a uniform electric field. (b) Asymmetric electric field creates an attraction force.

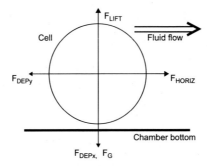

Figure 3.23: Forces Acting on a Cell in a Medium.
Forces affecting cells within the dielectric affinity column: gravity F_G, dielectrophoresis F_{DEPx}, F_{DEPy}, fluid drag F_{HORIZ}, and hydrodynamic lift effects F_{LIFT}. *From [67].*

The horizontal drag force in a contact with a wall is

$$F_{HORIZ} = 6\pi c r\, \eta(v_m - v_p) \tag{3.45}$$

where η is the dynamic viscosity, v_p is the cell velocity, v_m is the medium velocity and $c = 1.7$ (Fig. 3.23).

The y component of $\mathbf{F}_{DEP}$ and F_{HORIZ} defines the horizontal cell motion [65].

Applications

Gascoyne et al [67] show that breast cancer cells, MDA231, are dielectrically different from erythrocytes and leukocytes and demonstrate that cancerous cells can be removed from a blood sample because of dielectric differences. Each cell is dielectrically different based on its different threshold frequency. Separation chambers were constructed above electrodes. The electrode array sized 17.6 mm × 55 mm. The electrode element width and spacing were 80 μm. The electrode has a comb-shaped structure to create non-uniformity in the electric field to induce larger ∇E for the attraction force. An example of the electrode is shown in Fig. 3.24.

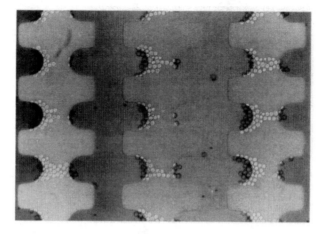

Figure 3.24: Cancer Cells Captured by Dielectrophoresis.
From [67].

Sinusoidal voltages of fixed or swept frequencies were applied to the electrodes. A suspension of MDA231 breast cancer cells were spiked into blood to test separation efficiency. First, buffer was filled into a chamber, and a 30 μl cell mixture was then injected. A 200-kHz signal of 5 V peak-peak voltage was applied to the electrodes to collect cells around the electrode.

After the initial collection at 200 kHz, a swept signal from 80 to 20 kHz was added to shake entrapped blood cells. A buffer solution was passed through the chamber to flush blood cells into the outlets. After 20 min, blood cells were further released by cross-flowing buffer from additional ports. Figure 3.24 shows captured cancer cells around the electrode tips. The purity of cancer cells remaining at the electrodes was better than 95%.

Wang et al reported on the separation of colorectal cancer cells using dielectrophoresis [68]. They designed a DEP colorectal cell separation system with two electrodes at the bottom surfaces of the microchannel. The electrode pair is in parallel and has a 45° angle to the stream-wise direction in the main channel. Alternating current electric signals are applied to the two electrodes. The operation principle was that negative DEP force acts on the target cells and repels them against the main stream. There is a side channel placed near the electrode gap at one side of the main channel. The channel carries the target particles that are separated from other particles in the main channel by the negative DEP force. The important part is finding a frequency that gives a negative DEP force specifically for cancer cells. Figure 3.25 shows the DEP forces on the HCT116 cells measured for different signal frequencies. The cancer cells experience a negative force in the frequency band of 1 Hz−6 MHz and 31−75 MHz and experience a positive DEP force in the frequency range of 6−31 MHz.

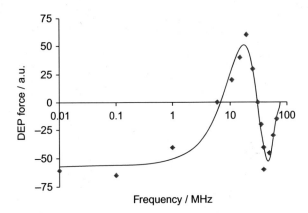

Figure 3.25: DEP Spectrum of the HCT116 Cells.
From [68].

Cell Separation Utilizing Mechanical Properties

Acoustic Separation

Acoustic separation, or acoustophoresis, is the method that separates particles using high intensity sound waves. Standing wave based cell separators have been promoted as a means for contactless handling and manipulation of cells [69–73]. Contactless manipulation eases the physical stresses and forces experienced by cells in the microfluidic channel thereby preserving the viability of the cells and allows for retrieval and post analysis.

Laminar flows in the microscale domain make the fractionation systems created by acoustic forces more efficient in continuous flow particle separation. Furthermore, microchannels require reduced resonator dimensions resulting in higher-resonance frequencies. Higher-resonance frequencies yield stronger radiation forces on particles because the primary radiation force is proportional to the frequency. Particle focusing and separation performance are both improved. However, standing wave based microfluidic separators are limited by channel geometry and particle geometry. As the particles become smaller, the primary radiation force rapidly decreases. Therefore, only particle sizes of about tenths to tens of micrometers can be manipulated by the primary radiation force within aqueous based solutions. Other physical parameters such as sedimentation and channel occlusion can also affect the performance of the primary radiation force [74].

Acoustic cell separation in a microfluidic channel begins with suspended particles in medium entering the rectangular cross-section channel, through an inlet. A piezoceramic actuator is used to generate a half-wavelength acoustic standing wave between the side walls of the microchannel which creates an acoustic force field that is perpendicular to the flow direction (see Fig. 3.26 (a)). The acoustic force moves the particles or cells laterally across the channel towards the center as they flow down the channel length. The rate at

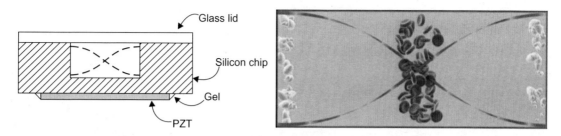

Figure 3.26: Acoustic Cell Separation. (a) Schematic of an Acoustic Separation Chip. (b) Particles Are Separated Based on Size and Density.
From [74].

which they laterally translate is dependent on the particle size, density, compressibility, and acoustic pressure amplitude. Most rigid particles and cells are moved to pressure nodes. Liquid elements or bubbles are moved to anti nodes.

The key principle of acoustic separation is to create a particle or cell gradient across the channel at the channel end in order to correctly sort and separate various particles and cells (see Fig. 3.26 (b)).

Various parameters that were of major consideration were channel length, the power applied to the piezoelectric element, and the axial acoustic primary radiation force (PRF). The channel length was determined by the desired flow rate, particle mixture used, and the acoustic input power. The power applied to the piezoelectric element was determined based on the size of the particles and the desired lateral displacement of the particle type. The axial acoustic PRF was the main driving force of the separation process. The magnitude of the axial PRF, F_r, is dependent on the properties of the medium and the particles, and the amplitude and wavelength of the standing wave. These dependencies can be described in equations given by the following:

$$F_r = - \left(\frac{\pi p_o^2 V_p \beta_m}{2\lambda} \right) \phi(\beta, \rho) \sin (2kx), \tag{3.46}$$

$$\phi = \frac{5\rho_p - 2\rho_m}{2\rho_p + \rho_m} - \frac{\beta_p}{\beta_m} \tag{3.47}$$

where φ is the particle's acoustic contrast factor, ρ_m is the density of the medium, ρ_p is the density of the particles, β_m is the compressibility of the medium, β is the compressibility of the particles, p_o is the pressure amplitude, V_p is the volume of the particle, λ is the ultrasonic wavelength, k is $2\pi/\lambda$, and x is the distance from a pressure node. The magnitude of the axial PRF is the primary factor in determining the rate of particle movement perpendicular to the flow direction.

The first term of the right side of Eq. 3.46 states that the force is proportional to the volume of the particle and the squared acoustic amplitude, which is related to the power supplied to the piezoelectric ceramic. Along with axial PRF, the opposing viscous drag force (see Eqs. 3.40 and 3.45) also plays a role in determining the rate of particle movement [17,69,75,76].

Figure 3.27 shows an top-view illustration of microchip-based acoustic separation reported by Laurell et al. The separation chip is composed in the same way as illustrated in Fig. 3.26. Particles with negative φ factors are collected in the central part of the flow, while particles with positive φ factors are moved to the sides of the channel to be separated. Separation efficiencies of more than 95% were demonstrated with a flow rate of 0.3 mL/min, an actuation voltage: 10 Vpp, and an actuation frequency of ~ 2 MHz. One application they studied is removal of human lipid (fat) particles from a suspension of red blood cells, simulating blood washing needed in open-heart surgery. They used a sonicated emulsion of tritium-labeled trioleine in saline solution as test lipid particles. With the optimized operation condition described above, they demonstrated that 85% of the lipid particles added to 1% by volume were removed independent of the concentration of erythrocytes which ranged 2.5−10.0% by volume. They used the concept of increasing the throughput by using eight parallel channels.

The microchip designed by Kapishnikov et al [77] was used to perform particle size sorting in a size spectroscope, rather than just a separator. The device operates on the same working principle as the previous device [74]. The use of the standing wave field can cause particles to accumulate in the nodes of the acoustic force and is dependent on particle properties and acoustic frequency and wavenumber. The microchannels were produced via soft lithography techniques using SU-8, a UV-sensitive epoxy. The microfluidic chip itself was made with silicone elastomer. Two designs were investigated. In the first design, separation of particles occurs between two parallel transducers located in the inlet and produces the standing wave (see Fig. 3.28). Filtered pure solvent is removed via two side outlets while the concentrated solution with particles is removed via the central middle channel. The second design is a three stage device that works in a similar way with particle separation occurring on each stage of the separator. Particles affected by the standing waves are removed via the two side outlets while the concentrated solution flows into the central channel and into the next stage for more dilution and separation. At each stage, the flow

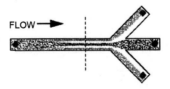

Figure 3.27: Separation of Particles with Negative (Black) and Positive (Gray) φ Factors.
From [69].

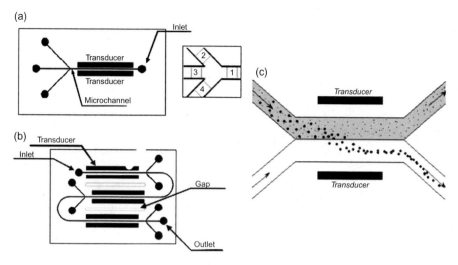

Figure 3.28: Schematic of the Acoustic Cell Separators by Kapishnikov et al. (a) The One-stage Microchannel. (b) The Three-stage Microchannel. (c) The Cell Size Sorter [77].

rate is three times less than the previous stage, which improves the filtering efficiency. The cell size sorting component is another microchannel and operates on the idea that particle velocity depends strongly on size and therefore, larger particles will move faster through an interface between a pure solvent and solution of different size particles [77].

Despite successful separation of particles and cells within the laboratory settings, microfluidic acoustic cell separation has seen little advancement in clinical settings. Compared to more developed techniques, such as fluorescence activated cell separation, microfluidic acoustic cell separation is less reliable [78]. Because the method is extremely reliant on physical parameters, such as particle or cell size and density, that are hard to control, the piezoceramic actuation must be precise and actuate at the correct fundamental frequency in order to effectively separate the particles and cells. The piezoceramic actuation must actuate at the right frequency to generate the gradient across the microfluidic channel which the separation is based upon. In large part, quantifying particle separations performed by acoustic separation remains challenging and validating methodology performance is equally as difficult. Especially when working with particles of micron sizes, the ability to effectively distinguish between different particle sizes dramatically decreases and the noise level rises [78].

Separation Based on Stiffness

In the next two sections, we discuss cell separation methods that utilize direct mechanical interaction between cells and the device. This section discusses separation based on stiffness of the cell. Zhang et al [79] developed a microfluidic chip to separate and enrich cells based on their deformability. The cells were also analyzed using fluorescence microscopy, flow cytometry, and assays to determine the qualities of their phenotype [79]. The separation chip, which was fabricated based on PDMS soft lithography, utilizes artificial microbarriers and hydrodynamic force to separate deformable cells from stiff cells (Fig. 3.29).

The separation occurred due to differences in deformability, while the cells also differ in their metastatic potential. Their work was applied to separate breast cancer cells MDA-MB-436 and MCF-7 that have distinct deformabilities and metastatic potential. More importantly, they separated a heterogeneous cancer cell line SUM149 into flexible and stiff subpopulations and analyzed their gene expression. The flexible population is associated with overexpression of genes which contribute to cancer formation. Their results suggested that tumor-initiating cells are more deformable and they are less differentiated in terms of cell biomechanics [79].

Another group has employed microfluidics to perform passive sorting and separation of cells by deformability [80]. Their technology employs microfluidic filters in series with pore sizes decreasing incrementally. As the cells flow through the device, the cells experience stresses that are similar to when they pass through capillary beds [80].

Separation Based on Size

There are three main techniques which are used to sort cells based on size: hydrodynamic sorting, microfilters, and centrifugation.

Davis et al separated white blood cells, red blood cells, and platelets from blood plasma at flow velocities of 1000 μm/s and volume rates up to 1 μL/min [81]. They utilized a size-

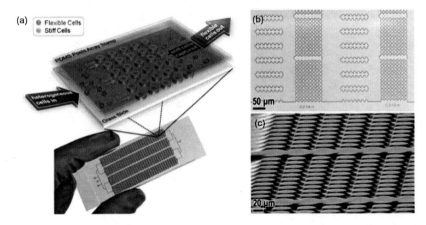

Figure 3.29: Cancer Cell Sorting Based on Deformability. (a) Scheme (upper) and photograph (lower) of the separation chip. (b) Optical microscopic image of the separation chip (c) SEM image of the separation chip.
From [79].

based hydrodynamic separation method previously introduced by Huang's group [82]. The method is based on a flow channel which they call a continuous-flow deterministic array. A particle smaller than a critical hydrodynamic diameter D_c follows streamlines cyclically through the gaps, moving in an average downward flow direction, staying within a flow stream. A particle larger than the critical diameter do not fit into the first streamline and are moved by hydrodynamic lateral drag and displaced into the next streamline at each obstacle. Streamlines for both a small particle and a large particle is shown in Fig. 3.30.

Microfilters can separate cells based on a combination of size and cell deformability. Some microfilters use micropores which act as sieves [83−85]. Figure 3.31 is an example of cancer cell separation from blood, reported in [84].

Circulating tumor cells are typically in the size range of 12−25 μm. Leukocytes, which compose most of the blood cell population, are typically in the size range of 7−15 μm. Size-

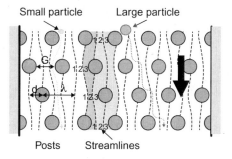

Figure 3.30: Hydrodynamic Separation Based on Size.
From [81].

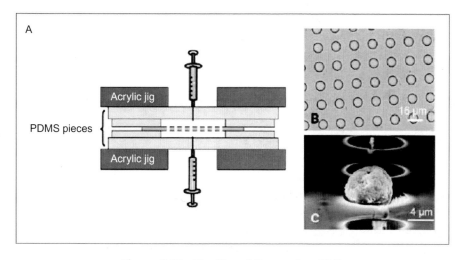

Figure 3.31: Size Based Separation [84].

based sorting is attractive because there is no sample preparation necessary and no labels are needed. The largest problem is the specificity. Size-based selection leaves types of cells that are morphologically similar to CTCs, and fails to find cancer cells that are as small as leukocytes. Clogging of blood cells is a practical problem associated with this method.

Centrifugation uses centrifugal force as the force of separation and can separate cells based on their density. For an example of blood cell separation, leukocytes (white cells), thrombocytes (platelets), and cancer cells are sorted in a thin layer called the buffy coat between the layers of erythrocytes (red cells) and plasma. There are different types of density centrifugation media, including Ficoll and OncoQuick [86].

Methods for Microfluidic Cell Separation

Immunoassay-based methods
- Flow cytometry, fluorescence-activated cell sorting (FACS)
- Magnetic separation, magnetic-activated cell sorting (MACS)
- Affinity mediated separation
  Utilization of electrical properties
- Electrophoresis
  Utilization of mechanical properties
- Acoustic separation
- Size-based separation (hydrodynamics microfilters, centrifugation)

3.3.3 Microfluidic Devices for DNA Analysis

DNA analysis is important for detecting diseases and mutations. It is now possible to determine if a patient carries a genetic disorder or has a certain gene or even to perform a paternity test using DNA analysis. There are several fundamental techniques used for DNA analysis including: detection, amplification (polymerase chain reaction (PCR)), and separation (electrophoresis). Furthermore, there is a continuous search for more efficient ways to perform these and other key steps in DNA analysis. Many steps can now be carried out in microfluidic devices. Microfluidics has the potential to make DNA analysis faster, more efficient, less expensive, and more accurate.

For DNA detection, gene microarrays and other microsystems are becoming more widely available. Microarrays allow for very large numbers of reactions to occur simultaneously within a very minute area, eliminating the need for large amounts of materials, equipment, and manpower. There have also been many advances in miniaturizing PCR as well as different electrophoretic methods. The aim for microfluidic devices in DNA analysis is to create a complete system in which DNA can be analyzed on a single chip. Ideally one device would be able to perform separation of samples, PCR, and detection of DNA.

Microfluidic Devices for DNA Amplification

DNA amplification is an essential first step for DNA analysis. PCR is generally used to perform DNA amplification, but the conventional PCR method is often time consuming, expensive, and carries high risks for contamination [88]. With the advent of microfabrication technology, the development of microscale PCR devices has been possible. Micro-PCR offers advantages including less use of reagents, rapid cooling/heating rates, decreased power consumption, and portability. The idea of a lab-on-a-chip (LoC) system is becoming increasingly popular and incorporating PCR in these devices will greatly improve efficiency and will greatly develop microfluidic diagnostic devices. PCR Microfluidic devices allow for multiple reactions to occur simultaneously while decreasing the risk of contamination and increasing efficiency [87].

Principles

PCR, invented by Kary Mullis in 1983, is a tool used to exponentially amplify a DNA sequence. To amplify the target DNA segment, two primers, a forward and reverse primer, are designed to complementary base pair with the leading and lagging parts of the target DNA. Primers, which are generally 18 to 30 bases long, are mixed into an aqueous solution containing the target DNA, reaction buffer, heat-stable DNA polymerase, deoxynucleoside triphosphates (dNTPs), and magnesium. PCR amplification is separated into three steps: (1) denaturation, (2) annealing, and (3) extension. The three steps are illustrated in Fig. 3.32.

1. The target DNA sequence is denatured to two single strands by heating the sample to about 98 °C for 30 s to 2 min.
2. The temperature is typically decreased to between 55 and 65 °C for 20 s to allow the primers to anneal to the target DNA. The annealing temperature is about 5° below the melting temperatures of the primers.
3. The temperature is increased to 70 to 80 °C for about 25 s per kilobyte of target DNA length and the DNA polymerase extends the primers by adding the appropriate dNTP.

The three-step amplification process is then repeated for 20 to 30 cycles to obtain high quantity of target DNA. Issues associated with PCR include amplification of nonspecific PCR products if the primers hybridize to non-complementary sequences. Potential by-products of PCR are primer dimers, primer molecules that hybridize to each other instead of matching with the target DNA.

There have been several approaches to realize microfluidics cased PCR systems. As described above, the key technologies required for microfluidic PCR is a capability of repeated precise temperature control and handling of small amount of liquid volumes without contamination.

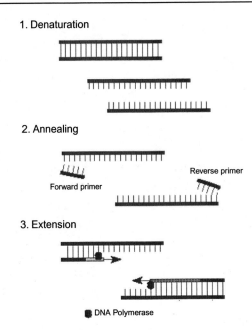

Figure 3.32: Polymerase Chain Reaction (PCR).

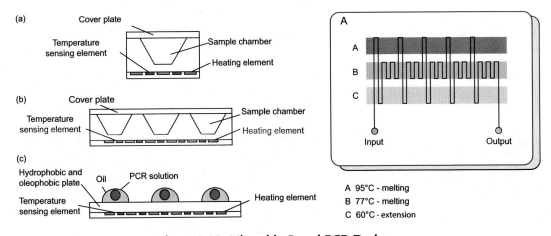

Figure 3.33: Microchip Based PCR Devices.
Left image is from [137] and the right image is from [92].

Figure 3.33 illustrates two types of PCR microfluidic devices. The one to the left represents the stationary chamber type and the one to the right represents the flow-through type.

During PCR in microfluidics devices the reaction mixture undergoes many cycles of varying temperature, and thermal cycling is implemented in two main ways. In the Stationary PCR device, there is a PCR inlet and outlet, where the solution flows through the device, and there are chambers in which the temperature can be controlled. In this case the PCR solution is stationary. An early stationary microfluidic PCR device was reported by Northrup et al [89,90]. This device contains heating elements with coils, which are heated by an electrical current. This type of device is generally made of silicon, glass, PDMS, PMMA, polycarbonate, polyimide, or epoxy.

In the flow-through PCR device, the PCR solution is pumped through the channel. The temperature goes through three temperature zones necessary for PCR amplification multiple times. The idea of flow-through PCR was first introduced by Nakano et al [91]. On-chip serpentine rectangular channel based flow-through PCR microfluidics is currently the most popular implementation. The device was first conceived by Kopp et al in 1998 [92].

There are various amounts of methods for fabrication of PCR microfluidics devices. Generally silicon/glass-based and polymer based microfabrication methods are used. Polymers such as PDMS and PMMA are also common. Silicon is the better choice for PCR devices because its higher thermal conductivity allows rapid thermal cycling. However, silicon's opacity prohibits optical measurements and is a significant drawback. PDMS has emerged as a cheaper, more biocompatible alternative than silicon, and is the current candidate for building disposable microfluidic PCR devices. Drawbacks from PDMS include the loss of sample due to its high permeability as well as bubble formation due to its inherent hydrophobicity. Microfluidic channels can be made using soft lithographic techniques. Silicon and glass-based devices are generally made by basic silicon micromachining techniques, which includes: photolithography, thermal growth of silicon oxide, chemical etching, electrochemical etching, ion etching, chemical vapor deposition, physical vapor deposition, epitaxy, and anodic bonding [93].

Applications

Conventional PCR techniques have been used to quantify viruses and bacteria including HIV-1 RNA levels in plasma [94], hepatitis B and C viruses [95], human papillomavirus [96], *Chlamydia trachomatis* infections from urine samples [97], *Neisseria gonorrhoeae* [98], and cytomegalovirus [99].

PCR microfluidic devices have a potential to expand the use of PCR into wider fields of research and clinical applications. One significant goal is efficient and accurate DNA detection. Compared to traditional PCR, PCR microfluidic devices are more effective at detecting DNA and are less prone to contamination. PCR microfluidic devices will amplify the DNA of the sample and detect target DNA for diagnosis of diseases such as human immunodeficiency virus, human papillomavirus, hepatitis virus, *Salmonella typhimurium*, *Mycobacterium tuberculosis*, and malaria [100]. The devices are also used to detect hereditary genetic disease, *E. coli*. microbial detection, and identification of various biological agents. Because the basic principle is the amplification and detection of specific DNA, there are many applications [101]. A microfluidic digital PCR device was created to enable multigene analysis of environmental bacteria [102]. The device allows for rapid separation and partitioning of single cells from a complex sample and represents a major breakthrough in environmental science, because microbial species can be easily identified.

Implementation

Several approaches have been reported on integrating PCR devices into micro total analytical systems (μTAS) [103−106]. An early, and excellent example of fully integrated PCR system is reported by Liu et al [103]. This μTAS can automatically perform every step of analysis including sample preparation, cell pre-concentration and purification, PCR, and electrochemical detection. The device has been proven to detect pathogenic bacteria in human blood samples and single nucleotide polymorphisms as well.

The device contains a plastic fluidic chip, a printed circuit board, and a sensor microarray chip, shown in Fig. 3.34. The chip is $60 \times 100 \times 2$ mm in size and contains channels and chambers that are from 0.5 to 1.2 mm deep and 1 to 5 mm wide. The chip is made out of polycarbonate using computer-controlled machining. Electrochemical pumping was achieved by adding 0.5 mm diameter platinum wires and placing them in contact with wells containing NaCl solutions. Thermally actuated micro-valves [107] were made by melting paraffin into the channels.

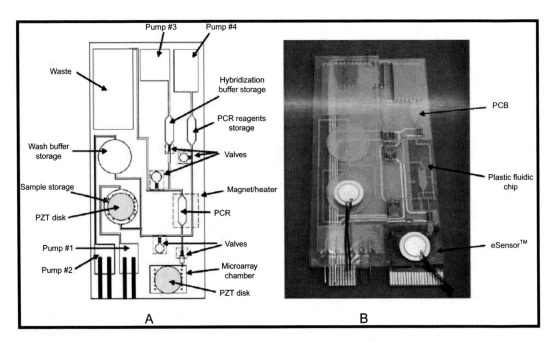

Figure 3.34: Integrated Biochip with PCR Functionality.

First, the sample is incubated with immunomagnetic capture beads so target cells can be labelled. Then, the sample is flowed into the PCR chamber, where the labeled cells will be trapped by a magnet, while the unlabeled solutes are washed away. The device is able to achieve complete mixing of a 50 µL chamber in 6 seconds and the target cell capture rate was 73%. The separated cells then undergo thermal lysis to extract the DNA and PCR is performed. After PCR, a hybridization buffer is added to hybridize the target DNA amplicon. A redox reaction occurs during hybridization which is detected by the platinum wires and measured voltammetrically. The device is able to detect *E. coli* from rabbit blood in 3.5 h. It was able to detect a single polymorphism in 2.7 h.

The Fluidigm BioMark System

The Fluidigm BioMark System was introduced in 2006 as the first commercial system for digital PCR based on integrated microfluidic microchips with integrated chambers and valves for partitioning samples. The underlying microfabrication technique was developed in 1998 by Quake et al at the California Institute of Technology, called Multilayer Soft Lithography (MSL). In this process, the soft silicone elastomer used to produce the microfluidic device deflects under pressure, creating an effective seal or valve that cuts off fluid flow within the channel (see Fig. 3.36). The valve forms the foundation behind the Fluidigm's many integrated microfluidic circuits, trademarked as the NanoFlex valve. The development of NanoFlex valves aided in the advancement in fluid manipulation as a precise, low-cost microscale platform capable of gentle and efficient handling of samples and reagents.

The Fluidigm BioMark System is one of the many microfluidic systems produced by Fluidigm that utilizes the NanoFlex valve on an integrated microfluidic microchip for single-cell gene expression profiling using digital PCR, genotyping, mutant detection, as well as real-time PCR. The automated system is capable of performing the essential steps of genotyping and PCR workflow including thermal cycling and fluorescence detection. It is ultimately a multiplexor, a combinatorial array of binary valve patterns, which controls the flow of fluids on the microchip. The number of control channels is logarithmically proportional to the number of flow channels, thus enabling high throughput through fewer controls. A device can contain thousands of channels and valves, but controlled through only a few inlets [108] (see Fig. 3.35). This multiplexing and high throughput capability enables handling of samples and primer-probes sets for up to 9216 PCR reactions. A typical digital PCR run using the BioMark system involves manual preparation, priming and pipetting samples into inlets on the chip, after which all mixing, thermal cycling, and fluorescence detection is performed automatically on the chip. As of the 2013, the system does not have FDA or other similar regulatory body approval, and the use is limited for research use only.

The chip is first primed using the NanoFlex integrated fluidic circuit (IFC) controller that pressurizes the control lines and closes the interface valves (IV) to prevent sample mixing.

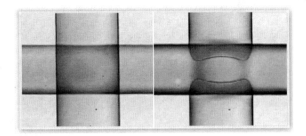

Figure 3.35: NanoFlex Valves Microfabricated Using a Multilayer Soft Lithography Technique.
Fluidigm http://www.fluidigm.com/technology.html. (a) http://molecular.roche.com/assays/Pages/ AmpliChipCYP450Test.aspx Courtesy of Roche.

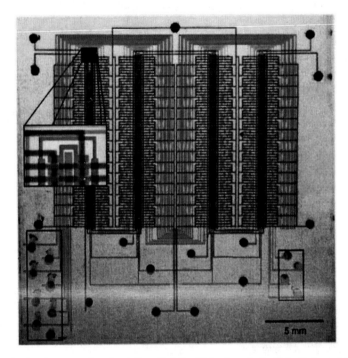

Figure 3.36: Optical Micrograph of a Microfluidic Chip Loaded with Colored Food Dye [108].

Samples of individually pipetted into sample inlets and gene expression assays are pipetted into detector inlets. The chip is then loaded onto the NanoFlex IFC controller for loading and mixing during which, pressure is applied to the fluid in the sample inlets and the fluid is pushed into fluid lines and into individual wells (Fig. 3.37). Simultaneously, the detector inlets are pushed into the fluid lines, prevented by IVs from mixing. Containment valves (CV) are then closed and IVs are opened, pushing reagents in the detector inlets into individual reaction chambers for mixing. After mixing, the IVs are closed and thermal cycled, a process that takes about 55 min [109].

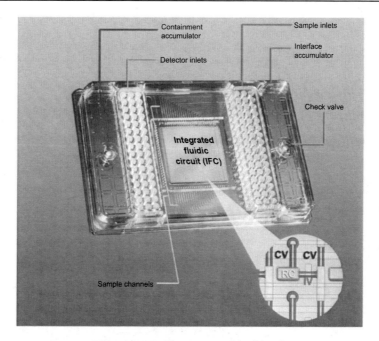

Figure 3.37: The Array Chip [109].

The BioMark System has been used by many as an integrative part of research and experimentation, especially for high throughput gene expression measurements that can be used for diagnostics. Tsui et al utilized the system as a tool for noninvasive prenatal diagnosis of the sex-linked hemophilia disease. Here, the BioMark system is used to perform digital PCR on DNA extracted from maternal plasma. The system served to automate the digital reaction setup by channeling nanoliter aliquots of PCR mixture into thousands of amplification chambers, thus making the process low-cost and high throughput [110]. Another example that utilizes the BioMark System was demonstrated by Yung et al. Here, it was used in the quantitative detection of two common epidermal growth factor receptor (EGFR) mutations in plasma and tumor tissues of non-small cell lung cancer patients. Use of the system enabled 9180 PCRs to be simultaneously performed at a nanoliter scale, whereby a single-mutant DNA molecule could be detected and quantities of mutant and wild-type sequences were determined [111]. Technologies like the Fluidigm BioMark System allow for low-cost, high-throughput single molecule amplifications that enable researchers to further explore and understand basic fundamental biology that underlie complex diseases of the human body.

DNA Microarray

DNA microarrays are popular microfluidic devices for genotyping [112–114]. The development of systems that can detect DNA is motivated by applications in the fields of gene profiling, disease diagnosing and forensic analysis. DNA microarrays surfaced in the mid-1990s and have since created successful companies such as Affymetrix. Currently, efforts in DNA microarray technology are realizing fully automated systems that can perform every single step starting from loading the analyte onto the microarray to automatically determining the presence of multiple genes of interest.

Working Principle

DNA microarrays work on the principle of DNA hybridization. In this technique, a sequence that is complementary to a DNA sequence of interest is immobilized on a substrate (the chip). If the complimentary sequence is present in the analyte, it will hybridize onto its complement located on the chip. Different methods can be used to detect the hybridization event. Techniques include optical detection methods, usually involving fluorescently tagged DNA molecules, as well as surface plasmon resonance, colorimetric, and surface-enhanced Raman spectroscopy [138].

Implementation

DNA microarrays are usually made from glass, plastic or silicon substrates. They contain hundreds of thousands of small oligonucleotide sequences. Oligonucleotides can be synthesized in situ or synthesized before being immobilized on a chip. The oligonucleotides are arranged spatially so that thousands of genes can be probed on a single chip. As a result, DNA microarrays have tremendous throughput because many different DNA probes can be placed on a single chip. Details of the implementation techniques are described in Chapter 2. Section 2.5.3.

Applications

DNA microarrays, or gene chips, are used in DNA sequencing, gene expression, doxological research, and single nucleotide polymorphism (SNP) detection. A SNP is a DNA sequence variation with a single nucleotide (A, T, C, or G) different between different individuals (for example, TAAATAC and TAAGTAC). Genetic variations may cause differences in the susceptibility to disease. Some SNPs are used to predict an individual's response to certain drugs, and susceptibility to toxins or diseases.

The massive parallelism of DNA arrays makes them attractive in applications that require screening large amounts of DNA for specific sequences. Although DNA microarrays have tremendous potential to be used as diagnostics devices in the clinic, most DNA microarrays are only used for research and development purposes due to lack of FDA-approval. This is not surprising, as there are many technical, medical and even marketing challenges to FDA approval. Only two DNA microarray-based diagnostic devices, MammaPrint and AmpliChip CYP450, are FDA approved as of 2013. The MammaPrint is a test that assesses the risk of women having recurring breast cancer, after the initial tumor has been surgically removed or treated with chemotherapy. The test was developed by Agendia, and was FDA-approved in 2007.

The AmpliChip CYP450 is the first FDA-approved (Class II device) DNA microarray-based test developed by Roche (Fig. 3.38a). It is a combination of the Affymetrix GeneChip technology and Roche's patented polymerase chain reaction amplification technology. It is used to screen for mutations in the *CYP2D6* and *CYP2C19* genes, which code for the production of CYP50 enzymes. These enzymes catalyze the oxidation of organic substances. They are responsible for the metabolism of about 25% of prescription drugs including antidepressants, antipsychotic, antiepileptics, and beta-blockers [115]. A mutation to any of these genes will affect how effective prescription drugs are. Detecting mutations to the *CYP2D6* and *CYP2C19* genes provides information helpful to decide on a course of treatment.

Each AmpliChip CYP450 microarray has a 20×20 μm^2 square grid that contains 15 129 probes, with each probe having about 10^7 identical copies of a specific oligonucleotide [116]. The probes of the AmpliChip are synthesized using photo-directed synthesis (see discussions in Chapter 2, Section 2.5.3 for Affymetrix gene chip). Probes are grouped in sets of four, with each probe being different by a single nucleotide: A, C, G, or T. Only one of the four probes in a set is considered a "perfect" match; the other three are considered "mismatches". The probes on the AmpliChip CYP450 can differentiate between wild-type or mutant sequences.

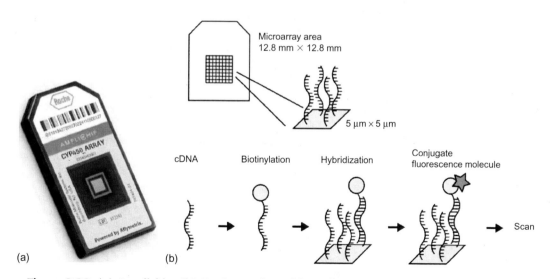

Figure 3.38: (a) Amplichip. (b) Design and Working Principle of the Affymetrix GeneChip.
(a) From http://molecular.roche.com/assays/Pages/AmpliChipCYP450Test.aspx. © 2008-2013 PerkinElmer, Inc. All rights reserved. Printed with permission http://www.perkinelmer.com/Catalog/Product/ID/122000.

Figure 3.35(b) shows the protocol with the AmpliChip CYP450 microarray: First, patient DNA is purified from blood by PCR. The DNA is then fragmented by adding DNAase. The fragmented DNA is labeled with biotin. Next, the DNA is placed in contact with the microarray to allow patient DNA to hybridize with the DNA on the microarray. Next, the unbound DNA is washed away and the DNA that is bound is stained with streptavidin-conjugated fluorescent dye. The AmpliChip CYP450 is scanned using the Affymetrix GeneChip Microarray platform, which has a laser that excites the fluorescent dye. The amount of fluorescence is proportional to bond DNA on each probe of the microarray.

Microfluidic Devices for DNA Separation

Electrophoresis is a method of separating macromolecules such as DNAs, proteins, and cells by charge and size. Electrophoresis used for cell separation is described in section 3.2.2. It is very useful in separating DNA fragments and sorting, and is a critical step in DNA analysis and sequencing. **Gel electrophoresis** [117,118] is a basic technique that separates analytes prepared in a porous gel medium. A gel is loaded with the analyte at one end. Electrodes are placed at opposite ends of the gel and the DNA fragments migrate towards the negative electrode. The rates of migration vary based on the size of fragments (Fig. 3.39). After a period of time the potential difference is removed and the migrated DNA can be imaged. After electrophoresis, molecules can be stained for visualization. DNA may be visualized under ultraviolet light after the DNA is intercalated with ethidium bromide, a UV fluorescent agent. Proteins may be visualized using silver stain or Coomassie Brilliant Blue dye. Gel separations are used for characterization of DNA. The applications are numerous and can be used for hereditary analysis, DNA sequencing, immunology and toxicology to name a few.

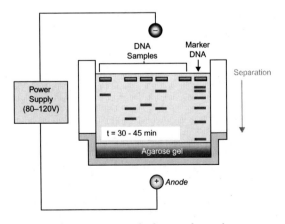

Figure 3.39: Gel Electrophoresis.

Gel electrophoresis can be performed on samples on a microfluidic chip. Advantages of using microfluidic analysis systems include accuracy and performance on a very small scale. Also many samples can be done in succession without having to reload samples, expediting DNA sequencing and analysis to provide instant lab results in a clinical environment.

Microfluidic Electrophoresis

The conventional method of gel electrophoresis is an inexpensive method of analyzing proteins and DNA, but is very time consuming and requires skills for operation. Capillary electrophoresis on microfluidic chip devices [119,120] can speed up and streamline the process. A microfluidic electrophoresis device requires a small sample to be placed in a reservoir. The entire process could be automated. Many samples can be processed in succession without having to reload samples. This can expedite the process of DNA sequencing and analysis and make it more viable for instant lab results in a clinical environment.

Capillary electrophoresis separates macromolecules by size and electrical polarity. A microfluidic channel ranging from five to two hundred microns in diameter is filled with a conducting buffer solution. A potential difference is applied to the ends of the capillary via electrodes placed in the sample and waste chambers (see Fig. 3.40). The potential difference creates a current through the channel. Once current begins to flow, charged particles (i.e., DNA fragments) begin to flow. Because the rate of their flow is dependent on their charge, the DNA fragments are distributed by charge along the microchannel in a way similar to a conventional gel electrophoresis.

Each well is connected to an electrode. A voltage is selectively applied to each pair of wells to perform sample injection and separation. The applied voltage ranges 100–3000 V. Generally, about ~100 nL of sample is taken to be analyzed. Capillary electrophoresis is most often performed in glass micro-machined structures. The commonly used material is Corning 7740 Pyrex glass. Photolithographic techniques are employed to fabricate these devices. Channels of varying diameters can be etched into the glass with buffered HF.

Commercial Products

Some products currently exist on the market such as the LabChip GX from PerkinElmer. The LabChip GX bench top device streamlines the DNA analysis process and only requires the technician to place the sample in its reservoir.

Figure 3.41 shows the steps involved in the LabChip electrophoresis system. Fluorescent dye and sieving polymer is used in the separation procedure. An automated vacuum system pulls the sample into the separation channel with the proper mixture of buffer and then applies the voltage. A plug is created to prevent backflow from the separation channel.

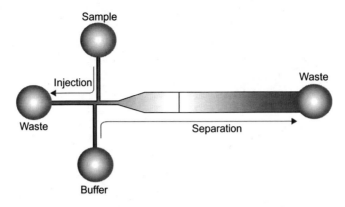

Figure 3.40: Microchannel-Based Capillary Electrophoresis.

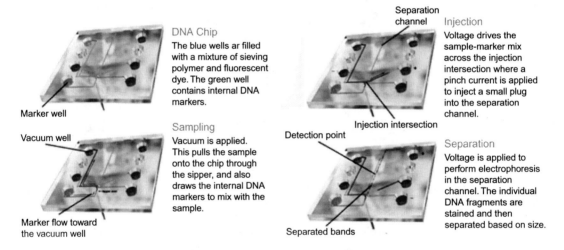

DNA Chip
The blue wells ar filled with a mixture of sieving polymer and fluorescent dye. The green well contains internal DNA markers.

Marker well

Vacuum well

Sampling
Vacuum is applied. This pulls the sample onto the chip through the sipper, and also draws the internal DNA markers to mix with the sample.

Marker flow toward the vacuum well

Separation channel

Injection
Voltage drives the sample-marker mix across the injection intersection where a pinch current is applied to inject a small plug into the separation channel.

Injection intersection

Detection point

Separation
Voltage is applied to perform electrophoresis in the separation channel. The individual DNA fragments are stained and then separated based on size.

Separated bands

Figure 3.41: PerkinElmer LabChip GX.

Then voltage is applied between the prepared sample chamber and the waste or end chamber to run the electrophoresis. Because very high voltages are used, electrophoresis only takes a few seconds. Longer times are not necessary because small amounts of sample are required. Once the separation is completed, the imaging step can now occur.

Agilent 2100 device is one of the most widely used commercial chip-based DNA analysis devices (Fig. 3.42). These microfluidic chips are manufactured from PDMS using soft lithography. The system uses interchangeable chips that interface with a benchtop device that acts as the power supply for electrophoresis and also contains the optical detection system. The chip works largely in the same way as the PerkinElmer system [121−123].

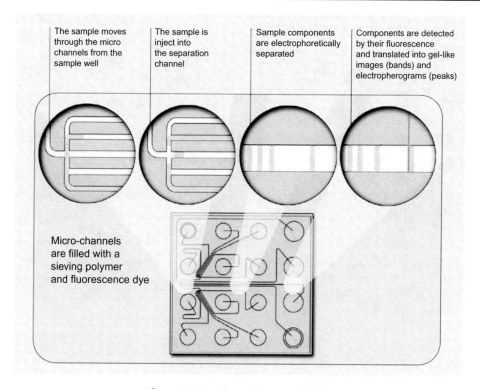

Figure 3.42: The Agilent LabChip.
The Agilent LabChip can perform tasks such as sizing and quantitation of DNA based on the principle of electrophoresis. *http://www.osa.sunysb.edu/udmf/Agilent_2100_Bioanalyzer.pdf*

DNA Sequencing

DNA sequencing is the process of determining the order of the four nucleotide bases, adenine (A), guanine (G), cytosine (C), and thymine (T), in a DNA strand.

Sanger Sequencing

Sanger sequencing, also known as the Sanger method is a widely used sequencing method. The Sanger method has been modified in several ways since it was developed by Sanger [124] and colleagues in 1977. As of 2008, the advanced Sanger method has achieved read-lengths of up to ~1000 bp, and per-base accuracies as high as 99.999%. It costs on the order of $0.50 per kilobase [125].

DNA fragments are synthesized and terminated with one of the four nucleotide bases. The lengths of the fragments are measured by the following process:

The DNA sample is divided into four separate tubes containing deoxyribonucleotides (dNTPs) which are precursors to DNA synthesis and dideoxynucleotides (ddNTPs) which are chain-terminating inhibitors of DNA polymerase. Each tube contains all four of the standard deoxynucleotides (dATP, dGTP, dCTP and dTTP) and one of the four dideoxynucleotides (ddATP, ddGTP, ddCTP, or ddTTP). The concentration of each ddNTP is about one-hundreth of those of the normal precursors (dNTPs).

During DNA synthesis, DNA polymerase adds nucleotides to the growing chain. A ddNTP is incorporated into the chain in place of a normal nucleotide with a certain probability, creating a DNA fragment terminated at that point. As a result, each tube produces DNA fragments with different lengths, but always terminated at one of the four nucleotide bases (A,G,C,T) that corresponds to the ddNTP added in the tube (see the following example). The lengths of these DNA fragments can be found by sorting them with the gel electrophoresis or capillary electrophoresis technique.

Let us use an example sequence: GAAACATG. What we will see from each tube is the lengths of the fragments that are synthesized, and we know at which base they are terminated.

Tube 1 (with ddATP: terminates at A)

?A	(GA – AACATG)
??A	(GAA – ACATG)
???A	(GAAA –CATG)
?????A	(GAAACA –TG)

Tube 2 (with ddGTP: terminates at G)

G	(G – AACATG)
???????G	(GAAACATG)

Tube 3 (with ddCTP: terminates at C)

????C	(GAAAC-ATG)

Tube 4 (with ddTTP: terminates at T)

??????T	(GAAACAT-G)

When we sort these fragments based on the lengths:

G
?A

??A
???A
????C
?????A
??????T
???????G

Now we find that the original DNA sequence was GAAACATG.

Microfluidic Systems for Sanger Sequencing

One approach for microfluidic DNA sequencing is miniaturizing Sanger sequencing. Mathies et al. developed a microfluidic device integrating all three Sanger sequencing steps, namely Sanger extension, purification, and electrophoretic analysis [126]. Figure 3.43 illustrates a flow chart of conventional DNA sequencing. DNAs are extracted from cultured cells and amplified. Sanger extension is then performed and purified to be analyzed on a capillary array electrophoresis (CAE) sequencer. Figure 3.43 (b) shows microfabricated sequencing lab on chip. Each step of the conventional protocol is miniaturized into a 15 cm-diameter microdevice, where 96 lanes are integrated into 48 doublet structures.

The same group performed complete Sanger sequencing from a 1 fmol DNA template. Up to 556 continuous bases were sequenced with 99% accuracy [127].

Nanopore Based Sequencing

A miniaturization technique very different from conventional Sanger sequencing is nanopore based sequencing [128]. It is based on translocation of a DNA molecule through a nanometer-sized pore that can accommodate only a single strand of RNA or DNA. Kasianowics et al used a 2.6 nm diameter ion channel in a lipid bilayer membrane. The membrane was formed across a $\sim100\,\mu$m diameter orifice in a 25 μm thick Teflon partition that separates two buffer filled compartments. The passage of each polynucleotide molecule was measured as a transient decrease of ionic current whose duration is proportional to the length of the molecule [129].

Figure 3.44 illustrates a solid state nanopore [139]. Figure 3.44(a) and (b) illustrates schematics of the experiment. Two isolated reservoirs are insulated by a silicon nitride membrane containing a single nanopore. An ionic current through the nanopore is induced by a bias voltage applied between the two reservoirs. While DNA molecules pass through the nanopore due to their negative charge, nanopore current is recorded to assess the molecule's interaction with the nanopore. The nanopore shown in Fig. 3.44(c) was fabricated in a 280 nm thick silicon nitride membrane supported by a 380 μm thick silicon chip. It was made by focused ion beam (FIB) milling followed by feedback controlled ion beam sculpting [130].

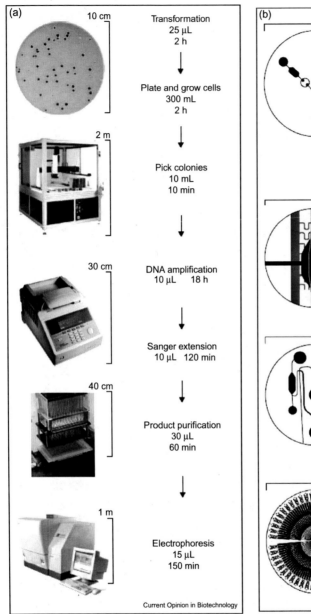
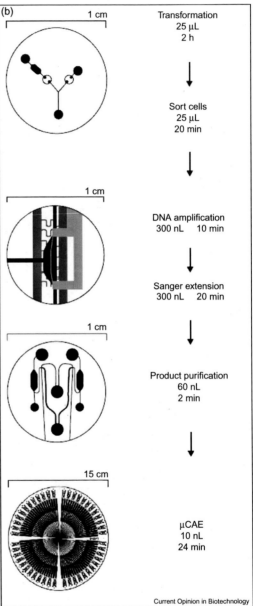

Figure 3.43: (a) A Flow Chart for Conventional DNA Sequencing. (b) Microfabricated DNA Sequencing Microdevice.

From [126].

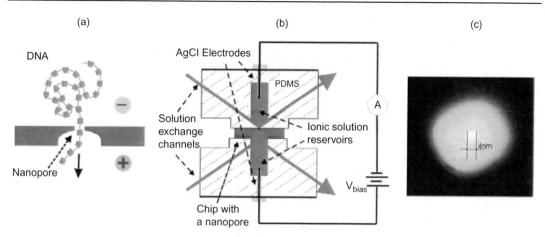

Figure 3.44: (a) Illustration of DNA Molecule Translocating Through a Solid-State Nanopore. (b) Experimental Setup for Single Molecule Measurements. (c) TEM of a 4 nm Silicon Nitride Nanopore in a 5–10 nm Thick Local Membrane.
From [139].

When a molecule passes through a nanopore, the ion current is reduced because the nanopore is partially blocked by the translocating molecule. One hypothesis for future nanopore based DNA sequencing is that the sequence of ionic current reduction may reflect that of bases in the polynucleotide molecule [128]. Manrao et al demonstrated the ability to resolve changes in current that correspond to a known DNA sequence [131]. They used a mutated form of the protein pore *Mycobacterium smegmatis* porin A (MspA) with phi29 DNA polymerase (DNAP). Phi29 DNAP controls the rate of DNA passage by acting like a motor to pull a single-stranded template through MspA as it synthesizes DNA. The authors recorded current sequences that match the known sequences using DNA sequences with ~50 nucleotides long readable regions (Fig. 3.45 (b) and (c)).

Problems

P3.1 Diffusion

1. If a molecule has diffused 10 nm in 1 s, how far does it travel in 4 s?
2. We measured that hemoglobin (diffusion coefficient $D = 7 \times 10^{-7}$ cm^2/s) in water takes 1 million seconds to diffuse 1 cm. A student designed a microfluidic channel with the width of 10 μm to mimic such a diffusion experiment at small scale.
   a. How long does it take for hemoglobin to diffuse across such a microfluidic channel?

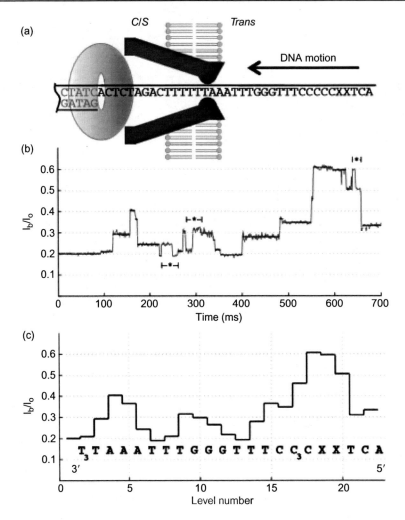

Figure 3.45: Nanopore-Based Single-nucleotide Resolution Sequencing [131].

b. Based on your calculation in (a), is it a good idea to design microfluidic diffusion-based mixers?

P3.2 Spindle Viscometer

1. A liquid is contained between a cylindrical spindle and a cylindrical container (Fig. P3.1). The height and the radius of the spindle is 15 mm and 10 mm, respectively. The gap between the spindle and the container is 1 mm. Because of the air trapped at the bottom of the spindle, viscous drag arises only on the sidewall of the spindle. If the torque of $T = 6.2 \times 10^{-6}$ N m is needed to rotate a spindle at 50 rpm, what is the viscosity of the liquid?

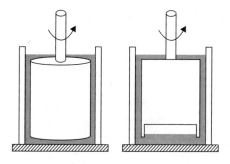

Figure P3.1

2. Figure P3.2 shows another type of viscometer where a liquid is contained between a cone-shaped spindle with an angle of 3° and a stationary substrate. The radius of the spindle is 20 mm. Calculate the torque needed to rotate the spindle at a rotational speed of 20 rpm when the viscosity of the liquid is 30 cP.

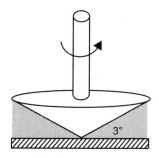

Figure P3.2

P3.3 Hagen—Poiseuille Equation

A micro capillary tubing can be used as a simple viscometer. The viscosity can be found from the flow rate, pressure drop and the tube geometry. Here is a measurement of a certain liquid in a capillary viscometer. Find the viscosity of the liquid.

Flow rate: 880 mm^3/s
Tube diameter: 500 μm
Tube length: 100 mm
Pressure drop: 0.1 Mpa

P3.4 Reynolds Number

The accepted transition Reynolds number for flow in a circular pipe is Re = 2300. For a 100 μm diameter pipe, at what velocity will the transition occur at 20 °C for (a) airflow and

(b) water flow? Assuming the kinematic viscosity for air at 20 °C is 1.5×10^{-5} m^2/s, for water is 1.0×10^{-6} m^2/s.

P3.5 Reynolds Number

Estimate the Reynolds numbers associated with honey bee, hummingbird, condor, and airplane. Use the information in the table. The chord length is usually taken as the characteristic length to find the Reynolds number of a wing. Kinematic viscosity of air is 1×10^{-5} m^2/s. Discuss the differences in the way they fly.

	Honey bee	Hummingbird	Condor	Airplane
Wing chord length	5 mm	3 cm	30 cm	3 m
Fly parameters	Wing flaps 200 times/second	Wing flaps 50 times/second	Velocity: 50 km/h	Velocity: 1000 km/h
Reynolds numbers				

P3.6 Laminar Flow

Consider a flow between two parallel plates, as shown in Fig P3.3. The viscosity of the fluid is μ, the gap between the two plate is H, and the average flow velocity is U.

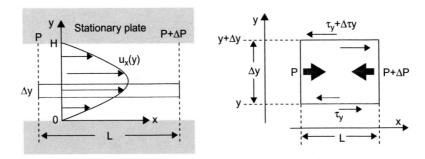

Figure P3.3

1. Form a differential equation of motion for a thin layer from y to $y + \Delta y$.
2. Show that the same equation as (1) can be derived from the Navier–Stokes equation.
3. Find the velocity profile $u_x(y)$ by solving the differential equation formed in (1).

P3.7 Laminar Flow

There is a flat microchannel with dimensions of width $w = 5$ mm, thickness $t = 0.5$ mm, and length $l = 20$ mm (Fig. P3.4). Assume the flow is laminar and uniform in the directions of

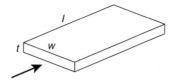

Figure P3.4

the width and the length. A liquid flows in the channel at a flow rate of $Q = 2.5$ mL/h. Find the velocity profile in z direction which follows a simple parabola defined by the flow rate Q.

P3.8 Microfluidic Pump

A microdevice engineer is designing a passive microfluidic "pump" to manipulate blood droplets. As shown in Fig. P3.5, two droplets are deposited at each end of a microfluidic channel, with the larger droplet on the left end. The channel is filled with blood. Without using any external force, which way would you think the blood is going to move towards?

Figure P3.5

a. From left to right, the larger droplet will disappear and the small droplet will grow bigger.
b. From right to left, the small droplet will disappear and the larger droplet will grow bigger.
c. No moving action involved, the pump does not work.

Choose among (a), (b), (c), and justify your choice.

P3.9 Hagen−Poiseuille Equation

A liquid with the viscosity μ is flowing in a circular microchannel with radius r and length L at a flow rate of Q pumped with a pressured of ΔP. The flow can be treated as laminar.

Now you need to design another microchannel for a liquid with the viscosity of 2μ. You have to use the same pumping pressure ΔP to keep the same flow rate Q.

1. If you can only change the length, what will be the length?
2. If you can only change the radius, what will be the radius?

P3.10 Hagen–Poiseuille Equation

1. Describe the pressure drop ΔP along the micro channel shown in Fig. P3.6 (left). You may use average velocity U, channel radius R, viscosity η and flow rate Q.

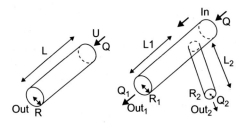

Figure P3.6

2. A flow introduced from the inlet is divided into two in a micro fluidic system shown in Fig. P3.6 (right). Calculate output flow rates Q_1 and Q_2. You may use incoming flow rate Q, radii R_1 and R_2 and channel lengths L_1 and L_2. Neglect the pressure drop at the connection. The fluid pressures at the two outlets are the same.

P3.11 Immunomagnetic Separation

Figure P3.7 shows a top view of a typical implementation of immunomagnetic cells separation, where a test tube with a diameter of 10 mm is surrounded by four magnets. The contour map shows a simulated magnetic field intensities. The magnitude of the field changes from $T = 0$ at the center to $T = 0.03$ at the wall.

1. Estimate the value of $\partial B^2/\partial r$ in the tube. Assume that the value of B^2 changes linearly along the radial direction.
2. Contained in the tube is blood that includes ~ 200 cancer cells labeled with magnetic nanoparticles. Calculate how long it takes to collect all the labeled cells to the side wall. Use the radius of the cells $R_C = 7.5\ \mu m$, effective magnetic volumetric susceptibility of the cell $\Delta\chi_C = 3 \times 10^{-3}$, the magnetic permeability of vacuum $\mu_0 = 4\pi \times 10^{-7}\ T \cdot m \cdot A^{-1}$, and Blood viscosity $\eta_B = 7.5 \times 10^{-3}$ kg·m^{-1}·s^{-1}.

$$\frac{kgm}{As^2}\left(\frac{kg}{ms}\right) = \frac{kg^2}{A^2s^3}$$

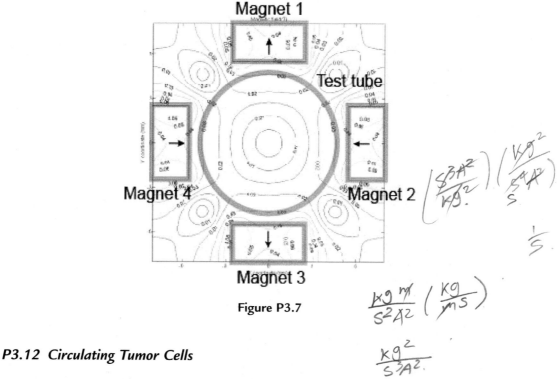

Figure P3.7

P3.12 Circulating Tumor Cells

1. Name three benefits of a system that can detect cancer cells from blood.
2. Name three difficulties in commercialization of such a system.
3. Name six methods of separating cancer cells from blood. Which method do you think is the most promising method? Explain why.

P3.13 Gel Electrophoresis

In gel electrophoresis, the amount of electrical charge is the (larger, smaller, same) for the longer DNA fragment. The viscous force acting on a DNA fragment is the (larger, smaller, same) for the longer DNA fragment. The rates of migration is the (larger, smaller, same) for the longer DNA fragment.

P3.14 PCR

Design a forward and reverse primer suitable for amplifying the underlined segment of "GACCTCCCAGGCCAGTGCAGTCGGGCC". In a practical case, primers are 10–25 bp long and the amplicon is 500 bp long, but for this exercise, design primers that are 5 bp long.

P3.15 PCR

We have prepared cDNAs from 10 cells and 10 000 cells. If we want to obtain the same amount of amplified DNA using PCR, what is the expected difference between the cycle times for the two samples?

References

[1] Eckhardt B. Introduction. Turbulence transition in pipe flow: 125th anniversary of the publication of Reynolds' paper. Philosophical Transactions of the Royal Society A: Mathematical, Physical and Engineering Sciences 2009;367:449−55.

[2] Schmidt-Nielsen K. Scaling: Why is Animal Size So Important? Cambridge: Cambridge University Press; 1984.

[3] Purcell EM. Life at low Reynolds number. American Journal of Physics 1977;45:3−11.

[4] Stone HA, Stroock AD, Ajdari A. Engineering flows in small devices: microfluidics toward a lab-on-a-chip. Annual Reviews of Fluid Mechanics 2004;36:381−411.

[5] Dittrich PS, Manz A. Lab-on-a-chip: microfluidics in drug discovery. Nature Reviews Drug Discovery 2006;5:210−8.

[6] Becker H, Gärtner C. Polymer microfabrication methods for microfluidic analytical applications. Electrophoresis 2000;21:12−26.

[7] Beebe DJ, Mensing GA, Walker GM. Physics and applications of microfluidics in biology. Annual Review of Biomedical Engineering 2002;4:261−86.

[8] Konrad R, Griebel A, Dörner W, Löwe H. Towards disposable lab-on-a-chip: poly (methylmethacrylate) microchip electrophoresis device with electrochemical detection. Electrophoresis 2002;23:596−601.

[9] Ahn CH, Choi J-W, Beaucage G, Nevin JH, Lee J-B, Puntambekar A, et al. Disposable smart lab on a chip for point-of-care clinical diagnostics. Proceedings of the IEEE 2004;92:154−73.

[10] Chin CD, Linder V, Sia SK. Lab-on-a-chip devices for global health: past studies and future opportunities. Lab on a Chip 2007;7:41−57.

[11] Whitesides GM. The origins and the future of microfluidics. Nature 2006;442:368−73.

[12] Anderson JR, Chiu DT, Wu H, Schueller OJ, Whitesides GM. Fabrication of microfluidic systems in poly (dimethylsiloxane). Electrophoresis 2000;21:27−40.

[13] Lorenz H, Despont M, Fahrni N, LaBianca N, Renaud P, Vettiger P. SU-8: a low-cost negative resist for MEMS. Journal of Micromechanics and Microengineering 1997;7:121.

[14] Becker H, Heim U. Hot embossing as a method for the fabrication of polymer high aspect ratio structures. Sensors and Actuators A: Physical 2000;83:130−5.

[15] McCormick RM, Nelson RJ, Alonso-Amigo MG, Benvegnu DJ, Hooper HH. Microchannel electrophoretic separations of DNA in injection-molded plastic substrates. Analytical Chemistry 1997;69:2626−30.

[16] Haneveld J, Jansen H, Berenschot E, Tas N, Elwenspoek M. Wet anisotropic etching for fluidic 1D nanochannels. Journal of Micromechanics and Microengineering 2003;13:S62.

[17] Nilsson A, Petersson F, Jönsson H, Laurell T. Acoustic control of suspended particles in micro fluidic chips. Lab Chip 2004;4:131−5.

[18] Jiang L, Mikkelsen J, Koo J-M, Huber D, Yao S, Zhang L, et al. Closed-loop electroosmotic microchannel cooling system for VLSI circuits. Components and Packaging Technologies, IEEE Transactions on 2002;25:347−55.

[19] Di Carlo D, Jeong K-H, Lee LP. Reagentless mechanical cell lysis by nanoscale barbs in microchannels for sample preparation. Lab Chip 2003;3:287−91.

x DNA = 16S S/Cycle   1Cycle/DNA

1000× DNA

[20]   Kaiser A, Klok CJ, Socha JJ, Lee W-K, Quinlan MC, Harrison JF. Increase in tracheal investment with beetle size supports hypothesis of oxygen limitation on insect gigantism. Proceedings of the National Academy of Sciences of the United States of America 2007;104:13198−203.

[21]   Liu RH, Stremler MA, Sharp KV, Olsen MG, Santiago JG, Adrian RJ, et al. Passive mixing in a three-dimensional serpentine microchannel. Journal of Microelectromechanical Systems 2000;9:190−7.

[22]   Nguyen N-T, Wu Z. Micromixers—a review. Journal of Micromechanics and Microengineering 2005;15:R1.

[23]   Stroock AD, Dertinger SKW, Ajdari A, Mezić I, Stone HA, Whitesides GM. Chaotic mixer for microchannels. Science 2002;295:647−51.

[24]   Wang H, Iovenitti P, Harvey E, Masood S. Optimizing layout of obstacles for enhanced mixing in microchannels. Smart Materials and Structures 2002;11:662.

[25]   Chang WC, Lee LP, Liepmann D. Biomimetic technique for adhesion-based collection and separation of cells in a microfluidic channel. Lab on a Chip 2005;5:64−73.

[26]   Nagrath S, Sequist LV, Maheswaran S, Bell DW, Irimia D, Ulkus L, et al. Isolation of rare circulating tumour cells in cancer patients by microchip technology. Nature 2007;450:1235−9.

[27]   Song H, Tice JD, Ismagilov RF. A microfluidic system for controlling reaction networks in time. Angewandte Chemie 2003;115:792−6.

[28]   Squires TM, Quake SR. Microfluidics: fluid physics at the nanoliter scale. Reviews of Modern Physics 2005;77:977.

[29]   Brody JP, Yager P. Diffusion-based extraction in a microfabricated device. Sensors and Actuators A: Physical 1997;58:13−8.

[30]   Shapiro HM. Practical Flow Cytometry. New York: Wiley-Liss; 2005.

[31]   Davey HM, Kell DB. Flow cytometry and cell sorting of heterogeneous microbial populations: the importance of single-cell analyses. Microbiological Reviews 1996;60:641−96.

[32]   Krishan A, Krishnamurthy H, Totey S. Applications of Flow Cytometry in Stem Cell Research and Tissue Regeneration. Oxford: Wiley-Blackwell; 2011.

[33]   Wolff A, Perch-Nielsen IR, Larsen U, Friis P, Goranovic G, Poulsen CR, et al. Integrating advanced functionality in a microfabricated high-throughput fluorescent-activated cell sorter. Lab on a Chip 2003;3:22−7.

[34]   Wang MM, Tu E, Raymond DE, Yang JM, Zhang H, Hagen N, et al. Microfluidic sorting of mammalian cells by optical force switching. Nature Biotechnology 2004;23:83−7.

[35]   Gurkan UA, Anand T, Tas H, Elkan D, Akay A, Keles HO, et al. Controlled viable release of selectively captured label-free cells in microchannels. Lab on a Chip 2011;11:3979−89.

[36]   Adams AA, Okagbare PI, Feng J, Hupert ML, Patterson D, Göttert J, et al. Highly efficient circulating tumor cell isolation from whole blood and label-free enumeration using polymer-based microfluidics with an integrated conductivity sensor. Journal of the American Chemical Society 2008;130:8633−41.

[37]   Moon S, Gurkan UA, Blander J, Fawzi WW, Aboud S, Mugusi F, et al. Enumeration of CD4+ T-cells using a portable microchip count platform in Tanzanian HIV-infected patients. PloS One 2011;6:e21409.

[38]   Chin CD, Laksanasopin T, Cheung YK, Steinmiller D, Linder V, Parsa H, et al. Microfluidics-based diagnostics of infectious diseases in the developing world. Nature Medicine 2011;17:1015−9.

[39]   Stott SL, Hsu C-H, Tsukrov DI, Yu M, Miyamoto DT, Waltman BA, et al. Isolation of circulating tumor cells using a microvortex-generating herringbone-chip. Proceedings of the National Academy of Sciences of the United States of America 2010;107:18392−7.

[40]   Wang S, Wang H, Jiao J, Chen KJ, Owens GE, Kamei Ki, et al. Three-dimensional nanostructured substrates toward efficient capture of circulating tumor cells. Angewandte Chemie 2009;121:9132−5.

[41]   Xu Y, Phillips JA, Yan J, Li Q, Fan ZH, Tan W. Aptamer-based microfluidic device for enrichment, sorting, and detection of multiple cancer cells. Analytical Chemistry 2009;81:7436−42.

[42]   Inglis DW, Riehn R, Austin R, Sturm J. Continuous microfluidic immunomagnetic cell separation. Applied Physics Letters 2004;85:5093−5.

[43] Talasaz AAH, Powell AA, Huber DE, Berbee JG, Roh KH, Yu W, et al. Isolating highly enriched populations of circulating epithelial cells and other rare cells from blood using a magnetic sweeper device. Proceedings of the National Academy of Sciences of the United States of America 2009;106:3970−5.

[44] Radisic M, Iyer RK, Murthy SK. Micro-and nanotechnology in cell separation. International Journal of Nanomedicine 2006;1:3.

[45] Miltenyi S, Müller W, Weichel W, Radbruch A. High gradient magnetic cell separation with MACS. Cytometry 1990;11:231−8.

[46] Taylor PA, Panoskaltsis-Mortari A, Swedin JM, Lucas PJ, Gress RE, Levine BL, et al. L-Selectinhi but not the L-selectinlo CD4+ 25+ T-regulatory cells are potent inhibitors of GVHD and BM graft rejection. Blood 2004;104:3804−12.

[47] Hansel TT, De Vries IJM, Iff T, Rihs S, Wandzilak M, Betz S, et al. An improved immunomagnetic procedure for the isolation of highly purified human blood eosinophils. Journal of Immunological Methods 1991;145:105−10.

[48] Zhao X, Deak E, Soderberg K, Linehan M, Spezzano D, Zhu J, et al. Vaginal submucosal dendritic cells, but not Langerhans cells, induce protective Th1 responses to herpes simplex virus-2. The Journal of Experimental Medicine 2003;197:153−62.

[49] Bauer M, Redecke V, Ellwart JW, Scherer B, Kremer JP, Wagner H, et al. Bacterial CpG-DNA triggers activation and maturation of human CD11c-, CD123+ dendritic cells. The Journal of Immunology 2001;166:5000−7.

[50] Burchill MA, Yang J, Vogtenhuber C, Blazar BR, Farrar MA. IL-2 receptor β-dependent STAT5 activation is required for the development of Foxp3+ regulatory T cells. The Journal of Immunology 2007;178:280−90.

[51] Paterlini-Brechot P, Benali NL. Circulating tumor cells (CTC) detection: clinical impact and future directions. Cancer Letters 2007;253:180−204.

[52] Mostert B, Sleijfer S, Foekens JA, Gratama JW. Circulating tumor cells (CTCs): detection methods and their clinical relevance in breast cancer. Cancer Treatment Reviews 2009;35:463−74.

[53] Allard WJ, Matera J, Miller MC, Repollet M, Connelly MC, Rao C, et al. Tumor cells circulate in the peripheral blood of all major carcinomas but not in healthy subjects or patients with nonmalignant diseases. Clinical Cancer Research 2004;10:6897−904.

[54] Fehm T, Sagalowsky A, Clifford E, Beitsch P, Saboorian H, Euhus D, et al. Cytogenetic evidence that circulating epithelial cells in patients with carcinoma are malignant. Clinical Cancer Research 2002;8:2073−84.

[55] Budd GT, Cristofanilli M, Ellis MJ, Stopeck A, Borden E, Miller MC, et al. Circulating tumor cells versus imaging—predicting overall survival in metastatic breast cancer. Clinical Cancer Research 2006;12:6403−9.

[56] Boyer TH. The force on a magnetic dipole. American Journal of Physics 1988;56:688.

[57] Pankhurst QA, Connolly J, Jones S, Dobson J. Applications of magnetic nanoparticles in biomedicine. Journal of physics D: Applied Physics 2003;36:R167.

[58] Xia N, Hunt TP, Mayers BT, Alsberg E, Whitesides GM, Westervelt RM, et al. Combined microfluidic-micromagnetic separation of living cells in continuous flow. Biomedical Microdevices 2006;8:299−308.

[59] Furdui VI, Kariuki JK, Harrison DJ. Microfabricated electrolysis pump system for isolating rare cells in blood. Journal of Micromechanics and Microengineering 2003;13:S164.

[60] Karl S, David M, Moore L, Grimberg BT, Michon P, Mueller I, et al. Enhanced detection of gametocytes by magnetic deposition microscopy predicts higher potential for Plasmodium falciparum transmission. Malar J 2008;7:66.

[61] Fang B, Zborowski M, Moore LR. Detection of rare MCF-7 breast carcinoma cells from mixtures of human peripheral leukocytes by magnetic deposition analysis. Cytometry 1999;36:294−302.

[62] Hoshino K, Huang YY, Lane N, Huebschman M, Uhr JW, Frenkel EP, et al. Microchip-based immunomagnetic detection of circulating tumor cells. Lab on a Chip 2011;11:3449−57.

[63] Huang Y, Hoshino K, Chen P, Wu C, Lane N, Huebschman M, et al. Immunomagnetic nanoscreening of circulating tumor cells with a motion controlled microfluidic system. Biomedical Microdevices 2012;1−9.

[64] Hoshino K, Chen P, Huang YY, Zhang X. Computational analysis of microfluidic immunomagnetic rare cell separation from a particulate blood flow. Analytical Chemistry 2012;84:4292−9.

[65] Choi JW, Oh KW, Han A, Wijayawardhana CA, Lannes C, Bhansali S, et al. Development and characterization of microfluidic devices and systems for magnetic bead-based biochemical detection. Biomedical Microdevices 2001;3:191−200.

[66] Pamme N. Magnetism and microfluidics. Lab Chip 2005;6:24−38.

[67] Becker FF, Wang XB, Huang Y, Pethig R, Vykoukal J, Gascoyne P. Separation of human breast cancer cells from blood by differential dielectric affinity. Proceedings of the National Academy of Sciences of the United States of America 1995;92:860−4.

[68] Yang F, Yang X, Jiang H, Bulkhaults P, Wood P, Hrushesky W, et al. Dielectrophoretic separation of colorectal cancer cells. Biomicrofluidics 2010;4:013204.

[69] Petersson F, Åberg L, Swärd-Nilsson AM, Laurell T. Free flow acoustophoresis: microfluidic-based mode of particle and cell separation. Analytical Chemistry 2007;79:5117−23.

[70] Lenshof A, Ahmad-Tajudin A, Järås K, Swärd-Nilsson AM, Åberg L, Marko-Varga G, et al. Acoustic whole blood plasmapheresis chip for prostate specific antigen microarray diagnostics. Analytical Chemistry 2009;81:6030−7.

[71] Bhagat AAS, Bow H, Hou HW, Tan SJ, Han J, Lim CT. Microfluidics for cell separation. Medical and Biological Engineering and Computing 2010;48:999−1014.

[72] Hultström J, Manneberg O, Dopf K, Hertz HM, Brismar H, Wiklund M. Proliferation and viability of adherent cells manipulated by standing-wave ultrasound in a microfluidic chip. Ultrasound in Medicine & Biology 2007;33:145−51.

[73] ORorke R, Wood C, Walti C, Evans S, Davies A, Cunningham J. Acousto-microfluidics: transporting microbubble and microparticle arrays in acoustic traps using surface acoustic waves. Journal of Applied Physics 2012;111: pp. 094911-094911-8.

[74] Laurell T, Petersson F, Nilsson A. Chip integrated strategies for acoustic separation and manipulation of cells and particles. Chemical Society Reviews 2007;36:492−506.

[75] Gossett DR, Weaver WM, Mach AJ, Hur SC, Tse HTK, Lee W, et al. Label-free cell separation and sorting in microfluidic systems. Analytical and Bioanalytical Chemistry 2010;397:3249−67.

[76] Bruus H. Acoustofluidics 10: scaling laws in acoustophoresis. Lab on a Chip 2012;12:1578−86.

[77] Kapishnikov S, Kantsler V, Steinberg V. Continuous particle size separation and size sorting using ultrasound in a microchannel. Journal of Statistical Mechanics: Theory and Experiment 2006;2006: P01012.

[78] Lenshof A, Magnusson C, Laurell T. Acoustofluidics 8: applications of acoustophoresis in continuous flow microsystems. Lab on a Chip 2012;12:1210−23.

[79] Zhang W, Kai K, Choi DS, Iwamoto T, Nguyen YH, Wong H, et al. Microfluidics separation reveals the stem-cell−like deformability of tumor-initiating cells. Proceedings of the National Academy of Sciences of the United States of America 2012;109:18707−12.

[80] Preira P, Grandné V, Forel JM, Gabriele S, Camara M, Theodoly O. Passive circulating cell sorting by deformability using a microfluidic gradual filter. Lab on a Chip 2013.

[81] Davis JA, Inglis DW, Morton KJ, Lawrence DA, Huang LR, Chou SY, et al. Deterministic hydrodynamics: taking blood apart. Proceedings of the National Academy of Sciences of the United States of America 2006;103:14779−84.

[82] Huang LR, Cox EC, Austin RH, Sturm JC. Continuous particle separation through deterministic lateral displacement. Science 2004;304:987−90.

[83] Zheng S, Lin H, Liu JQ, Balic M, Datar R, Cote RJ, et al. Membrane microfilter device for selective capture, electrolysis and genomic analysis of human circulating tumor cells. Journal of Chromatography A 2007;1162:154−61.

[84] Lin HK, Zheng S, Williams AJ, Balic M, Groshen S, Scher HI, et al. Portable filter-based microdevice for detection and characterization of circulating tumor cells. Clinical Cancer Research 2010;16:5011−8.

[85] Hosokawa M, Hayata T, Fukuda Y, Arakaki A, Yoshino T, Tanaka T, et al. Size-selective microcavity array for rapid and efficient detection of circulating tumor cells. Analytical Chemistry 2010;82:6629−35.

[86] Rosenberg R, Gertler R, Friederichs J, Fuehrer K, Dahm M, Phelps R, et al. Comparison of two density gradient centrifugation systems for the enrichment of disseminated tumor cells in blood. Cytometry 2002;49:150−8.

[87] Gomez FA. Biological Applications of Microfluidics. Hoboken, N.J.: Wiley-Interscience; 2008.

[88] Zourob M, Elwary S, Turner A. Principles of Bacterial Detection: Biosensors, Recognition Receptors, and Microsystems. New York: Springer; 2008.

[89] Xia Y, McClelland JJ, Gupta R, Qin D, Zhao XM, et al. Replica molding using polymeric materials: a practical step towards nanomanufacturing. Advances in Materials 1997;9:147−9.

[90] Woolley AT, Hadley D, Landre P, deMello AJ, Mathies RA, Northrup MA. Functional integration of PCR amplification and capillary electrophoresis in a microfabricated DNA analysis device. Analytical Chemistry 1996;68:4081−6.

[91] Nakano H, Matsuda K, Yohda M, Nagamune T, Endo I, Yamane T. High-speed polymerase chain-reaction in constant flow. Bioscience Biotechnology and Biochemistry 1994;58:349−52.

[92] Kopp MU, de Mello AJ, Manz A. Chemical amplification: continuous-flow PCR on a chip. Science 1998;280:1046−8.

[93] Elwenspoek M, Jansen HV. Silicon Micromachining. Cambridge, England; New York: Cambridge University Press; 1998.

[94] Piatak Jr M, Saag M, Yang L, Clark S, Kappes J, Luk K, et al. High levels of HIV-1 in plasma during all stages of infection determined by competitive PCR. Science-New York Then Washington 1993;259:1749−54.

[95] Yeh S-H, Tsai C-Y, Kao J-H, Liu C-J, Kuo T-J, Lin M-W, et al. Quantification and genotyping of hepatitis B virus in a single reaction by real-time PCR and melting curve analysis. Journal of Hepatology 2004;41:659−66.

[96] Walboomers JM, Jacobs MV, Manos MM, Bosch FX, Kummer JA, Shah KV, et al. Human papillomavirus is a necessary cause of invasive cervical cancer worldwide. The Journal of Pathology 1999;189:12−9.

[97] Quinn TC, Welsh L, Lentz A, Crotchfelt K, Zenilman J, Newhall J, et al. Diagnosis by AMPLICOR PCR of Chlamydia trachomatis infection in urine samples from women and men attending sexually transmitted disease clinics. Journal of Clinical Microbiology 1996;34:1401−6.

[98] Palmer H, Mallinson H, Wood R, Herring A. Evaluation of the specificities of five DNA amplification methods for the detection of Neisseria gonorrhoeae. Journal of Clinical Microbiology 2003;41:835−7.

[99] Boeckh M, Huang M, Ferrenberg J, Stevens-Ayers T, Stensland L, Nichols WG, et al. Optimization of quantitative detection of cytomegalovirus DNA in plasma by real-time PCR. Journal of Clinical Microbiology 2004;42:1142−8.

[100] Ohno K, Tachikawa K, Manz A. Microfluidics: applications for analytical purposes in chemistry and biochemistry. Electrophoresis Nov 2008;29:4443−53.

[101] Khandurina J, McKnight TE, Jacobson SC, Waters LC, Foote RS, Ramsey JM. Integrated system for rapid PCR-based DNA analysis in microfluidic devices. Analytical Chemistry 2000;72:2995−3000.

[102] Ottesen EA, Hong JW, Quake SR, Leadbetter JR. Microfluidic digital PCR enables multigene analysis of individual environmental bacteria. Science 2006;314:1464−7.

[103] Liu RH, Yang JN, Lenigk R, Bonanno J, Grodzinski P. Self-contained, fully integrated biochip for sample preparation, polymerase chain reaction amplification, and DNA microarray detection. Analytical Chemistry 2004;76:1824−31.

[104] Legendre LA, Bienvenue JM, Roper MG, Ferrance JP, Landers JP. A simple, valveless microfluidic sample preparation device for extraction and amplification of DNA from nanoliter-volume samples. Analytical Chemistry 2006;78:1444−51.

[105] Nakayama T, Kurosawa Y, Furui S, Kerman K, Kobayashi M, Rao SR, et al. Circumventing air bubbles in microfluidic systems and quantitative continuous-flow PCR applications. Analytical and Bioanalytical Chemistry 2006;386:1327–33.

[106] Easley CJ, Karlinsey JM, Bienvenue JM, Legendre LA, Roper MG, Feldman SH, et al. A fully integrated microfluidic genetic analysis system with sample-in-answer-out capability. Proceedings of the National Academy of Sciences of the United States of America 2006;103:19272–7.

[107] Liu RH, Bonanno J, Yang J, Lenigk R, Grodzinski P. Single-use, thermally actuated paraffin valves for microfluidic applications. Sensors and Actuators B: Chemical 2004;98:328–36.

[108] Henry C. Microfluidic circuits. Chemical & Engineering News 2002;80.

[109] Spurgeon SL, Jones RC, Ramakrishnan R. High throughput gene expression measurement with real time PCR in a microfluidic dynamic array. PloS One 2008;3:e1662.

[110] Tsui NB, Kadir RA, Chan KA, Chi C, Mellars G, Tuddenham EG, et al. Noninvasive prenatal diagnosis of hemophilia by microfluidics digital PCR analysis of maternal plasma DNA. Blood 2011;117:3684–91.

[111] Yung TK, Chan KA, Mok TS, Tong J, To K-F, Lo YD. Single-molecule detection of epidermal growth factor receptor mutations in plasma by microfluidics digital PCR in non-small cell lung cancer patients. Clinical Cancer Research 2009;15:2076–84.

[112] Lipshutz RJ, Fodor SPA, Gingeras TR, Lockhart DJ. High density synthetic oligonucleotide arrays. Nature Genetics 1999;21:20–4.

[113] Gresham D, Dunham MJ, Botstein D. Comparing whole genomes using DNA microarrays. Nature Reviews Genetics 2008;9:291–302.

[114] Massie CE, Mills IG. ChIPping away at gene regulation. Embo Reports Apr 2008;9:337–43.

[115] de Leon J, Susce MT, Murray-Carmichael E. The AmpliChip (TM) CYP450 genotyping test–integrating a new clinical tool. Molecular Diagnosis & Therapy 2006;10:135–51.

[116] Gardiner SJ, Begg EJ. Pharmacogenetics, drug-metabolizing enzymes, and clinical practice. Pharmacological Reviews 2006;58:521–90.

[117] Southern EM. Detection of specific sequences among DNA fragments separated by gel electrophoresis. Journal of Molecular Biology 1975;98:503–17.

[118] Weber K, Osborn M. The reliability of molecular weight determinations by dodecyl sulfate-polyacrylamide gel electrophoresis. Journal of Biological Chemistry 1969;244:4406–12.

[119] Effenhauser CS, Bruin GJ, Paulus A, Ehrat M. Integrated capillary electrophoresis on flexible silicone microdevices: analysis of DNA restriction fragments and detection of single DNA molecules on microchips. Analytical Chemistry 1997;69:3451–7.

[120] Dolnik V, Liu S, Jovanovich S. Capillary electrophoresis on microchip. Electrophoresis 2000;21:41–54.

[121] Y. Cheng, W. Lin, Y. Yen, L. Chen, P. Hwu, Q. Liu, et al., Ratio quantification of gene dosage by Agilent 2100 Bioanalyzer for detection of somatic gene deletions, Journal of Biophysical Chemistry, 2.

[122] Hathaway LJ, Brugger S, Martynova A, Aebi S, Mühlemann K. Use of the Agilent 2100 bioanalyzer for rapid and reproducible molecular typing of Streptococcus pneumoniae. Journal of Clinical Microbiology 2007;45:803–9.

[123] Lu CY, Tso DJ, Yang T, Jong YJ, Wei YH. Detection of DNA mutations associated with mitochondrial diseases by Agilent 2100 bioanalyzer. Clinica Chimica Acta 2002;318:97–105.

[124] Sanger F, Nicklen S, Coulson AR. DNA sequencing with chain-terminating inhibitors. Proceedings of the National Academy of Sciences of the United States of America 1977;74:5463–7.

[125] Shendure J, Ji H. Next-generation DNA sequencing. Nature Biotechnology 2008;26:1135–45.

[126] Paegel BM, Blazej RG, Mathies RA. Microfluidic devices for DNA sequencing: sample preparation and electrophoretic analysis. Current Opinion in Biotechnology 2003;14:42–50.

[127] Blazej RG, Kumaresan P, Mathies RA. Microfabricated bioprocessor for integrated nanoliter-scale Sanger DNA sequencing. Proceedings of the National Academy of Sciences of the united States of America 2006;103:7240–5.

[128] Branton D, Deamer DW, Marziali A, Bayley H, Benner SA, Butler T, et al. The potential and challenges of nanopore sequencing. Nature Biotechnology 2008;26:1146–53.

[129] Kasianowicz JJ, Brandin E, Branton D, Deamer DW. Characterization of individual polynucleotide molecules using a membrane channel. Proceedings of the National Academy of Sciences of the United States of America 1996;93:13770−3.

[130] Li J, Stein D, McMullan C, Branton D, Aziz MJ, Golovchenko JA. Ion-beam sculpting at nanometre length scales. Nature 2001;412:166−9.

[131] Manrao EA, Derrington IM, Laszlo AH, Langford KW, Hopper MK, Gillgren N, et al. Reading DNA at single-nucleotide resolution with a mutant MspA nanopore and phi29 DNA polymerase. Nature Biotechnology 2012;30:349−53.

[132] Fu AY, Spence C, Scherer A, Arnold FH, Quake SR. A microfabricated fluorescence-activated cell sorter. Nature Biotechnology 1999;17:1109−11.

[133] Sia SK, Linder V, Parviz BA, Siegel A, Whitesides GM. An integrated approach to a portable and low-cost immunoassay for resource-poor settings. Angewandte Chemie International Edition 2004;43: 498−502.

[134] Liberti PA, Rao CG, Terstappen LWMM. Optimization of ferrofluids and protocols for the enrichment of breast tumor cells in blood. Journal of Magnetism and Magnetic Materials 2001;225:301−7.

[135] Miller MC, Doyle GV, Terstappen LWMM. Significance of circulating tumor cells detected by the CellSearch system in patients with metastatic breast colorectal and prostate cancer. Journal of Oncology 2009;2010.

[136] Yung CW, Fiering J, Mueller AJ, Ingber DE. Micromagnetic−microfluidic blood cleansing device. Lab Chip 2009;9:1171−7.

[137] Zhang CS, Xing D. Miniaturized PCR chips for nucleic acid amplification and analysis: latest advances and future trends. Nucleic Acids Research 2007;35:4223−37.

[138] Sassolas A, Leca-Bouvier BD, Blum LJ. DNA biosensors and microarrays. Chemical Reviews 2008;108:109−39.

[139] Fologea D, Gershow M, Ledden B, McNabb DS, Golovchenko JA, Li J. Detecting single stranded DNA with a solid state nanopore. Nano letters. 2005;5(10):1905−9.

Further Reading

Panton RL. Incompressible Flow. third ed. Chichester: John Wiley & Sons; 2005.

Batchelor GK. An Introduction to Fluid Dynamics. Cambridge: Cambridge University Press; 2000.

Cemal Eringen A. Mechanics of Continua. second ed. Malabar FL: Krieger Pub Co; 1980.

Flügge S. Handbuch der Physik Encyclopedia of Physics. Berlin: Springer-Verlag; 1956.

CHAPTER 4

Electrical Transducers

Electrochemical Sensors and Semiconductor Molecular Sensors

Chapter Outline

4.1 Introduction

In this chapter, we describe the fundamental principles and important applications of molecular sensors based on electrical transduction. A variety of electrochemical devices, such as pH or other ion and gas sensors, have been widely used for chemical analysis. Electrical transduction is utilized to detect extraneous changes in chemical species. First, we describe principles and applications of conventional electrochemical measurements. We discuss the theories of how chemical reactions are related to electrical signals in an electrolyte–electrode system, followed by introduction of important practical applications.

Second, we will introduce molecular sensors based on semiconductors. Advances in silicon microfabrication have certainly expanded the scope of microelectronics based sensors towards a wide range of applications. The most important concept for integrated microelectronics-based sensors is the principle of the field effect transistor (FET). Basic theories of semiconducting materials and working principles of FETs are described and related to the design of molecular sensors. The utilization of biological recognition processes allows for the development of molecular sensors that are highly sensitive to specific biomolecules. Following the basics of FETs, we describe microelectronics based sensors that utilize immunoassay and DNA hybridization. The important aspect of sensor development is the advent of functional nanomaterials. Various materials have been used to create the electrochemical part of the sensor. The use of the nanomaterials in the FET sensor allow for a sensitive, real-time detection of a wide range of biological and chemical species.

At the end of the chapter we will introduce fundamentals and practice of nanomaterials used for electrical transduction. We describe sensors based on silicon and organic semiconductors, silicon nanowires, carbon nanotubes, and graphene, which are actively being tested and discussed within nanomaterial and nanosciences communities.

4.2 Electrochemical Sensors

4.2.1 Principles of Electrochemical Measurements

In this section we discuss basic theories of electrochemical measurements. Many chemical reactions involve transfer of electrons which can be measured as electrical potentials. We start from the fundamentals of **reduction−oxidation** (redox) reactions and how they are quantified. We also describe important examples of electrochemical sensors that are commonly used in daily life as well as laboratory based chemical analysis.

Redox Reaction

A **redox reaction** is a chemical reaction that has a transfer of at least one electron from one molecule to another. Substances that are capable of removing electrons from another substance are said to be oxidative or oxidizing and are called **oxidants** or **oxidizers**. When an oxidant accepts electrons from other substances, the oxidizer is reduced. Substances that are capable of transferring electrons to other substances are said to be reductive or reducing and are called **reductants** or **reducers**. When a reductant donates electrons to other substances, the reductant is oxidized.

Here is an example of a redox reaction.

$$Cu^{2+}(aq) + Zn(s) \rightarrow Cu(s) + Zn^{2+}(aq)$$

In this reaction, zinc transfers two electrons to copper. In this example, zinc is oxidized (i.e., acts as the reductant) and loses two electrons while copper is reduced (i.e. acts as the oxidant) and gains two electrons. **Half reactions** are used to clearly focus on either oxidation or reduction alone.

The above example can be rewritten as two half reactions:

$$Cu^{2+}(aq) + 2\ e^- \rightarrow Cu(s)$$

$$Zn(s) \rightarrow Zn^{2+}(aq) + 2e^-$$

An oxidant and its corresponding reductant which appear on the other side of a half-equation constitute a **redox couple**. In this example, Cu^{2+}/Cu and $Zn^{2+}/Zn(s)$ are redox couples.

Another example is simple salt solubilization:

$$AgCl(s) \rightarrow Ag^+(aq) + Cl^-(aq)$$

This is rewritten as

$$Ag(s) \rightarrow Ag^+(aq) + e^-$$
$$Cl(s) + e^- \rightarrow Cl^-(aq)$$

Redox potential or reduction potential is the tendency of a chemical species to acquire electrons and be reduced. Reduction potential is measured in volts (V), or millivolts (mV). This will be discussed in greater detail later.

Electrochemical Cell

When an electrode is placed in a salt solution (electrolyte) as shown in Fig. 4.1(a), a charge separation is generated at the metal-solution interface. This setup is called a **half-cell**. The charge separation in a half-cell cannot be measured directly, but we can quantify relative tendencies between two different materials by introducing another electrode–electrolyte combination (Fig. 4.1(b)).

The two half-cells compose an **electrochemical cell**. Each side of the cell is called a half-cell, and is electrically connected to each other by a conductive salt bridge.

A galvanic cell or voltaic cell is a type of electrochemical cell where spontaneous redox reactions take place when deriving electrical energy. The two electrodes of a galvanic cell constitute a pair of an anode and a cathode.

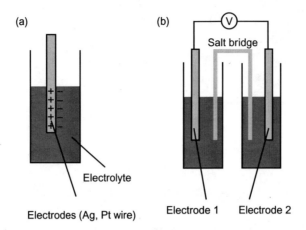

Figure 4.1: Electrochemical Cells. (a) Half-cell. (b) Electrochemical Cell Is Composed of Two Half-cells.

The **cathode** is defined as the electrode where **reduction** occurs. The reducing agent is the electron donor.

$$Ox_1 + n\,e^- \rightarrow Red_1 \quad [\text{Reduction of an oxidant } Ox_1]$$

The **anode** is defined as the electrode where **oxidation** occurs. The oxidizing agent is the electron acceptor.

$$Red_2 \rightarrow Ox_2 + n\,e^- \quad [\text{Oxidation of a reductant } Red_2]$$

These two reactions indicate that electrons move from the anode to the cathode, i.e. the current flows from the cathode to the anode. The potential difference between the anode and cathode measured in volts (V) is called the **cell potential** or **EMF** (cell electromotive force). Note that EMF is not considered a "force", which is measured in Newtons.

The **Daniell cell** is an example of galvanic cell which includes Zn and Cu electrodes. The Cu and Zn electrodes are immersed in copper (II) sulfate and zinc sulfate solutions, respectively, prepared in the following way:

$$CuSO_4(s) \rightarrow Cu^{2+}(aq) + SO_4^{2-}(aq)$$
$$ZnSO_4(s) \rightarrow Zn^{2+}(aq) + SO_4^{2-}(aq)$$

The two half reactions are written as:

$$Cu^{2+}(aq) + 2e^- \rightarrow Cu(s) \quad [\text{Reduction}]$$
$$Zn(s) \rightarrow 2e^- + Zn^{2+}(aq) \quad [\text{Oxidation}]$$

Here, copper works as the oxidant and zinc as the reductant. By adding two equations, we obtain

$$Cu^{2+}(aq) + Zn(s) \rightarrow Cu(s) + Zn^{2+}(aq)$$

which is the complete cell reaction of the Daniell cell. The Daniell cell can be also expressed in a following form depicting the structure of the cell:

$$Zn(s) \,|\, ZnSO_4(aq) \,|\, CuSO_4(aq) \,|\, Cu(s)$$

The Electrode Potential

When two metals are used as electrodes in an electrochemical cell, the direction of the current flow (i.e., which metal acts as the anode or the cathode) depends on the ionization tendencies of the metals. We can quantify them by measuring "relative" potentials of metals by introducing a reference potential. For this purpose, we determine the **electrode potential** for each half-cell relative to the hydrogen half-cell, where the reaction is written as:

$$H^+(aq) + e^- \rightarrow 1/2\ H(g) \quad [\textbf{Reduction}]$$

The ionization tendency is highly dependent on the temperature and pressure of the system. Here we use the standard state, namely:

- concentration of hydrogen 1 M
- hydrogen pressure 1 atm
- temperature: 25°C
- measured with inert electrode (Pt).

The electrode is called the standard hydrogen electrode (SHE).

Figure 4.2 shows the composition of a hydrogen electrode connected to another electrode which is to be measured. Hydrogen gas flows over the porous, or platinized, platinum which catalyzes the reaction between hydrogen molecules and hydrogen ions in solution.

Table 4.1 lists electrode potentials for some important materials at standard state, which is with solutes at an effective concentration of 1 M, and gases at a pressure of 1 atm. The **standard electrode potential**, or standard potential, is usually abbreviated as E°. (AgCl is often used as a reference electrode. Pt is often used as an inert electrode.)

By subtracting the potentials from each redox couple, we can find the EMF of an electrochemical cell that uses two metals as electrodes at standard state. For the Daniel cell,

$$E_{cell} = E_{Cu}{}^0 - E_{Zn}{}^0 = +0.34 - (-0.76) = 1.10\ V$$

is the voltage which will be found between the cathode (Cu) and the anode (Zn).

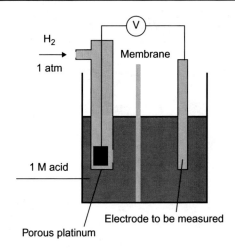

Figure 4.2: Standard Hydrogen Electrode for the Measurement of an Electrode Potential.
Adapted from [1].

Table 4.1: Electrode Potentials

Redox Couple		E [V]
$Ce^{4+}(aq) + e^-$	Ce^{3+} (aq)	$+1.61$
$Cu^{2+}(aq) + 2e^-$	Cu (s)	$+0.34$
AgCl (aq) $+ e^-$	Ag (aq) $+ Cl^-$ (aq)	$+0.22$
2 H (aq) $+ 2e^-$	H_2 (g)	$+0.00$
$Zn^{2-}(aq) + 2e^-$	Zn (s)	-0.76
$Na^+(aq) + e^-$	Na (s)	-2.71

Reference Electrode

In practice, routine SHE measurements using hydrogen gas is not convenient as hydrogen is an explosive gas. Types of reference electrodes for easier handling have been devised for practical applications.

Figure 4.3(a) illustrates the composition of a **silver–silver chloride (Ag|AgCl) electrode** with the following half-cell reaction:

$$AgCl(s) + e^- \rightarrow Ag(s) + Cl^-$$

The **saturated calomel electrode (SCE)** is a reference electrode based on the reaction between mercury and mercury(I) chloride. The half-cell reaction is:

$$Hg_2Cl_2(s) + 2\,e^- \rightarrow 2\,Hg(s) + 2\,Cl^-$$

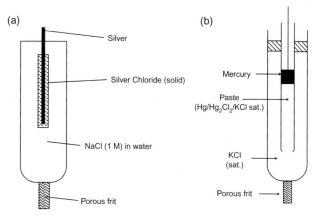

Figure 4.3: Reference Electrodes. (a) Silver-Silver Chloride Electrode. (b) Saturated-Calomel Electrode.

Figure 4.3(b) is the schematic of the electrode. A saturated water solution of KCl (potassium chloride) is in contact with the mercury and the mercury(I) chloride (also known as "calomel"). Linking the electrode to the solution in which the other electrode is immersed is a porous frit, which is a salt bridge. Both the Ag|AgCl electrode and the SCE are very commonly used types of reference electrodes. They can be easily prepared, inexpensive, and stable. Ag|AgCl may be preferred in some cases where use of mercury, which is toxic, should be avoided. However, Ag|AgCl tends to react more with other ions. Because reference systems based on Ag/AgCl can be fabricated in a small scale, they are often used in integrated micro or miniature systems. An example of Ag/AgCl electrode in a miniature system is described in Chapter 7, Section 7.3.6.

The Nernst Equation

The electrode potential is dependent on the temperature and concentration of the analyte in the system. The EMF E of a cell is related to the **Gibbs free energy** ΔG, which is the work obtainable from the thermodynamic system at a constant temperature and pressure. When we consider a half-cell reaction

$$Ox + n\,e^- \rightarrow Red$$

The Gibbs free energy ΔG is given by EMF E in the following way:

$$\Delta G = -nFE$$

Where F is the Faraday constant (96 485 C/mol). The value of ΔG tells us which way the reaction will shift to achieve equilibrium and the magnitude tells us how far the reaction is

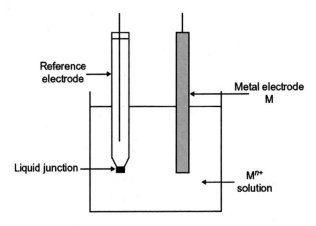

Figure 4.4: Electrode Potential Measurement.

from reaching equilibrium. The **Nernst equation** relates the electrode potential to the temperature and the standard-state electrode potential:

$$E = E^0 + \frac{RT}{nF} \ln\left(\frac{a_{Ox}}{a_{Red}}\right) \tag{4.1}$$

where a_{Ox} and a_{Red} are the chemical activities, or the effective thermodynamic concentration, of the oxidant and the reductant, respectively. For dilute solutions, they can be approximated simply as the concentrations:

$$E = E^0 + \frac{RT}{nF} \ln \frac{[Ox]}{[Red]} \tag{4.2}$$

Let us consider a realistic case of electrode potential measurement, as shown in Fig. 4.4. The metal electrode has the half-cell reaction of:

$$M^{n+} + ne^- = M$$

The direction of reaction (either from left to right or from right to left) depends on the potential.

The measured potential E_{cell} is given by the electrode potential $E_{M/M^{n+}}$, the liquid junction potential E_{lj} (the potential difference along the liquid junction) and the reference electrode potential E_{ref}:

$$E_{cell} = E_{M/M^{n+}} - E_{ref} - E_{lj} \tag{4.3}$$

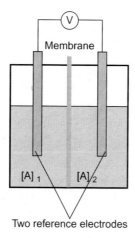

Figure 4.5: Concentration Cell.

From the Nernst equation, $E_{M/M^{n+}}$ is related to E_M^0 and the temperature and concentration.

$$E_{M/M^{n+}} = E_M^0 + \frac{RT}{nF} \ln \frac{[M^{n+}]}{[M]} \tag{4.4}$$

In this case, the reduced substance is a solid metal and $[M] = 1$.

$$E_{M/M^{n+}} = E_M^0 + \frac{RT}{nF} \ln [M^{n+}] \tag{4.5}$$

From Eqs. 4.3 and 4.5, the standard potential of the metal E_M^0 can be found.

$$E_M^0 = E_{cell} + E_{ref} + E_{lj} - \frac{RT}{nF} \ln [M^{n+}] \tag{4.6}$$

Another example is the concentration cell, where the difference of the concentration is measured as the potential. The electrochemical cells are linked by a selective membrane and contain the same half-cell electrode in each half of the cell, differing only in the concentration of the analyte, as shown in Fig. 4.5.

Assuming that the solution is dilute and $a_A = [A]$,

$$E_{cell} = \left(E^0 + \frac{RT}{nF} \ln [A]_1 \right) - \left(E^0 + \frac{RT}{nF} \ln [A]_2 \right) = \frac{RT}{nF} \ln \frac{[A]_1}{[A]_2} \tag{4.7}$$

This equation is important to help understand how concentration gradients behave in numerous biological, chemical, and physiological systems. The Nernst equation can be used to help model the equilibrium changes in a cell membrane and the same ideas are applied to model the Fermi level of a semiconductor, as described later.

The electrical double layer

When an electrode is immersed in an electrolyte solution, an interfacial region called the double layer is formed. The double layer consists of two parallel layers of charges. The first layer is made of ions adsorbed onto the electrode surface due to chemical interactions, and can be either positive or negative. The second layer is formed by ions attracted to the charge of the first layer via the Coulomb force. The electrical properties of this layer affect the electrochemical measurements. When the potential of a working electrode is changed, the double layer acts as a capacitor which must be charged. While the layer is being charged, a capacitive current, which is not related to the reduction or oxidation of the substrates, flows in the electrical system and interferes electrochemical measurements. The effect of capacitive current needs to be separated from the measured signals. Several methods are used to reduce or isolate the capacitive current.

4.2.2 Applications of Electrochemical Sensors

Based on the nature of the electrical signal, most of electrochemical sensors fall into the following three major categories:

1. **Potentiometric**: The electrostatic potentials (volts [V]) are measured. Little or practically no current is involved in the measurement.
2. **Voltammetric** or **amperometric**: Current measurement (amperes [A]) is involved. Potentials (volts [V]) of electrode are held constant or used as a variable input during measurements.
3. **Conductometric**: Resistance (ohms [Ω]) or conductance (siemens [S] = [$1/\Omega$]) is measured by alternating current.

We will describe the details of these methods and related techniques, followed by introduction of examples of sensors in practical applications.

Potentiometric Sensors

In potentiometric sensors, the potential difference between the working electrode and reference electrode is measured. Ion selective electrodes are typically used for obtaining the potential signal for specific ions. Potentiometric sensors use the potential of an electrode when there is no current. As we discussed earlier, the potential is proportional to the activity or relative concentration of species generated or consumed in the reaction.

A **pH meter** is one of the most commonly used applications of potentiometry. By definition, pH is related to the activity of H^+ ions in the following way:

$$pH = -\log_{10}[H^+] \tag{4.8}$$

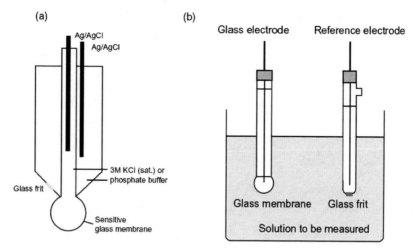

Figure 4.6: Composition of a pH Meter. (a) Schematic of a Typical pH Electrode. (b) pH Sensor Is Considered as a Specific Type of Concentration Cell Modeled as Illustrated.

For pure water or a liquid with the neutral pH, $[H^+] = [OH^-] = 10^{-7}$. A pH meter measures the potential difference between the liquid inside the reference glass electrode and the liquid to be measured. Figure 4.6(a) shows the schematic of a typical pH sensor. Highly hygroscopic soda-glass is commonly used for glass membrane. The surface of the glass must be hydrated for the membrane to be sensitive to H^+ ions. The sensitive glass membrane is highly selective and permeable only to hydrogen ions. The sensing electrode thus measures the potential of hydrogen ions in the outside liquid. The reference electrode is usually an Ag/AgCl electrode in a 3 M potassium chloride (KCl) solution contained in a glass tube. It is electrically connected to the outside liquid (i.e., the liquid to be measured) through a liquid junction formed by a glass frit. It is a specific type of concentration cell as modeled in Fig. 4.6(b). Applying Eq. 4.7 for this case, the measured potential $E = E_{cell}$ can be simply expressed as:

$$E = \frac{RT}{nF}(\ln[H^+] - \ln[H^+]_{ref}) \tag{4.9}$$

This can be simplified as:

$$E = K + S \log_{10}[H^+] \tag{4.10}$$

where

$$S = \frac{1}{\log_{10}e} \times \frac{RT}{nF} \tag{4.11}$$

Note that base 10 is used in Eq 4.10 to correlate Eq 4.9 with the value of pH.

The term K always includes an offset found in actual measurements. When the reference electrode is filled with a solution with a pH of 7,

$$E = E_0 + S \cdot (7 - \text{pH}) \tag{4.12}$$

The value of the offset E_0 needs to be found by calibration. Ideally, the measured potential equals zero when the value of pH is 7.

As seen in Eq. 4.11, the measured value in mV per pH unit is dependent on the temperature. The temperature has to be measured to find accurate pH values (see problem 4.2). For example, the actual values of pH at 25°C can be found from the following equation:

$$E = E_0 + 59.1 \cdot (7 - \text{pH}) \cdot (\text{mV}) \tag{4.13}$$

The same principle can be used for measurement of other ions by using proper ion-selective membranes. Other types of glass membranes that are sensitive to K^+, Na^+, Li^+, Ag^+ ions are available. Commercial ion sensitive sensors often use a polymer membrane that incorporates ion sensitive species.

One important application area of potentiometric ion sensing is trace-level analysis. These include detection of lead and copper ions found in drinking water. It has been also used to detect free copper ions present in seawater and how much cadmium ions are absorbed in plant roots [2].

Voltammetric and Amperometric Sensors

Voltametric and amperometric sensors measure current response between a working and reference electrode. In **voltammetric sensors**, the current response is measured as a function of applied potential. The potential is varied either step by step or continuously in order to determine the current as a function of the cells potential $I = f(E)$ which is called a voltammogram. Three electrodes (working, auxiliary and reference) are typically used for a voltammetric sensor (see Fig. 4.7). In principle, at least two electrodes are needed for the measurement. In practice, it is difficult to maintain a constant potential of an electrode when it is passing a current. The use of the three electrodes allows accurate application of potentials and the current measurement. The reference electrode acts as reference in measuring and stabilizing the potential of the working electrode. This is achieved by passing little or no current through the reference electrode. Thus the auxiliary electrode passes all the current through the working electrode.

The applied potential drives the electron transfer reaction of the analytes. The measured current directly indicates the rate of the electron transfer in an electrochemical reaction. See Section 4.4.2 on graphene-based glucose sensors for a specific example.

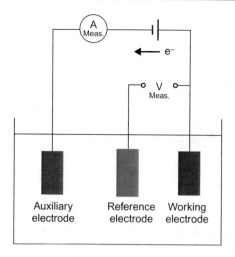

Figure 4.7: Three Electrode Setup for Voltammetry.

Amperometric sensors measure current response to detect the concentration of an analyte at a fixed potential. A simple form of amperometric measurement is single-potential amperometry or DC amperometry. A voltage is applied between two electrodes during measurement. A working potential is applied only for a short time to prevent damaging the electrode. The current is measured only while the potential is applied. A typical application of amperometry is the determination of glucose concentrations.

Conductometric Sensors

A **conductometric sensor** is based on conductive or semiconductive materials that change their conductivity upon interaction with chemical species. Figure 4.8 illustrates schematics of conductometric sensors. The sensing material that is specific to the analyte in question is sandwiched between two contact electrodes and the resistance of the entire device is measured.

Such an arrangement is typical for devices called chemiresistors or chemoresistors, mostly used for sensing in gases.

Conductometric sensors are made up of two basic components, electrodes and a sensitive layer. The resistance between the two sensing electrodes will change when the sensing layer comes in contact with a specific analyte. The change in resistance will increase as more analytes continue to bind to the sensing electrode. This resistance change could be measured and calibrated for actual measurements. Interdigital electrodes, as shown in Fig. 4.8(b), are commonly utilized to measure analyte concentration over a larger contact area. One of the main advantages of conductometric sensors is in their low manufacturing

(a)

(b)

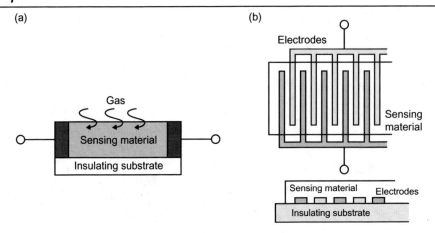

Figure 4.8: Conductometric Sensor. (a) Basic Concept. (b) Interdigital Electrodes.

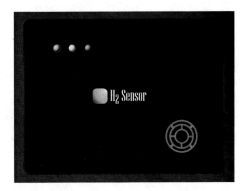

Figure 4.9: Conductometric Sensor Used as a Gas Monitor.

cost. Several preparation processes including screen printing, sputtering, and evaporation can be used to form the sensing layers. H_2 gas sensors usually installed on a wall, as shown in Fig. 4.9, are mostly conductometric sensors that utilize a reversible change in electrical resistance of a cerium oxide-based material.

W.Y. Lee et al. developed a conductometric urease biosensor using urease as the reacting agent. They used photolithography to create interdigital electrodes used in the sensor [3]. Urea would bind to the urease, which would result with a change in the conductivity between electrodes, thus allowing the sensor to detect a relative increase in urea concentration, as shown in Fig. 4.10.

The ability to measure the concentration of biomolecules at a high sensitivity and selectivity is of great importance. Kriz et al. used molecularly imprinted polymers to

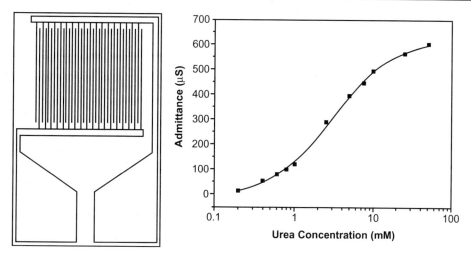

Figure 4.10: Conductometric Urease Biosensor. (a) Interdigital Electrodes of the Sensor. (b) Relationship Between Measured Conductivity and Urea Concentration.
From [3].

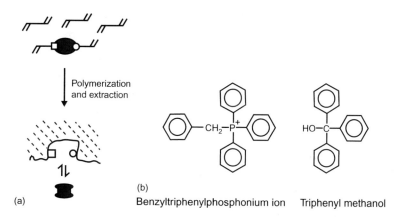

Figure 4.11: (a) Concept of Molecular Imprinting Technique. (b) Benzyltriphenylphosphonium Ion and the Reference Ion Used to Imprint the Reference Sensor.

construct the sensing element of a conductometric ion sensor. The imprinted polymers are polymers that have sensing sites available specifically to print molecules. Monomers were first cross-linked with print molecules (analytes) to form a polymer. The print molecules are then removed to create available sensing sites (see Fig. 4.11(a)). Benzyltriphenylphosphonium ions were used as a model (see Fig. 4.11(b)). When the sensor was exposed to the gas, the measured conductivity was significantly higher than that of a reference sensor which is printed for another type of molecule [4].

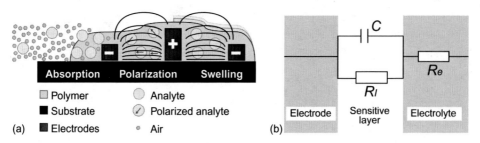

Figure 4.12: Capacitive Chemical Sensors. (a) Schematic Showing the Working Principle of a Capacitive Chemical Sensor (from [5]). (b) Simplified Equivalent Circuit.
From [6].

Capacitive Sensors

A **capacitive sensor** is based on measurement of changes in capacitance. In practice, sensor designs are very similar to those of conductometric sensors. Figure 4.12 shows a schematic of the sensing principle of a typical capacitive chemical sensor. Absorption of target molecules induces two relevant effects of changes in dielectric constant and swelling. Interdigitated electrodes as described for conductometric sensors are commonly used [5]. Usually measured for sensing is the impedance of the system [6], which includes the resistance of the sensitive layer as shown in the simplified circuit in Fig. 4.12(b). Capacitive sensors may be categorized as a special type of conductometric sensor. The impedance of the sensor is typically found indirectly by incorporating the sensor into an RC circuit. Changes in the resonant frequency of an oscillator, or the level of coupling (or attenuation) of an AC signal is used for measurement.

Glucose Sensors

Glucose sensors [7−9] are used to measure the blood glucose concentration of a patient and are an important part of managing diabetes mellitus. Type 1 and type 2 diabetes are the most common forms of diabetes. Type 1 diabetes is usually diagnosed in children and young adults and accounts for about 5% of all diagnosed cases of diabetes. Type 2 diabetes has been diagnosed in millions of Americans. According to diabetes report card 2012 issued by National Center for Chronic Disease Prevention and Health Promotion, 18.9% of US adults over 65 years old are diagnosed as diabetes in 2007−2009.

Patients with Type 1 diabetes may test their blood sugar five to ten times a day in order for them to effectively monitor their blood sugar levels. Type 2 diabetics may also consider monitoring their blood sugar levels daily based on their risk for future health complications due to the disease. Blood glucose testing may be also needed for patients with other diseases which may affect the pancreas such as cystic fibrosis. In sports medicine, it is used to monitor physical conditions of athletes. Normal blood glucose levels range between

80–120 mg/dL with spikes reaching up to 250 mg/dL after meals. The sensor should also be able to measure the extremes in blood sugar levels (between 20–500 mg/dL, or 1-30 mM) which a patient may experience during an episode of hyper or hypoglycemia and should have a resolution of ~ 1 mg/dL, or ~ 50 μM.

The majority of blood glucose sensors, or **glucose meters**, are categorized as **amperometric sensors**, which will be described in this chapter. In chapter 5, we will discuss techniques based on optical transduction such as **absorption spectroscopy** (see Section 5.5), **light scattering** and **Raman spectroscopy** (see Section 5.7). In chapter 7, we describe examples of implantable glucose sensors (see Section 7.3.6).

In amperometric glucose sensors, reducing property of glucose is measured as a current. Sensors contain electrodes to measure the current generated by an enzymatic reaction usually between glucose, an enzyme, and a mediator. Use of glucose oxidase (GOx or GOD) has become the gold standard for glucose sensing [10,11]. The initial concept of glucose enzyme electrodes, where a thin layer of GOx was entrapped via a semipermeable membrane, was introduced by Clark and Lyons [12]. Sensing was based on the measurement of the oxygen consumed by the enzyme-catalyzed reaction

$$\text{Glucose} + O_2 \xrightarrow{\text{GO}_x} \text{Gluconic acid} + H_2O_2$$
$$\text{Glucose} + \text{GOx}_{(ox)} \rightarrow \text{Gluconic acid} + \text{GOx}_{(red)}$$
$$\text{GOX}_{(red)} + 2M_{(ox)} \rightarrow \text{GOx}_{(ox)} + 2M_{(red)} + 2H^+$$
$$2M_{(red)} \rightarrow 2M_{(ox)} + 2e^-$$

In this method, glucose reacts with the enzyme $\text{GOx}_{(ox)}$. The reduced enzyme $\text{GOx}_{(red)}$ then reduces two mediator $M_{(ox)}$ ions to $M_{(red)}$, which is oxidized back to $M_{(ox)}$ at the electrode surface. The oxidation process $2M_{(red)} \rightarrow 2M_{(ox)} + 2e^-$ is measured as the current by the electrode. However, for this type of early glucose biosensors, a high operation potential is required to perform the amperometric measurement of hydrogen peroxide. Improved methods utilize artificial mediators instead of oxygen to transfer electrons between the GOx and the electrode [8]. Reduced mediators are formed and reoxidized at the electrode, providing an electrical signal to be measured.

A blood glucose test is typically performed by pricking the finger to draw blood, which is then applied to a disposable "test-strip". Figure 4.13 shows a typical glucose meter and a test strip. Each strip includes layers of electrodes, spacers, immobilized enzymes assembled in a small package. Continued research and development have worked to reduce the overall size of the sensor itself and reduce the amount of blood required for an accurate measurement ($\sim$ μL).

The advanced glucose electrodes do not use mediators and measures direct transfer between the enzyme and the electrode. The electrode directly transfers electrons using organic

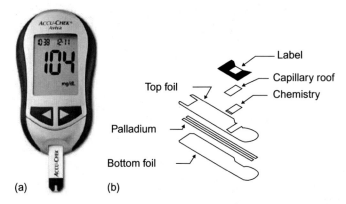

Figure 4.13: Blood Glucose Sensor. (a) Example of a Commercial Product. (b) Composition of a Test Strip Which Includes Electrodes.

(a) From https://www.accu-chek.com/us/; (b) from [13]. Courtesy of Roche.

conducting materials based on charge-transfer complexes. This type of electrodes have led to needle-type implantable sensors for continuous blood glucose monitoring [89, 90].

4.3 FET-Based Molecular Sensors

In this section we will discuss the basics of semiconductor physics in order to give the reader an understanding of the mechanism behind FET-based molecular sensors. We start from basics of semiconductor devices, followed by design and applications of FET based molecular sensors.

4.3.1 Semiconductor Basics

Electrical Conductivity

Here we describe basic parameters related to conductivity of solid state materials. Let us consider a material with a length L and a uniform cross section A. We assume a uniform flow of electric current within the material. When we apply a **voltage** (V), or an **electrical potential difference**, V as illustrated in Fig. 4.14, the **current** I(A) and the **electrical resistance** $R(\Omega)$ satisfy **Ohm's law**:

$$I = \frac{V}{R} \qquad (4.14)$$

The resistance R is proportional to the length L (cm or m) and inversely proportional to the cross section A (cm^2 or m^2):

$$R = \rho \frac{L}{A}, \qquad (4.15)$$

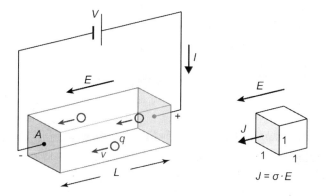

Figure 4.14: Electrical Conductivity.

where ρ is defined as the **electrical resistivity** (Ω cm or Ω m) of the material and is considered as the resistance of a material per unit length ($L = 1$ cm or 1 m) for unit cross section ($A = 1$ cm^2 or 1 m^2). Ohm $\cdot$ centimeter (Ω cm) is commonly used as the unit, although the SI unit of resistivity is ohm $\cdot$ meter (Ω m). In order to discuss conductive characteristics of a material, rather than the dimensions of the resistor, It is useful to describe Ohm's law in terms of a unit cubic volume. Plugging Eq. 4.15 into 4.14, we derive

$$\frac{I}{A} = \rho \frac{V}{L} \tag{4.16}$$

When we define the **electric field** (V/m) as $E = V/L$ and the **current density** (A/m^2 or A $\cdot$ cm^{-2}), or current per unit area, as $J = I/A$, Eq. 4.16 can be rewritten as:

$$E = \rho \cdot J \tag{4.17}$$

or

$$J = \frac{1}{\rho} \cdot E \tag{4.18}$$

Here we define the **conductivity** σ as the inverse of resistivity:

$$\sigma = \frac{1}{\rho} \tag{4.19}$$

The SI unit of conductivity is siemens per meter (S/m), where 1 S is the reciprocal of 1 Ω, i.e., S $= \Omega^{-1}$.

Using the conductivity, Eq. 4.18 can be rewritten as:

$$J = \sigma \cdot E, \tag{4.20}$$

This is a very important formula to describe electrical characteristics of semiconductors.

Charge Carriers and Mobility

An electric current is a flow of electric charges transported by **charge carriers**. Let us consider the current induced by transportation of charge carriers with an electric charge q (see Fig 4.14). The current density J is given as:

$$J = n \cdot q \cdot v, \tag{4.21}$$

where n and v are the concentration (i.e., the number of particles per unit volume) and the average velocity of the carriers, respectively.

If we assume n is independent of the electric field E, we find that the average velocity v has to be proportional to E, since we know from Eq. 4.20 that the current density J should be proportional to E. We write the average velocity as

$$v = \mu E, \tag{4.22}$$

where μ is the **mobility** of the charge carriers and indicates how quickly the charge carriers can move through the material when exposed to an electric field. The commonly used unit for mobility is (cm^2/V/s), while the SI unit is (m^2/V/s). Plugging Eq. 4.22 into 4.21, we find

$$J = n \cdot q \cdot \mu \cdot E \tag{4.23}$$

In metals, the charge carriers are electrons and because metals are conductors, charge can easily pass through the material. In an insulator, electrons do not flow freely in the material and do not conduct an electric current. In semiconductors, charge carriers are **electrons** and **holes**. An electron carries an electric charge of $-e$, and a hole carries a charge of $+e$. Since electrons and holes have different concentrations and mobilities, their respective current densities are considered independently, namely:

$$J_n = n \cdot (-e) \cdot \mu_n \cdot E \tag{4.24}$$

$$J_p = p \cdot e \cdot \mu_p \cdot E \tag{4.25}$$

where J_n and J_p are current densities of electrons and holes, n and p are electron and hole concentrations, and μ_n and μ_p are electron and hole mobilities, respectively.

The total current density J_{tot} is given as a summation of J_n and J_p:

$$J_{tot} = -J_n + J_p \tag{4.26}$$

The use of signs may be a little confusing. Electrons and holes have electric charges with opposite signs, which appears in Eqs 4.24 and 4.25. Under an electric field E, electrons and holes move in opposite directions, as seen in Eq. 4.26. As a result, both electrons and holes carry current in the same direction under an electric field E.

In practice, either electrons or holes are dominant and are considered the **majority carriers**. Either Eqs 4.24 or 4.25 instead of 4.26 gives sufficient approximation.

When electrons are the majority carriers, comparing Eqs 4.20 and 4.24 and neglecting the sign, we can find the conductivity and the resistivity:

$$\sigma_n = n \cdot e \cdot \mu_n \tag{4.27}$$

$$\rho_n = \frac{1}{n \cdot e \cdot \mu_n} \tag{4.28}$$

When holes are the majority carriers, comparing Eqs 4.20 and 4.25 we find the conductivity and the resistivity:

$$\sigma_p = p \cdot e \cdot \mu_p \tag{4.29}$$

$$\rho_p = \frac{1}{p \cdot e \cdot \mu_p} \tag{4.30}$$

The main idea behind semiconductor devices is the control of carrier concentrations by introducing impurities or by actively applying voltages or current. Changes in the carrier concentration are utilized as changes in the conductivity of an electrical device. For silicon, carrier concentration can be controlled between $\sim 10^{10} \, cm^{-3}$ and $\sim 10^{20} \, cm^{-3}$ by ion doping. On the other hand, carrier mobilities do not change as dramatically as the carrier concentrations. At room temperature, electron mobility changes between $100-1500 \, cm^2/V/s$, and hole mobility changes $50-500 \, cm^2/V/s$ depending on the concentration of doped ions.

Electrons and Holes

The states of electrons can be given by solution of Schrödinger's equation for the potential field of atoms. The energy levels of an electron around an atom have discrete quantum states called "orbitals", each of which corresponds to an amount of energy. Electrons wish to occupy low energy orbitals yet, due to the Pauli exclusion principle, no two electrons can occupy exactly the same orbital (quantum state) at the same time meaning additional electrons will tend to fill the next lowest orbital. The **energy band diagram** in solid-state physics of semiconductors is a representation of energy levels and the internal status of a

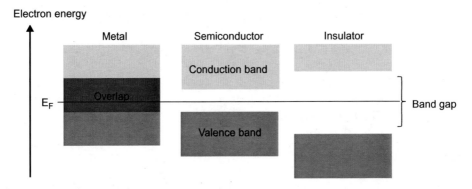

Figure 4.15: Energy Band Diagram of a Metal, a Semiconductor, and an Insulator.

material or a device. The same idea for visualization of energy levels is used to discuss optical properties of molecules. (See Chapter 5, Sections 5.6 and 5.7 for energy levels of molecules including fluorescent dyes, quantum dots, and other materials.

In the case of a crystal, atoms are closely packed together and more energy levels or orbitals are available for an electron. The difference in energy levels of the orbitals becomes increasingly small and considered to be continuous. However, the fact that the energy levels are continuous does not mean that an electron can take any energy level. The possible energy levels are still constrained by the crystal structure or the Schrödinger equation for the crystal. We consider two characteristic bands of energy levels, namely the valence band and the conduction band (see Fig 4.15).

- **Valence band:** The valence band is the band made up of the occupied molecular orbitals (filled shells). Electrons are tightly bound to the atom in the valence band.
- **Conduction band:** The conduction band corresponds to the orbitals that allow electrons to move within the crystal lattice.

Since an electron tends to fall to the lowest available energy state, it usually stays in the valence band. A certain amount of energy is needed to promote or move an electron from the valence band to the conduction band. In solid state, there are three different types of materials, namely a metal, a semiconductor and an insulator, which are characterized by the relative positions of the valence band and the conduction band in the energy band diagram.

1. Metal: In metals there are always free electrons without any excitation.
2. Semiconductor: electrons in semiconductors need to be excited to move to the conduction band.
3. Insulator: the band gap of an insulator is so large that very few electrons can jump the gap, and thus practically no current flows inside.

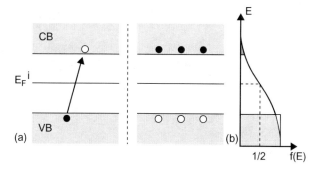

Figure 4.16: Electrons and Holes in an Energy Band Diagram. (a) Electrons Excited to the Conduction Band due to Thermal Excitation. (b) The Fermi–Dirac Distribution Function.

The energy difference between the top of the valence band and the bottom of the conduction band in an insulator or a semiconductor is called a **band gap**, or an **energy gap**. The band gap is a range where no electron states can exist.

The band gap is expressed in terms of electron volts (eV), which is a unit of energy equal to the amount acquired (or lost) by an electron moved across a potential difference of 1 V.

Semiconductors which do not contain impurities are called **intrinsic semiconductors**. For an intrinsic semiconductor, a small amount of electrons are excited to the conduction band due to thermal excitation, leaving the same amount of electron holes, or **holes,** in the valence band (see Fig. 4.16). An electron hole is a positively charged "imaginary" particle that can be used to describe the lack of an electron instead of analyzing the movement of many separate electrons in the valence band. Both electrons and holes are free to move and carry electric charges.

The Fermi–Dirac distribution function gives the probability that an electron occupies a state at energy E:

$$f(E) = \frac{1}{1 + \exp\left(\frac{E - E_f}{kT}\right)} \tag{4.31}$$

where k is the Planck constant, T is the temperature, and E_f is a hypothetical level of potential energy called the **Fermi level**. The Fermi level is an indication of how free electrons are distributed in the energy band diagram. The distribution function (Eq. 4.31) describes that the probability for a state S to be occupied by an electron is less than ½ in the region of $E > E_f$, and more than ½ if $E < E_f$. If the semiconductor is intrinsic (i.e., undoped), the Fermi level E_F^i is in the middle of the band gap.

As one can find from Eq. 4.31, the probability of occupancy becomes ½ at the Fermi level. The ½ probability may sound strange because there should be no electrons in the band gap or at the Fermi level (see Fig. 4.16). This happens because there is no available state at E_f. The density of electrons is given as the product of density of states and the probability for those sates to be occupied. More specifically, the density of electrons $n(E)$ at an energy level E is the product of density of states in the conduction band $g_c(E)$ and the probability for them to be filled $f(E)$.

$$n(E) = g_c(E) \cdot f(E) \tag{4.32}$$

The probability of a hole to exist in a state at E equals the probability for that state not to be occupied, i.e. $1 - f(E)$. The density of holes $p(E)$ is thus given as

$$p(E) = g_v(E)[1 - f(E)], \tag{4.33}$$

where $g_v(E)$ is the density of states in the valence band.

The density of states $g_C(E)$ can be found by solving the Schrödinger equation.

$$g_c(E) = \frac{8\sqrt{2\pi}}{h^3} m_e^{3/2} \sqrt{E - E_c} \quad (E \geq E_c), \tag{4.34}$$

where m_e is the effective mass of an electron and h is the Planck constant

For the density of states in valence band,

$$g_v(E) = \frac{8\sqrt{2\pi}}{h^3} m_h^{3/2} \sqrt{E_v - E} \quad (E \leq E_v), \tag{4.35}$$

where m_h is the effective mass of a hole. Now the electron density n can be calculated by integrating $n(E)$:

$$n = \int_{E_c}^{\infty} n(E) \cdot dE = \int_{E_c}^{\infty} g_c(E) f(E) \cdot dE \tag{4.36}$$

When the band gap of a semiconductor is large enough and the Fermi level is more than $3kT$ away from either band edge, or $E - E_f > 3kT$, we can use the following approximation and the semiconductor is called a non-degenerate semiconductor.

$$f(E) = \frac{1}{1 + \exp\left(\frac{E - E_f}{kT}\right)} \approx \exp\left(\frac{E_f - E}{kT}\right) \tag{4.37}$$

From Eqs 4.34, 4.36, and 4.37

$$n = \int_{E_c}^{\infty} \frac{8\sqrt{2\pi}}{h^3} m_e^{3/2} \sqrt{E - E_c} \exp\left(\frac{E_f - E}{kT}\right) \cdot dE = N_c \exp\left(\frac{E_f - E_c}{kT}\right) \tag{4.38}$$

$$N_c = 2\left[\frac{2\pi m_e kT}{h^2}\right]^{3/2}, \tag{4.39}$$

where N_c is called the effective density of states in the conduction band.

The hole density p can be found in a similar way using Eq. 4.33:

$$p = \int_{-\infty}^{E_v} p(E) \cdot dE = \int_{-\infty}^{E_v} g_v(E) \cdot [1 - f(E)] \cdot dE \tag{4.40}$$

For a non-degenerate semiconductor

$$1 - f(E) = \frac{\exp\left(\frac{E - E_f}{kT}\right)}{1 + \exp\left(\frac{E - E_f}{kT}\right)} = \frac{1}{1 + \exp\left(\frac{E_f - E}{kT}\right)} \approx \exp\left(\frac{E - E_f}{kT}\right) \tag{4.41}$$

The symmetry between Eqs 4.37 and 4.41 can be also seen in Fig. 4.16. From Eqs 4.33, 4.40, and 4.41, the hole density is calculated as:

$$p = N_v \exp\left(-\frac{E_f - E_v}{kT}\right) \tag{4.42}$$

$$N_v = 2\left[\frac{2\pi m_p kT}{h^2}\right]^{3/2} \tag{4.43}$$

where N_v is called the effective density of states.

In an intrinsic semiconductor, the hole and the electron concentrations are equal (see Fig. 4.16). When we define the Fermi level of an intrinsic material $E_f = E_i$, the career concentrations are

$$n_i = n = N_c \exp\left(\frac{E_i - E_c}{kT}\right) \tag{4.44}$$

$$n_i = p = N_v \exp\left(\frac{E_i - E_v}{kT}\right) \tag{4.45}$$

Where n_i is the intrinsic hole and electron concentration. Multiply Eqs 4.44 and 4.45 and take the square root, one will find

$$n_i = \sqrt{N_c \cdot N_v} \, \exp\left(\frac{E_v - E_c}{2kT}\right) \tag{4.46}$$

From Eqs 4.38, 4.42, 4.44 and 4.45, we can rewrite n and p as

$$n = n_i \, \exp\left(\frac{E_f - E_i}{kT}\right) \tag{4.47}$$

$$p = n_i \, \exp\left(\frac{E_i - E_f}{kT}\right) \tag{4.48}$$

Eqs. 4.47 and 4.48 give an important relationship

$$np = n_i^2, \tag{4.49}$$

Eqs. 4.47 and 4.48 can be also rewritten as:

$$E_f = E_i + kT \, \ln\frac{n}{n_i} \tag{4.50}$$

$$E_f = E_i - kT \, \ln\frac{p}{n_i} \tag{4.51}$$

These equations indicate that the Fermi level of a semiconductor at a known temperature can be found from the career concentration. Fermi level is somehow similar to the pH of an ionic solution we discussed earlier. This is because both measure the potential of a system in equilibrium. In fact, Eqs 4.50 and 4.51 are another form of the Nernst equation. The important difference to note is that discrete energy levels or band gaps are not considered in ionic solutions.

Fermi Level and Carrier Concentrations

Electron concentration n and hole concentration p are given as:

$$n = n_i \, \exp\left(\frac{E_f - E_i}{kT}\right)$$

$$p = n_i \, \exp\left(\frac{E_i - E_f}{kT}\right)$$

(Continued)

Fermi Level and Carrier Concentrations (Continued)

where E_f, E_i, and n_i are Fermi level, intrinsic Fermi level, and intrinsic carrier concentration, respectively. They can be also written as:

$$E_f = E_i + kT \ln \frac{n}{n_i}$$

$$E_f = E_i + kT \ln \frac{p}{n_i}$$

Carrier concentrations n, p, and n_i satisfy

$$np = n_i^2$$

Doping

Doping is an introduction of impurities into a pure semiconductor crystal. A small portion of silicon atoms in a silicon crystal, which have a filled valance shell, can be replaced with an atom which has one more or one less electron in its valance shell thus increasing or decreasing the materials conductivity.

p-Type Doping

When a crystal is doped with atoms that can accept electrons, such atoms are called **acceptors** and the doping is p-type. Boron, aluminum, and gallium (group III elements) are used as p-type dopants for silicon. Figure 4.17 illustrates an example of a boron atom doped

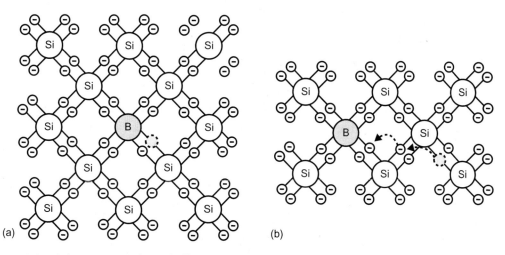

Figure 4.17: (a) p-type Doping of Silicon. (b) Electrons Moving from Hole to Hole, Resulting in a "Hole" Moving in the Other Direction.

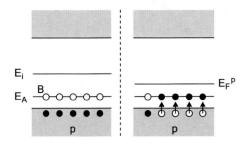

Figure 4.18: Energy Band Diagram of a p-type Semiconductor.

in a silicon crystal. The three valence electrons of boron are used to form covalent bonds with three of the neighboring Si atoms, but the bond with the fourth neighbor is unoccupied. The unoccupied bond can accept an electron from neighboring bonds, creating a new unoccupied bond. In this way, the unoccupied bond created by the acceptor can move freely in the crystal lattice (just like an opening slot in a sliding tile puzzle). This empty bond can be considered as a **hole**, which is a positively charged imaginary particle.

As shown in Fig. 4.18, the energy level of unoccupied bonds (or acceptor orbitals) in a p-type semiconductor (E_A) is only slightly higher than the valence band, and it is much easier for an electron to be thermally excited to occupy one of them than to go up to the bottom of the conduction band. As a result, most of the excited electrons are trapped by the acceptors, and the Fermi level is lowered. If the hole concentration is known, the Fermi level E_F^P can be found using Eq. 4.51:

$$E_f = E_i - kT \ln \frac{p}{n_i} \tag{4.51}$$

Eq. 4.51 indicates that if there are more free holes than in an intrinsic silicon, the Fermi level $E_f (= E_F^p)$ is lower than E_i.

n-Type Doping

When a crystal is doped with atoms that can provide one free electron, such atoms are called donors and the doping is n-type. Phosphorus, arsenic, antimony, and bismuth (group V elements) can be used as n-type dopants for silicon. Figure 4.19 illustrates an example of a phosphorus atom in a silicon crystal. Four of the valence electrons form covalent bonds with neighboring silicon atoms. The fifth electron is only weakly bonded to the phosphorus atom, and can move around larger orbitals of silicon atoms which are in the conduction band of the silicon crystal. Note that this electron is "relatively" free but is still somehow attracted by the donor phosphorus atom, because phosphorus has one more proton than silicon and is positively charged after donating an electron. See the discussion in p-n diode.

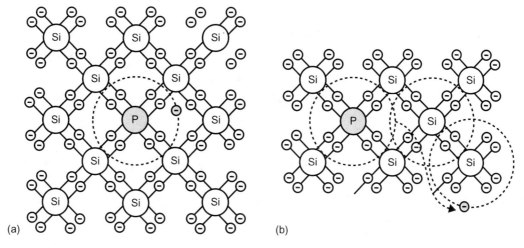

Figure 4.19: (a) n-type Doping of Silicon. (b) The Fifth Valence Electron Moving to Orbitals of Silicon.

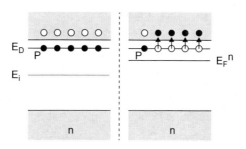

Figure 4.20: Energy Band Diagram of a n-type Semiconductor.

The additional electrons from donors can move to the conduction band, and raises the Fermi level as in the Figure 4.20. E_D is the energy level of the donor orbitals. The Fermi level $E_F{}^n$ can be found from Eq. 4.50 if the electron concentration is known.

$$E_f = E_i + kT\ln\frac{n}{n_i} \tag{4.50}$$

Eq. 4.50 indicates that if there are more free electrons than in an intrinsic silicon, the Fermi level $E_f(= E_F^n)$ is higher than E_i.

p−n Junction

When bringing an n- and a p-doped semiconductor together, electrons and holes begin to diffuse into the other side (see Fig. 4.21). Because of the difference between the Fermi

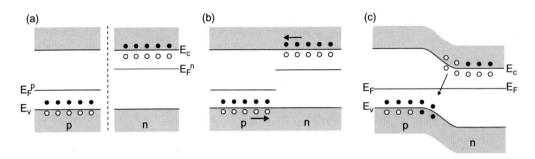

Figure 4.21: Band Bending at the p—n Junction.

energies as shown in Fig. 4.21(a), more electrons tend to travel from the n-region to the p-region than the other way, and more holes tend to travel from the p-region to the n-region than the other way (Fig. 4.21(b)). This unbalanced diffusion of electrons and holes continues until the Fermi energies at both sides are aligned and reaching thermal equilibrium (Fig. 4.21(c)). The diffusion mainly occurs in the region close to the junction. Carrier concentrations in the areas far enough from the junction stay the same as before the contact. Thus $E_c - E_f$ on the right side and $E_f - E_v$ at the left side in Fig. 4.21(c) should be the same as those in Fig. 4.21(a). This local change in energy is called **band bending**.

We can analytically find the band bending described above. Electrons and holes leave the charged core atoms behind, which then leads to a potential drop that stops further diffusion. Figure 4.22 shows how charges move from one region to the other. When a free electron moves from the n-region to the p-region, it leaves the core phosphorus atom, which donated the free electron and is now positively charged. The result is a negatively charged thin layer in the p-region at the interface and a positively charged area in the n-region (Fig. 4.22(a) and (b)). The electric field can be found by integrating the charges along the x-axis (Fig. 4.22(c)). The integration of the electric field then gives the "built-in" voltage distribution in the device (Fig. 4.22(d)). The energy levels or voltages for electrons are analogous to a gravitational potential for a moving mass. The slope in the voltage keeps electrons from moving up to the p region and prevents further distribution of electrons.

p—n Diode

A p—n diode can be constructed based on a p—n junction that allows current to flow in only one direction. On application of a reverse voltage, as in Fig. 4.23, the slope in the voltage distribution becomes larger compared with the case in Fig. 4.22(d). Note that application of a positive voltage means lowering energies in the energy diagram. Diffusion of holes and electrons are prevented and no current flows the device.

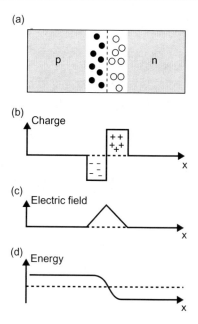

Figure 4.22: Analysis of a p—n Junction. (a) Charge, (b) Electric Field and (c) Energy Distributions.

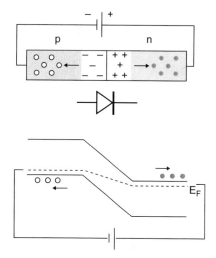

Figure 4.23: p—n Diode Applied with a Reverse Voltage.

On application of a forward voltage (Fig. 4.24), the voltage slope becomes smaller and allows more electrons to diffuse from the n-region to the p-region. The circuitry allows electrons to circulate through the wire and the diode.

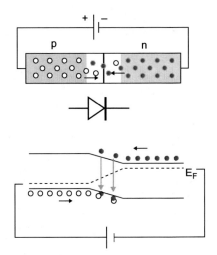

Figure 4.24: p—n Diode Applied with a Forward Voltage.

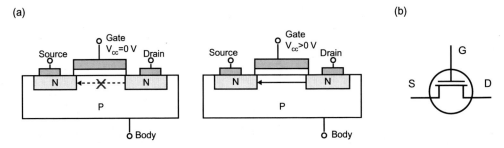

Figure 4.25: nMOSFET. (a) Diagram of nMOSFET. (b) Electronic Symbol of nMOSFET.

MOSFET

The **MOSFET** (metal oxide semiconductor field effect transistor) is a transistor used for amplifying or switching electronic signals. Figure 4.25(a) shows a diagram of nMOSFET (n-channel MOSFET). It typically consists of three terminals, namely source (S), gate (G), and drain (D). The source and drain are n-type wells made on p-type silicon. The gate is a poly-silicon layer deposited on a thin insulating layer between the substrate and the gate. The MOSFET utilizes silicon oxide (SiO_2) as the insulator, and that is the main reason why it is called MOSFET. Metal layers are used for terminals and wirings.

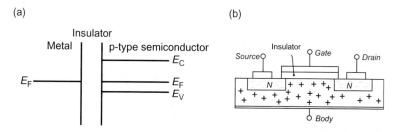

Figure 4.26: Flat Band Condition. (a) Energy Band Diagram. (b) nMOSFET with no Voltage Applied to the Gate.

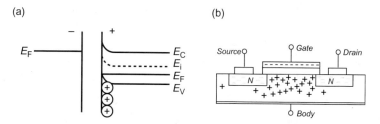

Figure 4.27: Accumulation. (a) Energy Band Diagram. (b) Accumulation of Charges due to a Smaller Voltage Applied to the Gate.

The body (B) or substrate, which serves as a reference, is a fourth terminal sometimes seen on MOSFETS but is normally grounded and not shown in the device diagram. Figure 4.25(b) is the commonly used symbol for nMOSFET.

At a normal state, the impedance between the source and the drain is large enough to prevent conduction. When a certain positive bias is applied to the gate, it allows current flow between the source and the drain.

Flat Band

When no voltage is applied to the gate, Fermi levels are equal at contact. The p−n junctions at the source-body and drain-body interfaces prevents current flow in either direction (Fig. 4.26).

Accumulation

Fermi level in the body shifts lower if a voltage smaller than the flatband voltage is applied to the gate (Fig. 4.27(a)). Accumulation of charges occurs on both sides as shown in Fig. 4.27(b) where the electrons move to the metal side and holes move to the

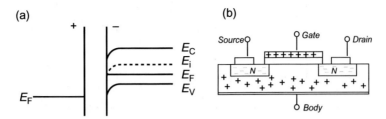

Figure 4.28: Depletion. (a) Energy Band Diagram. (b) Depletion of Charges due to a Voltage Applied to the Gate Accumulation.

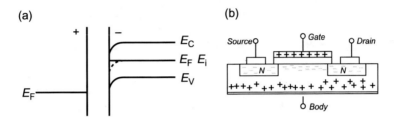

Figure 4.29: Inversion. (a) Energy Band Diagram. (b) Electrons Attracted to the Interface on the Gate Insulator.

semiconductor side. During accumulation, no current flows between the source and the drain.

Depletion

If a voltage slightly larger than the flatband voltage is applied to the gate, the Fermi level in the body further increases (Fig. 4.28(a)). Charges are depleted on both sides of the metal and the semiconductor (electrons on the metal side, holes on the the semiconductor side). In this state, still no current flows.

Inversion

If a large voltage is applied to the gate, the Fermi level in the body shifts higher (Fig. 4.29(a)). Electrons in the body are attracted to the interface on the gate insulator, even in a p-type semiconductor, free electrons exist as minority carriers. Semiconductor becomes n-type at the interface, forming a thin layer called an n-channel (Fig. 4.29(b)). At the semiconductor-insulator interface, the Fermi level E_F is close to the conduction band edge E_C. Now conduction between the source and the drain is allowed.

The pMOSFET is a MOSFET where the polarities of the semiconductors are opposite from those of the nMOSFET (Fig. 4.30). For the operation of pMOSFET, we can do the same

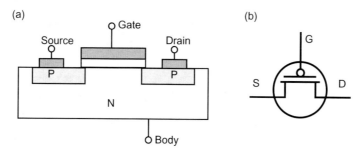

Figure 4.30: pMOSFET. (a) Diagram of a pMOSFET. (b) Electronic Symbol of a pMOSFET.

discussion as for the nMOSFET. Since the polarities are opposite, the pMOSFET conducts current when a lower voltage with reference to the body voltage V_B is applied to the gate. For a pMOSFET, Positive bias V_{cc} is often applied to the source and the body, i.e., $V_S = V_B = V_{cc}$. The gate voltage does not need to be the absolute negative voltage. It needs to be relatively lower than V_B by a certain voltage. The electronic symbol of a pMOSFET is shown in Fig. 4.30(b).

A silicon device that includes both pMOSFETs and nMOSFETs on a single chip is called a complementary metal−oxide−semiconductor (CMOS) device. The words "complementary" refer to the fact that the typical digital devices use symmetrical pair of pMOSFET and nMOSFET for use in logic operations. A CMOS process, or simply CMOS, refers to the process that integrates both p-type and n-type FETs, and could almost mean a "standard silicon process." The fabrication processes of analog devices that are compatible with CMOS may also be called CMOS processes. Figure 4.31 is a typical CMOS process. Details of techniques used in each steps are described in Chapter 2.

4.3.2 Silicon FET-Based Molecular Sensors

Silicon FET-based molecular sensors can be used for a variety of applications, such as chemical and biomolecular sensing. FET sensors have sensing elements immobilized on their gate insulators and use the degree of electrical activity at that interface to determine species concentration. Popular areas of chemical sensing include pH sensing and hydrogen sensing. By identifying the relationship between the pH level of an environment to the electrical activity of a sensor, FETs can be used to monitor acidity or alkalinity of various environments. These sensors can also be modified to identify the presence of certain chemical species. The term IGFET (insulated-gate field-effect transistor) works similarly to MOSFET and is often considered synonymous. However, IGFETs could imply that their gates are made of materials other than metal and insulated by materials other than oxide. In this section, we discuss conventional silicon FET-based molecular sensors including both

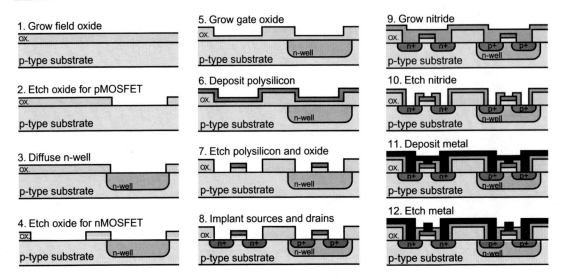

Figure 4.31: Standard CMOS Process.

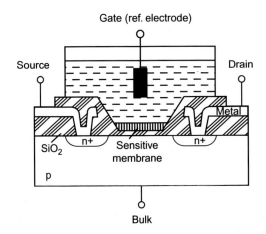

Figure 4.32: Ion-sensitive Field Effect Transistor.
From [15].

MOSFETs and IGFETs. Molecular sensors that utilize other types of FETs will be discussed in later sections.

Ion Selective FET Sensors

An ion-sensitive field effect transistor (ISFET) is used to measure ion concentrations in a solution and was first introduced by Bergvold in 1970 [14]. The structure and principle of ISFETs are based on those of MOSFETs or IGFETs. The main difference is the replacement of a voltage gate with a sensing electrode (Fig. 4.32). When the ion concentration at the gate electrode changes, the Fermi level in the sensor body changes (see later) and the source-drain current changes. One can find several types of commercially available ISFETs designed for sensing ions such as H^+, K^+, and Ca^{2+}.

Fabrication and Implementation

Structurally, an ISFET is fabricated in a similar manner with MOSFETs. The fabrication process of an ISFET is based on CMOS fabrication and can be summarized into the following steps [15]:

1. Field oxide growth
2. Photolithography for gate oxide definition
3. Gate oxide growth
4. Sensitive inorganic membrane deposition
5. Photolithography of the sensitive membrane
6. Photolithography of contacts
7. Metal deposition
8. Photolithography for metal patterning
9. Passivation deposition
10. Photolithography for passivation opening over bonding pads and ISFET gates [15].

The transistor is built upon a p-type silicon substrate. The source and the drain are separated by a channel overlain by silica and a metal gate. In this scenario, the silica acts as an insulator for the substrate. The gate voltage to be applied between the substrate and the gate is chosen based on its polarity and magnitude so that an n-type inversion layer forms in the channel between the source and drain regions. The magnitude of the drain current can be found based on the effective electrical resistance of this n-type surface inversion layer and the source-drain voltage difference. A relationship between the drain current and the electrochemical properties of the desired chemical species are then found. The ion selectivity comes from the ion-selective membranes deposited on the gate. Membranes of glass or polymers such as poly vinyl chloride (PVC) modified for ion selectivity are commonly used [16–18].

Silicon FET Sensors for Biosensing Applications

FET-based molecular sensors can be used in numerous of applications including an abundant number of biosensing applications (BioFETs). FETs have been used as immunologically modified field effect transistors (ImmunoFETs) [19]. For this type of sensor, antibodies are immobilized on the gate insulator, as shown in Fig. 4.33. The sensor is then placed in the presence of antigens, either in vivo or in vitro, as an array of sensors, for the purpose of them binding to the antibodies on the gate. This leads to a detectable change in the charge distribution, which then modulates the drain current of the transistor.

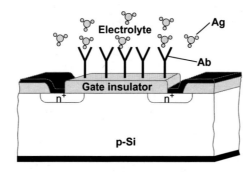

Figure 4.33: Immunologically Modified Field Effect Transistor.
From [19].

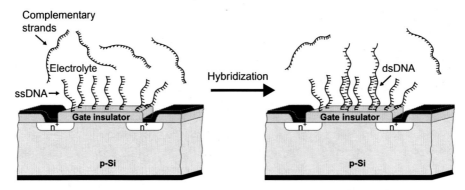

Figure 4.34: DNA-Modified Field Effect Transistor.
From [19].

Another application of a FET based sensor is a DNA-modified field effect transistor (DNA-FET) also called a gene-modified field effect transistor (GenFET) [19−22]. GenFETs use immobilized single stranded DNA (ssDNA), which binds to complementary strands in the solution. Figure 4.34 illustrates the working principle of a DNA-FET. By using ssDNA that is complementary to certain areas of the genome that code for specific genetic disorders, GenFETs are a useful tool in disease detection. The recognition process of the hybridization of two ssDNA strands produces a measurable signal that is detected by the transducer. Immobilization of thiol DNAs onto a thin gold electrode is commonly utilized [20].

FET-based molecular sensors have also been used in the detection of whole cell entities [19,23]. A "cell-transistor" is made by direct coupling of a single cell (or cell system) to the gate insulator of a FET. This sensor is potentially sensitive to all sorts of cellular events like metabolism, growth, or toxicity [19].

4.3.3 Organic Transistor-Based Sensors

Development of molecular sensors based on organic and polymeric FET is a new region of research. Organic chemicals can form chemical interactions with various analytes. Just as with the silicon based FETs, the semiconducting properties of the organic materials allows for the transduction of chemical information to electronic information [24]. Organic semiconductors include single molecules, short chain (oligomers), and organic polymers.

Organic thin-film transistors (OTFT) comprise a thin film of organic semiconducting material as the active layer of the transistor [25]. The source and the drain are in contact with the organic semiconductor. The gate electrode is usually located in the substrate, and can be used to modulate the source-drain current. Just like silicon based transistors, organic semiconductors can be either n-type or p-type depending on the semiconductor material and the doped impurities. Commonly used organic semiconducting materials include naphthalene tetracarboxylic derivatives, pentacene, oligothiophenes, and polythiophenes.

OTFTs can be categorized into two types, namely organic field effect transistors (OFET) and organic electrochemical transistors (OECTs). OFETs utilize field-effect doping, where the career density in the organic semiconductor is regulated by the electric field from the gate electrode across the insulating layer (Fig. 4.35(a) and (b)). OECTs utilize electrochemical doping, where the conductivity of the organic semiconductor is controlled by ions from an electrolyte (Fig. 4.35(c)). The operating voltages of OECTs are usually lower, but their switching times are much slower (on the scale of seconds or longer) because movement of ions are involved for the sensing.

Varieties of configurations have been introduced for OFET sensors. Figure 4.35(a) is a conventional OFET design, where the active layer is exposed to gases or liquids. Figure 4.35(b) is an example of ion-sensitive FET (ISOFET), where electric field across the insulating gate dielectric is modulated by ions at the electrolyte/insulator interface.

Analytes that have been tested include oxygen, iodine, bromine [26], NO_2 [27], ozone [28], and other organic gases [29]. Glucose sensing [30,31] may be one of most important future applications.

A specific analyte produces different patterns of responses to different semiconductors [29,32]. Utilization of a large number of semiconductors may facilitate the construction of an electronic-nose-type sensor that can identify a variety of analytes. Figure 4.36 shows responses of several organic/polymeric semiconductors to different analytes.

Among organic semiconductive materials, conductive polymers (which, despite their name, are considered semiconductive and are not as conductive as metals) have been especially focused on as materials for molecular sensors [33,34]. Polymer FETs can be used for many purposes, including chemical sensing [33,35], humidity sensing [36], and organic vapor

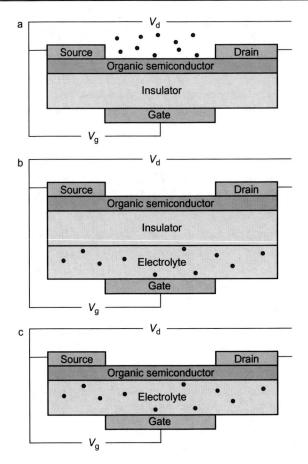

Figure 4.35: Three Types of Molecular Sensors Based on Organic Thin Film FETs (OFETs).
From [25].

sensing [37]. Conducting polymers have found use in electronics by acting as selective layers for chemical sensors [33]. When exposed to a gas, the conductance of a conducting polymer changes and can function as the transducer itself. Additionally, by employing different active layers, polymeric FETs can be used to detect various analyte molecules with good stability and sensitivity [35]. This is accomplished by means of direct semiconductor−analyte interactions and specific receptor molecules percolated throughout the semiconductor layer.

4.3.4 Nanowire Based FET Sensors

FET sensors based on nanowires are emerging as a powerful class of sensitive, electrical sensors for the direct detection of biological and chemical species. The working principle of nanowire-based FET sensors is similar to that of conventional silicon FET-based sensors.

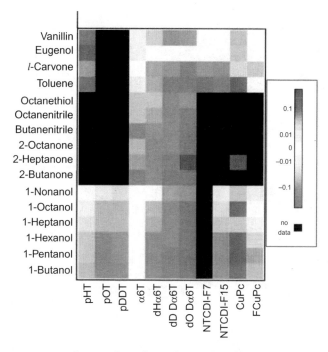

Figure 4.36: Responses of Organic/polymeric Semiconductors to Different Analytes.
From [32].

Typically, silicon nanowires or carbon nanotubes are used as the main material for the FETs. The nanowire surface, which works as the sensing gate, is modified to attach molecular receptors for the analyte of interest. The binding of charged species will result in the depletion or accumulation of charge carriers within the nanowire FET. The binding of molecules can then be monitored by a change in electrical properties such as conductance. In comparison to conventional planar FETs, silicon nanowire sensor devices have a one-dimensional morphology [38]. The nanometer-scale cross-section of the nanowire allows for the depletion or accumulation of charge carriers in the bulk material while the chemical species bind to the surface. The high surface to volume ratio of these nanoscale structures enable nanowire-based FET sensors to overcome the sensitivity limitations of planar FET sensors (see discussion in Chapter 2, Section 2.2). This allows a nanowire based sensing FET to have sufficient sensitivity to detect single particles [39].

Silicon Nanowire-Based FET Sensor

Silicon is advantageous as the materials for molecular sensors in many aspects. Semiconducting properties of silicon have been well studied. The dopant type and concentration can be precisely controlled. Abundant fabrication techniques available for silicon microdevices can be easily utilized for FET-based nanowire sensors. They can be

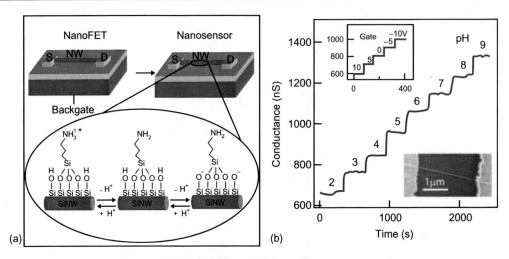

Figure 4.37: Silicon Nanowire Sensor. (a) Composition of the Sensor. (b) pH Measurement with the Nanowire Sensor.

From [38].

modified with receptors for many applications and demonstrate their potential for highly sensitive real-time detection of chemical and biological species. A dense array of sensors could be prepared to create a highly sensitive and label-free sensor for screening and in-vivo diagnostics. Building blocks of silicon nanowire sensors may also be more readily integrated into silicon-based industrial processing and fabrication than sensors based on other nanowire materials.

Yi Cui et al. proposed a boron-doped silicon nanowire based molecular sensor used to detect biological and chemical species electrically using an FET [38]. The sensors versatility was demonstrated by attaching various receptor molecules to the surface of the nanowires.

Figure 4.37(a) is a diagram showing the conversion of a nanowire FET into a pH sensing sensor. The nanowire is connected with two electrodes to measure conductance. In one of the nanowire applications, 3-aminopropyltriethoxysilane (APTES, see Chapter 2, Section 2.5.3) was immobilized on the nanowire to provide a surface that can undergo protonation and deprotonation. The change in surface charge can chemically gate the nanowires allowing the FET to detect the change in pH based on the change in surface charge. Figure 4.37(b) shows the result of real-time conductance detection for an APTES-modified silicon nanowire for a pH range between 2 and 9. The top plot shows the time-dependent conductance of the nanowire as a function of back-gate voltage and the bottom shows an SEM image of a typical silicon nanowire device.

The silicon nanowires were prepared using a nanocluster-mediated vapor-liquid-solid growth method [40]. The nanowires could then be aligned using a flow aligning technique on oxidized silicon substrates which could then be attached to electrical contacts through electron-beam lithography. The sensor is fabricated by modifying the SiO_2 surface of a solid state FET device. To detect the presence of a protein in a solution, biotin can be bound to the oxide surface of the nanowires. Biotin binds with high selectivity to the protein streptavidin, which will trigger an increase in conductance measured by the FET in real-time. The device is able to detect protein concentrations of at least 10 pM [38]. These experiments showed the device had very low nonspecific binding of the protein, the binding interaction is highly specific, and that nanowire devices are ultrasensitive detectors. This shows a clear approach could lead to the development of sensor devices of real value. The developed nanowire-based molecular sensor showed that the conductance of biotin-modified nanowires increased with a constant value when introduced to a 250 nM streptavidin solution. The change in conductance is consistent with binding of a negatively charged species to the p-type nanowire surface and the fact that streptavidin is negatively charged at pH of the solutions used. These experiments show the nanowire based molecular sensors ability to monitor protein concentration in real-time. Future research could lead to devices to monitor protein expression, and medical diagnostics in real-time [41].

Figure 4.38 shows the structure of the FET sensor developed by Kim et al. [42]. It shares several fundamental principles with the sensor shown Fig. 4.37. The FET is first surface functionalized with 3-aminopropyltriethoxysilane (APTES), and modified with glutaraldehyde to obtain aldehyde groups. Antibodies for PSA (prostate-specific antigen) were then linked to the surface. They demonstrated the evidence of the real-time detection of 1 fg/mL PSA.

The significance of this sensor is that the silicon nanowire FET is fabricated based on the conventional "top down" approach, namely the conventional lithography based approach, while most of other nanowire based sensors are made by bottom up approach. Silicon nanowire based sensors are interesting examples that induces discussion on bottom up approach and top down approach (see Chapter 2, Section 2.6 for details of top down and bottom up approaches). A similar top down approach was used in sensors reported by Elfström et al. [43]. In 2011, Gao et al. reported silicon nanowire FETs fabricated with a CMOS compatible process with the gate width less than 20 nm. They demonstrated detection of 1 fM of target DNA and single-nucleotide polymorphism discrimination [44].

One thing to note is that this approach was possible mainly because the sensors were made of silicon, which is the building block of the contemporary nano/microfabrication devices. We will introduce FET sensors based on nanomaterials of bottom up approaches in the following sections.

Polymer Nanowire FET Sensors

Conductive polymers are often chosen as fabrication materials over other nanomaterials because of their electrical properties, ease of fabrication and potential low cost production.

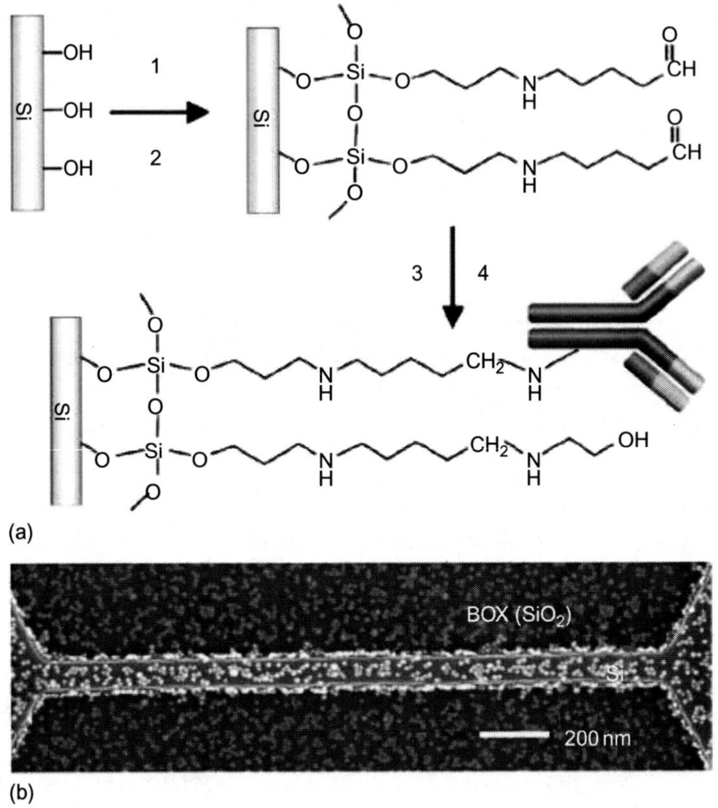

(a)

(b)

Figure 4.38: Silicon Nanowire Sensor Fabricated by Top Down Approach.
From [42].

Conductive polymer nanostructures are rapidly emerging as FET-based bio/chemical sensors. In this section, we discuss the working principle and fabrication methods for a conducting polymer nanowire FET-based molecular sensor [24,45,46]. As we discussed for the silicon nanowire FET sensors, conducting polymers undergo changes in their electrical conductivity when exposed to electrochemical activity. The amount of change varies with the amount of interactions of ions or charge transfer between molecules. Using this principle, these types of FETs can be used for multiple chemiresistive and biomolecule sensing applications. While on its own, a conducting polymer lacks the specificity/selectivity to function as a biosensor, biological recognition molecules can be incorporated into the sensor to give it that specificity and selectivity. The most important part of the fabrication of these types of biosensors is thus the immobilization of the biorecognition molecule to the polymer substrate or nanowire.

An example of a polymer nanowire FET-based sensor is shown in Fig. 4.39. In this design, the polypyrrole nanowire was formed by the electropolymerization of an aqueous solution

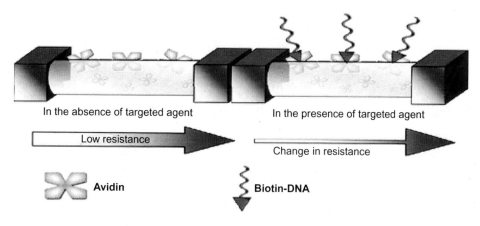

In the absence of targeted agent   In the presence of targeted agent

Low resistance

Change in resistance

Avidin   Biotin-DNA

Figure 4.39: Polymer Nanowire FET Sensor.
From [24].

of pyrrole monomer. The nanowire channel measures 100 or 200 nm wide and 3 μm long between gold electrodes on prefabricated silicon substrate. Avidin- or streptavidin-conjugated ZnSe/CdSe quantum dots are incorporated into the nanowire as the model biomolecule during the nanowire synthesis [24]. When exposed to biotin − DNA, the avidin − and streptavidin − polypyrrole nanowires generated a rapid change in resistance to as low as 1 nM. Advantages of the method include direct incorporation of functional biological molecules, site-specific positioning, built-in electrical contacts, and scalability to large-scale arrays over the silicon nanowire and carbon nanotube biosensors.

The polymer nanowires can be grown either template-free or with a template [45]. Template techniques include the use of soft-templates and hard templates. Soft-template fabrication involves the use of liquid crystals and micelles to form polymer nanotubes. Hard-template fabrication utilizes a substrate with a particular morphology for one directional growth of polymer chains [47]. An example of scaffold upon which to grow these chains is anodized aluminum oxide with track-etched polycarbonate membranes. One face of the template is coated with a conductive metal, like gold or silver, to function as an anode and the electrochemically conducting polymer is then deposited into the pores of the scaffold membrane via electrophoretic deposition. Conducting polymer nanowire length is dependent on current density and deposition time. This method can also be altered by replacing the electrolyte and depositing different layers at a time for multi-layered fabrication. The template-free synthesis process relies on preferential growth of nanowires in the presence of a driving force. Examples of template-free synthesis include self-assembly, mechanical stretching, dip-pen lithography (DPN), electrospinning and electrochemical synthesis. Via self-assembly, nanoribbons are fabricated by dropping solution containing a polymer derivative onto substrate and then allowing that solution to evaporate. This creates nanoribbons of all different lengths to be selected from depending on your need [45]. Electrospinning has recently emerged as an alternative method of preparing nanowires of conducting polymers for applications in electronic devices. The high voltage used to create a charged fiber may cause the polymer chains efficiently packed within the nanowires. Single electrospun polymer nanowire-based transistors have been demonstrated with mobilities as high as $0.03cm^2/(Vs)$ [91].

4.4 Carbon Nanotubes and Graphene-Based Sensors

Carbon emerged as the key element in materials engineering through the development of carbon nanotubes (CNTs) and graphene. The discovery of these two materials has made a substantial impact with their unique properties which have advanced research and potential product development in molecular and biological sensors. In this section, we discuss synthesis and basic properties of CNTs and graphene, followed by introduction of important applications in molecular sensing.

4.4.1 CNT-Based Molecular Sensors

The first discovery of carbon nanotubes were by Sumio Iijima in 1991 when he was carrying out transmission electron microscopy studies [48]. Iijima specifically observed multi-walled carbon nanotubes (MWCNTs), and two years later Iijima produced single-walled carbon nanotubes (SWNTs) [49]. Since their discovery, CNTs have created hype in the materials science world on their exceptional bulk properties that show promise for numerous applications [50−53]. We describe chemical, mechanical and electrical characteristics of carbon nanotubes and how they are utilized in electrochemical sensors.

Carbon Nanotubes

Carbon nanotubes are a unique material due to the properties it contains mainly due to their unique geometry. It is important to note that the material contains a high length to diameter aspect ratio on the order of 10^7 as CNTs can be nanometer sized in their diameter but can contain a length on the scale of centimeters [52−56].

In the CNT crystal structure, all carbon bonds are fulfilled, which create strong chemical bonds and provide stability to the material. This means that CNTs are chemically inert and are advantageous for biomedical applications because the material is biocompatible. Furthermore, CNTs are about twice as strong as steel and due to its geometry is more stable when placed under compression in comparison to graphene [56]. The crystal structure allows for one-dimensional electron carrier transport along the carbon lattice which creates ballistic conduction. Conduction is able to occur at superior rates and yet able to keep

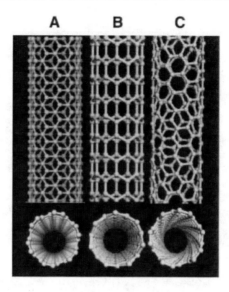

Figure 4.40: Three Types of Single Walled Carbon Nanotubes (SWNTs): (A) Armchair, (B) Zig-zag, (C) Chiral CNTs.
From [57].

power dissipation at a minimum which reduces any heat loss to the environment. The strong covalent bonding between carbon atoms also increases resistance to electro-migration which contributes to the material's ability to transport electrons so quickly along its surface [53].

Carbon nanotubes are divided into two groups: single-walled carbon nanotubes (SWNTs) and multi walled carbon nanotubes (MWNTs).

SWNTs typically have a diameter of about 1 nm, and can contain different lattice geometries dependent upon the lattice direction. Figure 4.40 shows the three different lattice patterns observed in SWNTs: (A) armchair, (B) zig-zag, (C) and chiral CNTs [57].

The properties of SWNTs change significantly with the lattice directions. All armchair SWNTs are metals (i.e., conductive). Other SWNTs are semiconductors with the band gap varying from zero to about 2 eV, depending on their lattice directions and the diameters. Since a MWNT comprises an array of different SWNTs, most of MWNTs can be considered to be metals.

SWNTs are likely candidates for key sensing element in molecular sensors. Bachtold et al. reported on a SWNT-based FET [58], which suggests basic designs for SWNT-based molecular sensors (Fig. 4.41).

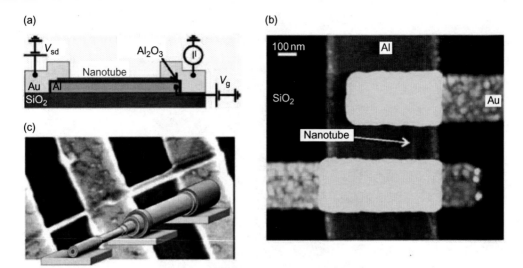

Figure 4.41: Carbon Nanotube-Based FET. (a) Schematic of a CNT-Based FET. (b) Scanning Tunneling Microscope Photograph of a SWNT FET. (c) Image of a MWNT Where Electrical Pulses Are Used to Remove Unwanted Carbon Layers.
From [57].

Synthesis of Carbon Nanotubes

Various methods have been demonstrated [59] to produce carbon nanotubes, and all of them require high temperatures in their processes. SWCNTs and MWCNTs can be made by laser ablation or pulsed-laser vaporization (PLV) of graphite [60], a method in which the material is made through etching of a solid graphite surface. Other methods include carbon arch (CA) discharge [48] and decomposition in an oxygen free environment [61]. The most commonly used has been chemical vapor deposition (CVD) which produces a more pure sample of the nanotubes [50]. Growth by CVD has allowed for nanotubes to be grown on a surface in an ordered, aligned fashion. One of the catalysts used in CVD is iron oxide along with a silica substrate on which the CNT growth occurs. During CNT growth, the outer surfaces of the nanotubes interact with one another via van der Waals forces, and these interactions are what promote the nanotubes to grow perpendicular to the silica surface and remain aligned [62].

It is not possible to generate a completely pure batch of CNTs, and the defects it contains can significantly alter its mechanical and electrochemical properties. Synthesizing SWCNTs can result in large concentrations of impurities. These defects are typically removed through acid washing [63], but this process can result in further defects such as shortening the nanotubes and increasing the cost for purification [57].

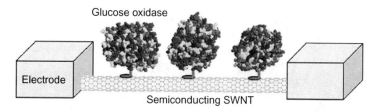

Figure 4.42: Single Walled Carbon Nanotube (SWCNT)-Based Molecular Sensor.
From [65].

Carbon Nanotube-Based Molecular Sensors

Kong et al. first demonstrated a SWNT-based chemical sensor. Upon exposure to gaseous molecules such as NO_2 or NH_3, the conductance of a semiconducting SWNT is found to dramatically increase or decrease [64]. CNTs have also made a significant contribution to the development of molecular sensors for biomedical applications. For example, electrochemical biosensors have been created to be enzyme electrodes by coating the external surface of individual SWNTs with specific biological enzymes that enable the altered CNTs to act as a single molecule biosensor (Fig. 4.42) [65]. One problem of using carbon nanotubes is difficulty of production. Typical production of carbon nanotubes creates a mixture of metallic and semiconducting nanotubes while only semiconducting nanotubes can be used for FET sensors.

Many researchers have also developed processing techniques to suit CNTs for glucose detection. Production of well-aligned MWNTs was found useful in creating nano-electrodes which also took advantage of the high electrocatalytic activity of CNTs [66]. Lin et al. attached enzymes to CNT tips and then submerged in a solution with glucose oxidase to finish construction of the biosensor for glucose detection [67]. CNTs were grown in a conductive substrate to guarantee its conductivity. Figure 4.43 illustrates the fabrication procedure.

One advantageous characteristic of carbon nanotubes for molecular sensing include the extremely small dimensions and the high surface volume rate. They allow for detection of single molecular events. There are several interesting reports on recent advancement of single molecular sensing with carbon nanotube FET-based sensors. Sorgenfrei et al. reported on single-molecule measurement of DNA hybridization with a carbon nanotube FET [68]. Choi et al. reported on a SWNT FET sensor attached with a single lysozyme molecule [69]. Lysozymes are enzymes that catalyze hydrolysis of a specific type of chemical bond found in bacterial cell walls. A schematic and an atomic force microscopy measurement of the lysozyme sensor are shown in Figure 4.44(a) and (b), respectively. First they fabricated SWNT FETs and functionalized with linker

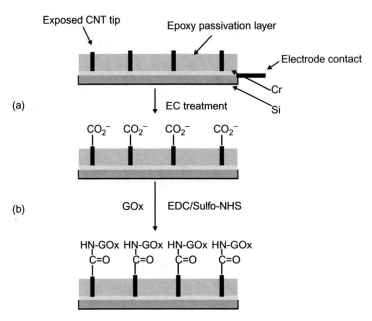

Figure 4.43: Glucose Sensor Based on Carbon Nanotube Electrode. (A) Electrochemical Treatment of the CNT NEEs for Functionalization. (B) Coupling of the Enzyme GOx to the Functionalized CNT NEEs.
From [67].

molecules, which strongly adhere to SWNT sidewalls. Next, the devices were incubated in a solution of the pseudo wild-type, single-cysteine variant of T4 lysozyme. The functional thiol of the lysozyme reacted with an SWNT-bound linker to produce covalent bonding. Reaction conditions were tuned to readily produce sensors with having only one lysozyme. Figure 4.44(c) shows the measured channel current as a function of applied gate voltages. Figure 4.44(d) shows response of current to the lysozyme substrate peptidoglycan, a polysaccharide found in bacterial cell walls, added to the solution at $t = 0$ with a concentration of 25 µg/mL. It was measured that about 100 chemical bonds on average are processively hydrolyzed at 15-Hz rates, before lysozyme returns to its nonproductive motion. Figure 4.44(c) is the measured channel current as a function of the gate voltage.

4.4.2 Graphene-Based Molecular Sensors

Graphene is a flat, one atom thick layer of repeating six-membered carbon rings (Fig. 4.45(a)). Graphene has great potential for use in a diverse variety of applications in a wide range of fields. It was recently brought to attention through the work of Andre Geim

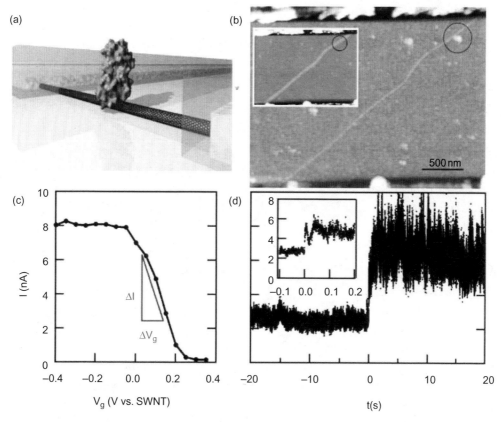

Figure 4.44: Single Molecule Measurement with a Carbon Nanotube FET Sensor.
From [69].

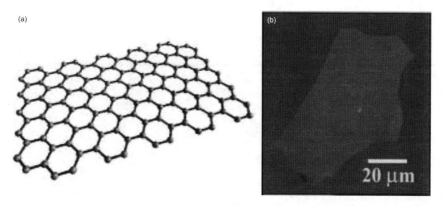

Figure 4.45: Graphene. (a) Atomic Structure of Graphene. (b) Graphene Film.
From [71].

and Konstantin Novoselov in 2004, for which they were awarded the Nobel Prize. The ultimately large surface-to-volume ratio and high carrier mobility of graphene make it an excellent candidate for the material for a future molecular sensor, but challenges stand in the way of their production at an industrial level. We first describe basic characteristics of graphene. We then discuss molecular sensors based on graphene.

Properties of Graphene

The atomic structure of graphite has been long known, and there have been attempts at mechanical exfoliation to produce thin samples [70]. Geim, Novoselov and colleagues at the University of Manchester reported the isolation of two-dimensional individual crystal planes from a graphite flake in 2004 [71]. They used a "peeling" method utilizing cellophane tape to successively remove single layers (Fig. 4.45(b)), and tested electrical properties.

Current integrated circuits are typically fabricated using silicon, but this technology is limited by a few key technical issues such as tunneling current and power dissipations. Carrier mobility is considered to be one of the factors to impede improvement of silicon devices. In graphene, charge transport is practically considered ballistic. Intrinsic mobility limit of graphene is inferred to be 2×10^5 cm^2/V/s, which exceeds that of indium antimonide, the inorganic semiconductor with the highest known mobility ($\sim 7.7 \times 10^4$ cm^2/V/s) and that of semiconducting carbon nanotubes ($\sim 1 \times 10^5$ cm^2/V/s) [72].

Bolotin et al. measured electron mobility in excess of 200 000 cm^2/V/s at electron densities of $\sim 2 \times 10^{11}$ cm^{-2} by suspending single layer graphene. They used a combination of electron beam lithography and etching to achieve suspension ~ 150 nm above a Si/SiO$_2$ gate electrode and electrical contacts to the graphene. Figure 4.46 is an SEM image of a typical device prepared for the four-probe measurement. The voltage between electrodes 2 and 3 was measured while the current is sent between electrodes 1 and 4. The resistance is

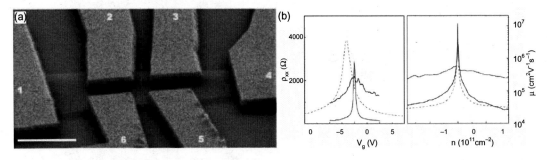

Figure 4.46: Measurements of Electrical Properties of Graphene. (a) SEM of a Four Probe Measurement. (b) Resistance and Mobilities of Graphene.
From [73].

shown in Fig. 4.46(b) as a function of gate voltage applied between graphene and the substrate.

Graphene also exhibits excellent mechanical properties. The breaking strength graphene was measured to be 200 times greater than that of steel [74].

Synthesis of Graphene

Difficulties in graphene fabrication are the primary rate-limiting factors in furthering graphene-based applications. To achieve large-scale application, the standardization of fabrication methods must produce a balance of high-quality, high-yield monolayer sheets [75,76]. The traditional top-down approach to mechanical exfoliation cannot produce high-quality samples at a high-throughput rate.

Chemical-based approaches have yielded some advancement in mass-fabrication. The most promising is the initial oxidation of graphite to graphite oxide, followed by successive mechanochemical and thermal exfoliation to eventually reduce the sheets to reduced graphene oxide layers. Single layer graphene was chemically derived by Ruoff et al. in 2006. This technique allows for some hydrophilic properties of graphene oxide to be retained as functional groups for biomedical applications. However, hydrazine or dimethylhydrazine used to reduce graphene is highly toxic and dangerously reactive, and the use of them remains a serious challenge for large-scale production.

CVD is also a promising method because of the compatibility with the current silicon process. Kim et al. reported the direct synthesis of large-scale graphene films using CVD on thin nickel layers [77]. The films were patterned based on lithography of the nickel film, and can be transferred to another substrate based on the PDMS stamping technique.

Graphene-Based Gas Sensors

In 2007, Geim et al. demonstrated single molecule sensitivity of a graphene-based device to NO_2 and NH_3 gases [78]. Changes in resistivity caused by exposure to various gases were measured for an about 10 μm-sized graphene film (Fig. 4.47(a)). In principle, the adsorbed gas molecules change the local carrier concentration in the graphene film. The opposite sign of carriers are generated by NO_2 (electron withdrawing species, p-type) and NH_3 (electron-donor, n-type), resulting in negative and positive step-like changes in resistance, respectively, as shown in Fig. 4.47(b).

Fowler et al. developed a chemical sensor based on chemically converted graphene. The sensitivity is measured in response to different concentrations, temperatures, and types of gaseous molecules to test its effectiveness as a molecular sensor [79]. The device is based on the principle of four-probe measurement. The resistance is measured by the four interdigitated electrodes arranged in a serpentine shape with 10 μm spacing to create larger

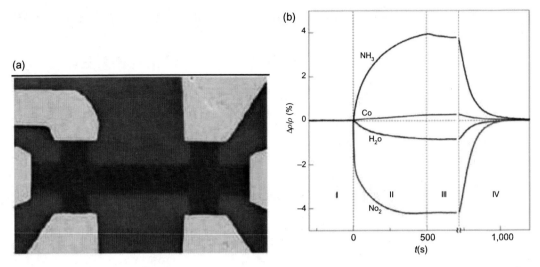

Figure 4.47: Graphene Based Gas Sensor. (a) SEM Image of the Gas Sensor. (b) Changes in Resistance in Responses to NO_2 and NH_3 Gases.
From [78].

contact area as shown in Fig. 4.48(a) and (b). Figure 4.48(c) and (d) are responses to 5 ppm NO_2 and NH_3 in dry nitrogen, respectively.

Graphene for Glucose Sensing

Carbon is one of the most widely used materials for electrochemical analysis. With the potential advantages of its unique physicochemical properties, graphene is expected to provide a good platform for electrochemical study of carbon materials [80].

Direct electrochemical measurements of glucose oxidase (GOD) based on graphene were reported independently by Shan et al. [81] and Kang et al. [82]. Shan et al. modified the surface of a glassy carbon (GC) electrode to construct a graphene-based electrochemical sensor. Polyvinylpyrrolidone (PVP)-protected graphene in a polyethylenimine-functionalized ionic liquid (PFIL) solution was first dropcasted onto the GC electrode. The GC electrode is then immersed in a GOD solution to construct the graphene modified electrode.

Figure 4.49 shows cyclic voltammetric measurements (see earlier) at the graphene-GOD-PFIL modified GC electrode in various concentrations of glucose PBS solution from 2 mM to 14 mM. The measurement was made with a conventional three-electrode cell with a platinum wire as the auxiliary electrode and an Ag/AgCl electrode as the reference electrode. A pair of well-defined redox peaks are visible. The inset is the calibration curve of the amperometric responses at -0.49 V. The excellent

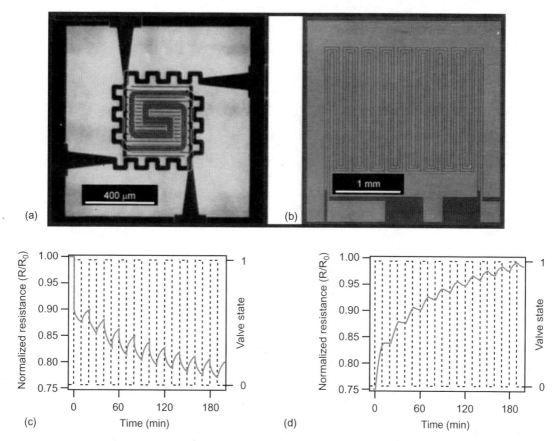

Figure 4.48: Chemical Sensor Based on Chemically Converted Graphene. (a) Micrograph of Four Interdigitated Electrodes. (b) Zoom in Image of the Interdigitated Electrodes. (c) Changes in Resistance (R/R_0) in Response to Pulsed 5ppm Flows of NO_2 and (d) NH_3.
From [79].

electronic properties and biocompatibility of graphene-based composites allowed the direct electron transfer of redox enzyme while maintaining its bioactivity.

Graphene-Based DNA Sensing

Electrochemical DNA sensors can potentially offer a compact and highly sensitive analysis platform for detecting specific DNA sequences associated with disease. Graphene-based electrochemical DNA sensors may allow miniaturization of devices as well as reduction of sample volumes needed for analysis. A simple but important step toward such sensors is the detection based on the direct oxidation of DNA. Zhou et al. reported an electrochemical DNA sensor based on chemically reduced graphene oxide (CR-GO) [83]. CR-GO modified

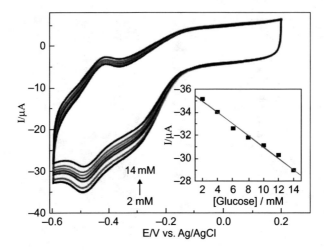

Figure 4.49: Electrochemical Measurements of Glucose Oxidase with a Graphene-Based Electrode. See Text for Details.
From [81].

GC electrode showed more favorable electron transfer kinetics than graphite modified GC (graphite/GC) and GC electrodes. The signals of the four free bases of DNA (guanine, adenine, thymine, and cytosine) on the CR-GO/GC electrode are measured in differential pulse voltammograms (DPV), and they are all more efficiently separated than with the other electrodes. In addition, they were able to detect a single-nucleotide polymorphism (SNP) site for short oligomers with a particular sequence without hybridization or labeling processes.

Functionalization of graphene has been also the topic of interest. The biocompatibility and recognition properties of the component will lead to the realization of a new type of molecular sensors. Adsorption between graphene and ssDNA [84,85] is based on the similarity between graphene and carbon nanotubes, which have been previously functionalized with a self-assembled layer of ssDNA [86]. Patil et al. first demonstrated stable aqueous suspension of a DNA coat on graphene [85]. Hydrogen bonding between the DNA sugar backbone and oxygen-containing graphene oxide may support the affinity of the DNA coat. Liu et al. deposited DNA coating on reduced graphene oxide using a stepwise procedure. Thiolated DNA oligos were adsorbed onto graphene oxide nanosheets via periodic sonication during incubation [87]. Lv et al. showed DNA-graphene binding was much more efficient than other tested dispersants (sodium dodecyl sulfate and cetyltrimethyl ammonium bromide) [88]. In addition, Lv et al. reported the formation of a monolayer of globular ssDNA on both sides of the graphene sheet. They reported on the potential of single-stranded DNA in combination with graphene sensors to create a potent electrochemical sensor.

4.5 Conclusion

In this chapter, we describe the fundamental principles and important applications of molecular sensors based on electrical transduction. We describe principles and applications of conventional electrochemical measurements, after which we introduce molecular sensors based on semiconductors. Basic theories of semiconducting materials and working principles of FETs are described and related to the design of molecular sensors. The utilization of biological recognition processes allows for the development of molecular sensors that are highly sensitive to specific biomolecules. The use of nanomaterials in the FET sensor allow for a sensitive, real-time detection of a wide range of biological and chemical species.

Problems

P4.1 Potential of Electrodes

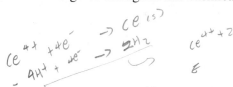

1. Calculate the standard potential for the redox couple Ce^{4+}/Ce, for which the reduction half-reaction is: $Ce^{4+}(aq) + 4e^- \rightarrow Ce(s)$.
2. Calculate the standard potential of the couple $Au^{3+}(aq)/Au^+(aq)$.

P4.2 pH Meter

The following data were obtained for the calibration of a pH meter at 25°C.

pH	5	6	7	8	9
E (mV)	112.3	53.2	−5.9	−65.0	−124.1

1. Find the relationship $E = E_0 + S\,(7 - pH)$ from the calibration data.
2. When $E = 63.4$ mV was obtained for a sample at 25°C, what is the pH of the sample?
3. If the pH meter does not have temperature compensation. What value of E is obtained for a sample with a pH of 6.5 at 35°C?

P4.3 Glucose Sensor

A glucose sensor consists of two electrodes micromachined in a cavity in a silicon wafer. One electrode is platinum and the other is silver/silver chloride. Describe the reactions taking place at the anode and the cathode when a potential is applied large enough for reduction and oxidation reactions to take place.

P4.4 Glucose Sensor

1. How many people in the US need blood glucose tests in the care of diabetes mellitus?
2. What sensitivity is usually required for a glucose meter?

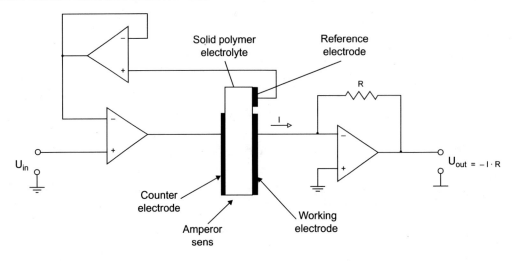

Figure P4.1

P4.5 Electrochemical Sensors

As shown in Fig. P4.1, an amperometric sensor was connected in a circuit consisting of three operational amplifiers:

1. Describe the difference between the amperometric electrodes and potentiometric electrodes.
2. Qualitatively draw the current I, as a function of voltage V applied on the amperometric sensor ($I-V$ curve). Describe the electrochemical process corresponding to each section on the $I-V$ curve.
3. Describe the function of the three operational amplifier circuitry.

P4.6 Glucose Sensor

A medical device engineer decides to use nanoporous silica as the membrane for designing an implantable glucose sensor. The silica membrane will act as the barrier between blood and the electrode.

1. What can be interpreted about the surface properties of the membrane from the figure showing a water droplet on the membranes surface?
2. Will the use of nanoporous silica help promote transport of glucose or oxygen across the membrane?
3. If 5 μM glucose is getting oxidized at the anode of such a sensor every second, what is the current that the sensor will show? (Faraday's constant, $F = 96\,485$ C/mol).

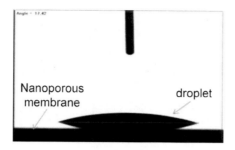

P4.7 Semiconductor Basics

Answer "True" or "False" for the following statements. Justify your answer.

a. Energy bands in crystals result from the periodicity of a crystal structure and describe the orbitals that are filled or can be filled if enough energy is provided.
b. Bands are not continuous but have gaps between them.
c. Phosphorus is the only n-type dopant used to dope silicon
d. Filled orbitals create bands that are able to conduct → conduction bands.
e. Not completely filled bands are able to conduct → valence bands.
f. The Fermi level is the energy level at which the probability to fill that level is 100%.
g. The difference between metals, semiconductors and insulators is the location of the Fermi level.
h. If the lattice constant of a semiconductor A is bigger than another semiconductor B, then B will have the larger band gap.
i. Boron doped silicon is a p-type semiconductor; therefore there is no negative charge carriers in the doped materials.

P4.8 Charge Carriers, Conductivity, Fermi level

We have a boron-doped silicon wafer with the hole concentration of 1.0×10^{15} cm^{-3} at 300 K.

1. Find the electron concentration of the wafer at 300 K. Use the intrinsic carrier concentration of silicon at 300 K is $n_i = 7.9 \times 10^9$ cm^{-3}.
2. Find the level energy (eV) E_f measured relative to the top of the valence band E_v.
3. Find the resistivity of the wafer. Use the electron's charge $e = 1.6 \times 10^{-19}$ C and the hole mobility $\mu_p = 350$ cm^2/V/s.

P4.9 Charge Carriers, Conductivity, Fermi level

An n-type wafer has a resistivity of 1 Ω cm at 300 K.

1. Find the electron concentration. Use the electron mobility $\mu_n = 1000 \text{ cm}^2/\text{V/s}$.
2. Find the Fermi level (eV) E_f of the wafer measured relative to the bottom of the conduction band E_c.

P4.10 Temperature Dependence

A piece of silicon with dimensions of $1 \times 1 \times 1$ mm is diced from an n-type wafer. It has a resistivity of 10 Ω cm at 300 K. Explain this piece can be used as a temperature senor which changes the resistivity as a function of temperature T. Find the responsivity $\Delta R/\Delta T$ at 300 K.

P4.11 FET-Based Molecular Sensors

1. Design a field effect transistor (FET)-based molecular sensor, which can detect a specific sequence of DNA. Draw a diagram showing the structure of the sensor.
2. Explain how the sensor can be fabricated.
3. Explain the principle of operation for the sensor you designed.

P4.12 Nanomaterials ⸺ thickness.

1. What are the typical size ranges of the following nanomaterials? Use the characteristic dimensions in the answer: MWNTs, SWNTs, colloidal QDs, gold nanoparticles, graphene.
2. Based on the International Technology Roadmap for Semiconductors, what is the smallest feature size of semiconductor manufacturing process in the year of 2012?

P4.13 Carbon Nanotubes

1. What is the typical diameter range of single walled carbon nantubes (SWNTs)?
2. Illustrate the structures of armchair, zigzag, and chiral SWNTs. Explain the electrical characteristics of each type.
3. Justify the statement that "Most of multi-walled carbon nanotubes (MWNTs) are electrically conductive". Is this true or false, and explain the reason.
4. List three advantages and three disadvantages of carbon-nanotube based FET biosensor. Suggest solutions to the listed disadvantageous points.

P4.14 Graphene-Based Molecular Sensors

1. Name four reasons why graphene is considered an excellent material for sensors or other active devices.
2. Name four difficulties in fabricating graphene-based molecular sensors.

References

[1] Eggins BR. Chemical Sensors and Biosensors. New York: Wiley; 2008.

[2] Bakker E, Pretsch E. Potentiometric sensors for trace-level analysis. TrAC Trends in Analytical Chemistry 2005;24:199−207.

[3] Lee WY, Lee KS, Kim TH, Shin MC, Park JK. Microfabricated Conductometric Urea Biosensor Based on Sol-Gel Immobilized Urease. Electroanalysis 2000;12:78−82.

[4] Kriz D, Kempe M, Mosbach K. Introduction of molecularly imprinted polymers as recognition elements in conductometric chemical sensors. Sensors and Actuators B: Chemical 1996;33:178−81.

[5] Kummer AM, Hierlemann A, Baltes H. Tuning sensitivity and selectivity of complementary metal oxide semiconductor-based capacitive chemical microsensors. Analytical Chemistry 2004;76:2470−7.

[6] Panasyuk TL, Mirsky VM, Piletsky SA, Wolfbeis OS. Electropolymerized molecularly imprinted polymers as receptor layers in capacitive chemical sensors. Analytical Chemistry 1999;71:4609−13.

[7] Wang J. Glucose biosensors: 40 years of advances and challenges. Electroanalysis 2001;13:983.

[8] Wang J. Electrochemical glucose biosensors. Chemical Reviews 2008;108:814.

[9] Oliver N, Toumazou C, Cass A, Johnston D. Glucose sensors: a review of current and emerging technology. Diabetic Medicine 2009;26:197−210.

[10] Ferri S, Kojima K, Sode K. Review of glucose oxidases and glucose dehydrogenases: a bird's eye view of glucose sensing enzymes. Journal of Diabetes Science and Technology 2011;5:1068−76.

[11] Marks V. An improved glucose-oxidase method for determining blood CSF and urine glucose levels. Clinica Chimica Acta 1959;4:395−400.

[12] Clark LC, Lyons C. Electrode systems for continuous monitoring in cardiovascular surgery. Annals of the New York Academy of Sciences 1962;102:29−45.

[13] Kuhn LS. Biosensors: blockbuster or bomb? The Electrochemical Society Interface 1998;7:26−31.

[14] Bergveld P. Development, operation, and application of the ion-sensitive field-effect transistor as a tool for electrophysiology. Biomedical Engineering, IEEE Transactions on 1972;:342−51.

[15] Cané C, Gràcia I, Merlos A. Microtechnologies for PH ISFET chemical sensors. Microelectronics Journal 1997;28:389−405.

[16] Moody G, Thomas J, Slater JM. Modified poly (vinyl chloride) matrix membranes for ion-selective field effect transistor sensors. Analyst 1988;113:1703−7.

[17] Moss SD, Janata J, Johnson CC. Potassium ion-sensitive field effect transistor. Analytical Chemistry 1975;47:2238−43.

[18] Karbue I, Tamiya E, Dicks JM, Gotoh M. A microsensor for urea based on an ion-selective field effect transistor. Analytica Chimica Acta 1986;185:195−200.

[19] Schoning MJ, Poghossian A. Recent advances in biologically sensitive field-effect transistors (BioFETs). Analyst 2002;127:1137−51.

[20] Kim DS, Jeong YT, Park HJ, Shin JK, Choi P, Lee JH, et al. An FET-type charge sensor for highly sensitive detection of DNA sequence. Biosensors and Bioelectronics 2004;20:69−74.

[21] Mascini M, Palchetti I, Marrazza G. DNA electrochemical biosensors. Fresenius' Journal of Analytical Chemistry 2001;369:15−22.

[22] Poghossian A, Cherstvy A, Ingebrandt S, Offenhäusser A, Schöning M. Possibilities and limitations of label-free detection of DNA hybridization with field-effect-based devices. Sensors and Actuators B: Chemical 2005;111:470−80.

[23] Errachid A, Zine N, Samitier J, Bausells J. FET-based chemical sensor systems fabricated with standard technologies. Electroanalysis 2004;16:1843−51.

[24] Ramanathan K, Mangesh A, Yun M, Chen W, Myung NV, Mulchandani A. Bioaffinity sensing using biologically functionalized conducting-polymer nanowire. Journal of the American Chemical Society 2005;127:496−7.

[25] Mabeck JT, Malliaras GG. Chemical and biological sensors based on organic thin-film transistors. Analytical and Bioanalytical Chemistry 2006;384:343−53.

[26] Laurs H, Heiland G. Electrical and optical properties of phthalocyanine films. Thin Solid Films 1987;149:129−42.

[27] Hu W, Liu Y, Xu Y, Liu S, Zhou S, Zhu D, et al. The gas sensitivity of a metal-insulator-semiconductor field-effect-transistor based on Langmuir−Blodgett films of a new asymmetrically substituted phthalocyanine. Thin Solid Films 2000;360:256−60.

[28] Bouvet M, Leroy A, Simon J, Tournilhac F, Guillaud G, Lessnick P, et al. Detection and titration of ozone using metallophthalocyanine based field effect transistors. Sensors and Actuators B: Chemical 2001;72:86−93.

[29] Crone B, Dodabalapur A, Gelperin A, Torsi L, Katz H, Lovinger A, et al. Electronic sensing of vapors with organic transistors. Applied Physics Letters 2001;78:2229−31.

[30] Someya T, Dodabalapur A, Gelperin A, Katz HE, Bao Z. Integration and response of organic electronics with aqueous microfluidics. Langmuir 2002;18:5299−302.

[31] Bartic C, Campitelli A, Borghs S. Field-effect detection of chemical species with hybrid organic/inorganic transistors. Applied Physics Letters 2003;82:475−7.

[32] Torsi L, Dodabalapur A. Organic thin-film transistors as plastic analytical sensors. Analytical Chemistry 2005;77:380−7.

[33] Janata J, Josowicz M. Conducting polymers in electronic chemical sensors. Nature Materials 2003;2 (1):19−24.

[34] Lange U, Roznyatovskaya NV, Mirsky VM. Conducting polymers in chemical sensors and arrays. Analytica Chimica Acta 2008;614:1−26.

[35] Wang L, Fine D, Sharma D, Torsi L, Dodabalapur A. Nanoscale organic and polymeric field effect transistors as chemical sensors. Analytical and Bioanalytical Chemistry 2006;384:310−21.

[36] Nilsson D, Kugler T, Svensson PO, Berggren M. An all-organic sensor−transistor based on a novel electrochemical transducer concept printed electrochemical sensors on paper. Sensors and Actuators B: Chemical 2002;86:193−7.

[37] Covington JA, Gardner JW, Briand D, de Rooij NF. A polymer gate FET sensor array for detecting organic vapours. Sensors and Actuators B: Chemical 2001;77:155−62.

[38] Cui Y, Wei Q, Park H, Lieber CM. Nanowire nanosensors for highly sensitive and selective detection of biological and chemical species. Science 2001;293:1289−92.

[39] Patolsky F, Zheng G, Hayden O, Lakadamyali M, Zhuang X, Lieber CM. Electrical detection of single viruses. Proceedings of the National Academy of Sciences of the United States of America 2004;101:14017−22.

[40] Cui Y, Duan X, Hu J, Lieber CM. Doping and electrical transport in silicon nanowires. The Journal of Physical Chemistry B 2000;104:5213−6.

[41] Ahn JH, Choi SJ, Han JW, Park TJ, Lee SY, Choi YK. Double-gate nanowire field effect transistor for a biosensor. Nano Letters 2010;10:2934.

[42] Kim A, Ah CS, Yu HY, Yang J-H, Baek I-B, Ahn C-G, et al. Ultrasensitive, label-free, and real-time immunodetection using silicon field-effect transistors. Applied Physics Letters 2007;91(10):103901−3.

[43] Elfström N, Karlström AE, Linnros J. Silicon nanoribbons for electrical detection of biomolecules. Nano letters 2008;8:945−9.

[44] Gao A, Lu N, Dai P, Li T, Pei H, Gao X, et al. Silicon-nanowire-based CMOS-compatible field-effect transistor nanosensors for ultrasensitive electrical detection of nucleic acids. Nano Letters 2011;11:3974−8.

[45] Bangar MA, Chen W, Myung NV, Mulchandani A. Conducting polymer 1-dimensional nanostructures for FET sensors. Thin Solid Films 2010;519:964−73.

[46] Hangarter CM, Bangar M, Mulchandani A, Myung NV. Conducting polymer nanowires for chemiresistive and FET-based bio/chemical sensors. Journal of Materials Chemistry 2010;20:3131−40.

[47] Hulteen JC, Martin CR. A general template-based method for the preparation ofnanomaterials. J. Mater. Chem. 1997;7:1075−87.

[48] Iijima S. Helical microtubules of graphitic carbon. Nature 1991;354:56−8.

[49] S. Iijima, T. Ichihashi, Single-shell carbon nanotubes of 1-nm diameter, 1993.

[50] Coleman JN, Khan U, Blau WJ, Gun'ko YK. Small but strong: a review of the mechanical properties of carbon nanotube—polymer composites. Carbon 2006;44:1624—52.

[51] Breuer O, Sundararaj U. Big returns from small fibers: a review of polymer/carbon nanotube composites. Polymer Composites 2004;25:630—45.

[52] Wang J. Carbon-nanotube based electrochemical biosensors: a review. Electroanalysis 2004;17:7—14.

[53] Avouris P, Appenzeller J, Martel R, Wind SJ. Carbon nanotube electronics. Proceedings of the IEEE 2003;91:1772—84.

[54] Jacobs CB, Peairs MJ, Venton BJ. Review: carbon nanotube based electrochemical sensors for biomolecules. Analytica Chimica Acta 2010;662:105—27.

[55] Mahar B, Laslau C, Yip R, Sun Y. Development of carbon nanotube-based sensors—a review. Sensors Journal, IEEE 2007;7:266—84.

[56] Jorio A, Dresselhaus G, Dresselhaus MS. Carbon Nanotubes: Advanced Topics in the Synthesis, Structure, Properties and Applications, vol. 111. New York: Springer; 2008.

[57] Baughman RH, Zakhidov AA, de Heer WA. Carbon nanotubes—the route toward applications. Science 2002;297:787—92.

[58] Bachtold A, Hadley P, Nakanishi T, Dekker C. Logic circuits with carbon nanotube transistors. Science 2001;294:1317—20.

[59] Kuzmany H, Kukovecz A, Simon F, Holzweber M, Kramberger C, Pichler T. Functionalization of carbon nanotubes. Synthetic Metals 2004;141:113—22.

[60] Guo T, Nikolaev P, Thess A, Colbert D, Smalley R. Catalytic growth of single-walled manotubes by laser vaporization. Chemical Physics Letters 1995;243:49—54.

[61] Nikolaev P, Bronikowski MJ, Bradley RK, Rohmund F, Colbert DT, Smith K, et al. Gas-phase catalytic growth of single-walled carbon nanotubes from carbon monoxide. Chemical Physics Letters 1999;313:91—7.

[62] Dai H. Carbon nanotubes: opportunities and challenges. Surface Science 2002;500:218—41.

[63] Chiang I, Brinson B, Smalley R, Margrave J, Hauge R. Purification and characterization of single-wall carbon nanotubes. The Journal of Physical Chemistry B 2001;105:1157—61.

[64] Kong J, Franklin NR, Zhou C, Chapline MG, Peng S, Cho K, et al. Nanotube molecular wires as chemical sensors. Science 2000;287:622—5.

[65] Besteman K, Lee JO, Wiertz FGM, Heering HA, Dekker C. Enzyme-coated carbon nanotubes as single-molecule biosensors. Nano Letters 2003;3:727—30.

[66] Ye JS, Wen Y, De Zhang W, Ming Gan L, Xu GQ, Sheu FS. Nonenzymatic glucose detection using multi-walled carbon nanotube electrodes. Electrochemistry Communications 2004;6:66—70.

[67] Lin Y, Lu F, Tu Y, Ren Z. Glucose biosensors based on carbon nanotube nanoelectrode ensembles. Nano Letters 2004;4:191—5.

[68] Sorgenfrei S, Chiu C-Y, Gonzalez Jr RL, Yu Y-J, Kim P, Nuckolls C, et al. Label-free single-molecule detection of DNA-hybridization kinetics with a carbon nanotube field-effect transistor. Nature Nanotechnology 2011;6:126—32.

[69] Zhang W, Kai K, Choi DS, Iwamoto T, Nguyen YH, Wong H, et al. Microfluidics separation reveals the stem-cell—like deformability of tumor-initiating cells. Proceedings of the National Academy of Sciences 2012;109:18707—12.

[70] Allen MJ, Tung VC, Kaner RB. Honeycomb carbon: a review of graphene. Chemical Reviews 2010;110:132.

[71] Novoselov K, Geim A, Morozov S, Jiang D, Zhang Y, Dubonos S, et al. Electric field effect in atomically thin carbon films. Science 2004;306:666—9.

[72] Chen JH, Jang C, Xiao S, Ishigami M, Fuhrer MS. Intrinsic and extrinsic performance limits of graphene devices on SiO2. Nature Nanotechnology 2008;3:206—9.

[73] Bolotin KI, Sikes K, Jiang Z, Klima M, Fudenberg G, Hone J, et al. Ultra high electron mobility in suspended graphene. Solid State Communications 2008;146:351—5.

[74] Lee C, Wei X, Kysar JW, Hone J. Measurement of the elastic properties and intrinsic strength of monolayer graphene. Science 2008;321:385—8.

[75] Chen D, Tang L, Li J. Graphene-based materials in electrochemistry. Chemical Society Reviews 2010;39:3157–80.

[76] Geim AK, Novoselov KS. The rise of graphene. Nature Materials 2007;6:183–91.

[77] Kim KS, Zhao Y, Jang H, Lee SY, Kim JM, Kim KS, et al. Large-scale pattern growth of graphene films for stretchable transparent electrodes. Nature 2009;457:706–10.

[78] Schedin F, Geim A, Morozov S, Hill E, Blake P, Katsnelson M, et al. Detection of individual gas molecules adsorbed on graphene. Nature Materials 2007;6:652–5.

[79] Fowler JD, Allen MJ, Tung VC, Yang Y, Kaner RB, Weiller BH. Practical chemical sensors from chemically derived graphene. Acs Nano 2009;3:301–6.

[80] Shao Y, Wang J, Wu H, Liu J, Aksay IA, Lin Y. Graphene based electrochemical sensors and biosensors: a review. Electroanalysis 2010;22:1027–36.

[81] Shan C, Yang H, Song J, Han D, Ivaska A, Niu L. Direct electrochemistry of glucose oxidase and biosensing for glucose based on graphene. Analytical Chemistry 2009;81:2378–82.

[82] Kang X, Wang J, Wu H, Aksay IA, Liu J, Lin Y. Glucose Oxidase–graphene–chitosan modified electrode for direct electrochemistry and glucose sensing. Biosensors and Bioelectronics 2009;25:901–5.

[83] Zhou M, Zhai Y, Dong S. Electrochemical sensing and biosensing platform based on chemically reduced graphene oxide. Analytical Chemistry 2009;81:5603–13.

[84] Lu Y, Goldsmith BR, Kybert NJ, Johnson ATC. DNA-decorated graphene chemical sensors. Applied Physics Letters 2010;97. pp. 083107-3

[85] Patil AJ, Vickery JL, Scott TB, Mann S. Aqueous stabilization and self-assembly of graphene sheets into layered bio-nanocomposites using DNA. Advanced Materials 2009;21:3159–64.

[86] Staii C, Johnson Jr AT, Chen M, Gelperin A. DNA-decorated carbon nanotubes for chemical sensing. Nano Letters 2005;5:1774–8.

[87] Liu J, Li Y, Li Y, Li J, Deng Z. Noncovalent DNA decorations of graphene oxide and reduced graphene oxide toward water-soluble metal–carbon hybrid nanostructures via self-assembly. J. Mater. Chem. 2009;20:900–6.

[88] Lv W, Guo M, Liang MH, Jin FM, Cui L, Zhi L, et al. Graphene-DNA hybrids: self-assembly and electrochemical detection performance. Journal of Materials Chemistry 2010;20:6668–73.

[89] Wang J. Electrochemical glucose biosensors chem. Rev 2008;108:814–25.

[90] Yoo E-H, Lee S-Y. Glucose biosensors: an overview of use in clinical practice. Sensors 2010;10:4558–76.

[91] Briseno AL, Mannsfeld SC, Jenekhe SA, Bao Z, Xia Y. Introducing organic nanowire transistors. Materials Today 2008;11:38–47.

Further Reading

Electrochemistry

Eggins BR. Chemical sensors and biosensors. New York: Wiley; 2008.

Bard AJ, Faulkner LR. Electrochemical methods: Fundamentals and applications. second ed. New York: Wiley; 2000.

Bond AAM, Sholz F. Electroanalytical methods: Guide to experiments and applications. New York: Springer; 2010.

Semiconductor physics

Sedra AS, Smith KC. Microelectronic circuits revised edition. Oxford: Oxford University Press, 2007.

Streetman B, Banerjee S. Solid State Electronic Devices, fifth ed. Englewood Cliffs: Prentice Hall, 2005.

Weste N, Harris D. CMOS VLSI design: A circuits and systems perspective. fourth ed. Boston: Addison-Wesley; 2010.

Baker RJ. CMOS circuit design, layout, and simulation. third ed. New York: Wiley-IEEE Press; 2010.

Razavi B. Design of analog CMOS integrated circuits. New York: McGraw-Hill; 2000.

CHAPTER 5

Optical Transducers

Optical Molecular Sensors and Optical Spectroscopy

Chapter Outline

5.1 Introduction

The study of optics is a vital part to all sciences because it is the basis of numerous technologies applied in the life sciences, chemistry, physics and engineering. More than 50 Nobel Laureates have contributed to optics research, advancing studies on key technologies including Raman scattering (awarded in 1930), laser (1950), fluorescent proteins (2008), optical fibers (2009), and CCD (2009).

The advantages of optical transducers include lack of direct contact, high spatial resolution, and relatively easy detection. In optical waveguides, there are several high contrast modes available for light transmission, sensing and imaging. Disadvantages include the need for the optical transducer to be transparent at the wavelength used for sensing, the use of labels, interferences, and issues arising from fluorophore decay in fluorescence imaging.

In this chapter, we will describe molecular sensors based on optical transduction. We introduce basic electromagnetic theory, study waveguides and sensors based on waveguide structures, and cover surface plasmon resonance-based (SPR) sensors. Lastly, we introduce optical spectroscopy techniques and describe theories and practices of optical absorption, scattering, and fluorescence.

5.2 Basic EM Theory

Light is a form of electromagnetic wave that consists of oscillating electric and magnetic fields. Figure 5.1 illustrates the wavelength and frequency range of visible and invisible light along with those of other types of electromagnetic waves. The wavelength of visible light ranges from about 400 nm to 650 nm, which defines several critical dimensions in optical sensing techniques.

The electric and magnetic fields of propagating light are perpendicular to each other and to the direction of propagation. Such fields are oscillating harmonically in temporal and spatial domains. Two electromagnetic waves can interfere, which leads to an interference term that contains a cosine of the phase difference that can be positive or negative. If an electromagnetic

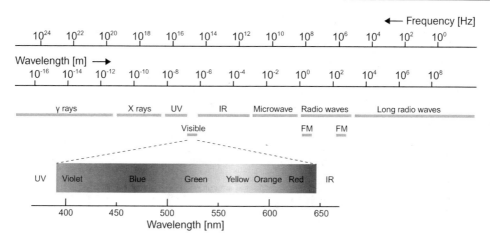

Figure 5.1: Light as an Electromagnetic Wave.

wave hits an interface to another medium, boundary conditions have to be fulfilled for the parallel and normal components. The surface defines a plane of incidence that contains the surface normal and the incoming beam. The electric and magnetic fields can be decomposed into separate polarization components parallel and perpendicular to the plane of incidence: p- and s-polarization or transverse-magnetic (TM) and transverse-electric (TE) polarization. The two polarizations behave differently at the interface. Only p or TM polarization can show a Brewster angle or can excite surface plasmons, since it is the only polarization with a field normal to the surface. Boundary conditions produce Snell's law, Fresnel coefficients for the reflectivity, and the evanescent field.

The behavior of light as an electromagnetic wave is governed by Maxwell's equations. We start from Maxwell's equations to find the wave equation of propagating light, followed by description of important principles that constitute the fundamentals of optical sensing.

5.2.1 Maxwell's Equations

The following equations are the differential form of Maxwell's equations:

$$\nabla \times \mathbf{E} = -\frac{\partial \mathbf{B}}{\partial t} \quad \text{(Faraday's Law)} \tag{5.1}$$

$$\nabla \times \mathbf{H} = \mathbf{J} + \frac{\partial \mathbf{D}}{\partial t} \quad \text{(Ampere's Law)} \tag{5.2}$$

$$\nabla \cdot \mathbf{D} = \rho_v \quad \text{(Gauss's Laws)} \tag{5.3}$$

$$\nabla \cdot \mathbf{B} = 0 \quad \text{(Gauss's Laws)} \tag{5.4}$$

where **E** is the electric field, **H** is the magnetic field, **D** is the electric flux density, **B** is the magnetic flux density, **J** is the electric current density, and ρ_v is the electric charge density.

The following equations show the conservative relation between the flux densities and the fields:

$$D = \varepsilon_0 E + P = \varepsilon E \tag{5.5}$$

$$B = \mu_0 H + \mu_0 M = \mu H \tag{5.6}$$

where **P** is the electric polarization, **M** is the magnetization, ε is the permittivity and μ is the permeability.

The following boundary conditions are applied for source free media where $\rho_v = 0$ and $J = 0$.

$$E_{\parallel} \text{ is continuous: } S \times (E_2 - E_1) = 0 \tag{5.7}$$

$$H_{\parallel} \text{ is continuous: } S \times (H_2 - H_1) = 0 \tag{5.8}$$

$$D_{\perp} \text{ is continuous: } S \cdot (D_2 - D_1) = 0 \tag{5.9}$$

$$B_{\perp} \text{ is continuous: } S \cdot (B_2 - B_1) = 0 \tag{5.10}$$

Here, **S** is the normal unit vector to the boundary surface, $A_{\parallel}$ and $A_{\perp}$ are tangential and normal components, respectively, of a field **A** at the boundary, A_1 and A_2 are field **A** in media 1 and 2, respectively, at the boundary $(A = E, H, D, B)$.

5.2.2 Wave Equation

To derive wave equation from the Maxwell equations, we take the curls of both sides of Faraday's Law.

$$\nabla \times (\nabla \times E) = \nabla \times \left(-\frac{\partial B}{\partial t} \right) \tag{5.11}$$

For the left side of Eq. 5.11, we use the vector identity $\nabla \times \nabla \times A = \nabla(\nabla \cdot A) - (\nabla \cdot \nabla)A$, which is true for any vector **A**, and an assumption that the divergence of the electric field is zero, namely $\nabla \cdot E = 0$.

$$\nabla \times \nabla \times E = \nabla(\nabla \cdot E) - (\nabla \cdot \nabla)E = -\nabla^2 E, \tag{5.12}$$

For the right side of Eq. 5.11, the curl operation and the differentiation operation can be switched since both operations are continuous and linear. If the light propagates non-conducting media, **J** is zero, which derives $\nabla \times B = \mu\varepsilon \frac{\partial E}{\partial t}$ from Eqs 5.2, 5.5 and 5.6.

$$\nabla \times \left(-\frac{\partial B}{\partial t} \right) = -\frac{\partial}{\partial t}(\nabla \times B) = -\frac{\partial}{\partial t}\left(\mu\varepsilon \frac{\partial E}{\partial t} \right) = -\mu\varepsilon \frac{\partial^2 E}{\partial t^2} \tag{5.13}$$

Plugging Eqs 5.12 and 5.13 into 5.11, we obtain

$$\nabla^2 \mathbf{E} - \mu\varepsilon \frac{\partial^2 \mathbf{E}}{\partial t^2} = 0 \tag{5.14}$$

Equation 5.14 is called the electric field wave equation.

Following the same procedure from Ampere's Law:

$$\nabla^2 \mathbf{B} - \mu\varepsilon \frac{\partial^2 \mathbf{B}}{\partial t^2} = 0 \tag{5.15}$$

Equation 5.15 is called the magnetic field wave equation.

Equation 5.14 has a solution known as a plane wave (Fig. 5.2):

$$\mathbf{E}(\mathbf{r}, t) = \mathbf{E}_0 e^{j(\omega t + \phi - \mathbf{k} \cdot \mathbf{r})} \tag{5.16}$$

When we use $e^{j(\ )}$ to express the phase of a wave, we only consider the real part of it. More specifically, we may use another form in sinusoidal expression:

$$\mathbf{E}(\mathbf{r}, t) = \mathbf{E}_0 \cos(\omega t + \phi - \mathbf{k} \cdot \mathbf{r}), \tag{5.17}$$

where $\mathbf{r}$ is a position vector, t is time, $\mathbf{E}_0$ is a vector perpendicular to the propagation, and $\mathbf{k}$ is a vector with the direction of the propagation. The vector $\mathbf{k}$ is called the **wave vector**, and the magnitude of the wave vector is given as:

$$|\mathbf{k}| = \frac{2\pi}{\lambda} \cdot n, \tag{5.18}$$

where λ and n are the wavelength and the refractive index, respectively.

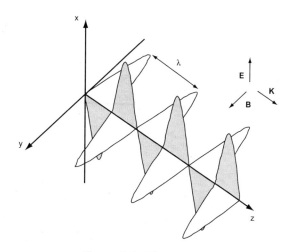

Figure 5.2: Plane Wave.

It should be noted that plane waves are just one form of solution to the wave equations, and there are many other possible solutions that describe different forms of light wave propagation.

Here, we use a coordinate system where the direction of $\mathbf{E}_0$ is chosen as that of x axis and direction of $\mathbf{k}$ is chosen as that of z axis. In this case,

$$E_x(z, t) = E_0 \exp j\left(\omega t + \phi - \frac{2\pi n}{\lambda} \cdot z\right) \tag{5.19}$$

in Eq. 5.19, we consider the magnitude of vector $\mathbf{E}_0$. Both E_x and E_0 are scalers. In this book, vectors are shown with a bold character, and scalars are italic. Another important thing to note is that the left side of Eq. 5.19 is a real number. We only consider the real part of the right side.

Eq. 5.19 may help better understand the meaning of each term in Eq. 5.16. Furthermore, plugging Eq. 5.19 into Eq. 5.1, we find that $\mathbf{B}$ is in the direction of the y axis.

$$B_y = \sqrt{\varepsilon_0 \cdot \mu_0} \cdot E_0 \exp j\left(\omega t + \phi - \frac{2\pi n}{\lambda} \cdot z\right) \tag{5.20}$$

We can rewrite Eq. 5.20 in a simplified form:

$$B_y = B_0 \exp j\left(\omega t + \phi - \frac{2\pi n}{\lambda} \cdot z\right), \tag{5.21}$$

where

$$B_0 = \sqrt{\varepsilon_0 \cdot \mu_0} \cdot E_0 \tag{5.22}$$

Now we find the important fact that the electric and magnetic fields of a plane wave are perpendicular to each other and to the direction of propagation. The wave vector can be redefined using the directions of $\mathbf{E}$ and $\mathbf{B}$:

$$\mathbf{k} = \frac{2\pi}{\lambda} \cdot n \cdot \frac{\mathbf{E} \times \mathbf{B}}{|\mathbf{E} \times \mathbf{B}|} \tag{5.23}$$

5.2.3 Phasor Notation

When the time variation of a field is sinusoidal, the field is called a time-harmonic field. Plane waves are an example of a time−harmonic field. For a time-harmonic field, we can separately consider the space term and the time term. For the case of harmonic plane wave in Eq. 5.16, it can be rewritten as:

$$\mathbf{E}(\mathbf{r}, t) = \mathbf{E}(\mathbf{r}) \cdot e^{j(\omega t + \phi)}, \tag{5.24}$$

where $\mathbf{E}(\mathbf{r})$ is the space term given as

$$\mathbf{E}(\mathbf{r}) = \mathbf{E}_0 e^{-j\,\mathbf{k}\cdot\mathbf{r}} \tag{5.25}$$

The term $\mathbf{E}(\mathbf{r})$ in Eq. 5.24 is called the **phasor**. It is a vector, each component of which has an amplitude and a phase. Equation 5.24 is identical to 5.16 but is convenient to calculate higher-order differentiations and integrations than instantaneous expression.

Plane Wave

Wave equations are derived from Maxwell's equations:

$$\nabla^2 \mathbf{E} - \mu\varepsilon \frac{\partial^2 \mathbf{E}}{\partial t^2} = 0$$

$$\nabla^2 \mathbf{B} - \mu\varepsilon \frac{\partial^2 \mathbf{B}}{\partial t^2} = 0$$

Plane wave is one of solutions to wave equations

$$\mathbf{E}(\mathbf{r}, t) = \mathbf{E}_0 e^{j(\omega t + \phi - \mathbf{k}\cdot\mathbf{r})}$$

The direction of the propagation is given by the wave vector $\mathbf{k}$

$$\mathbf{k} = \frac{2\pi}{\lambda} \cdot n \cdot \frac{\mathbf{E} \times \mathbf{B}}{|\mathbf{E} \times \mathbf{B}|}$$

When the direction of x and y axes are chosen as those of $\mathbf{E}$ and $\mathbf{B}$, respectively,

$$E_x(z, t) = E_0 \exp j\left(\omega t + \phi - \frac{2\pi n}{\lambda} \cdot z\right)$$

$$B_y(z, t) = B_0 \exp j\left(\omega t + \phi - \frac{2\pi n}{\lambda} \cdot z\right)$$

5.2.4 Interference

The electric field of a plane electromagnetic traveling wave is given as:

$$\mathbf{E} = \mathbf{E}_0 e^{j(\omega t + \phi - \mathbf{k}\cdot\mathbf{r})} \tag{5.26}$$

Let us consider multiple waves that are present at one place at the same time with the same wavelength and frequency.

$$\mathbf{E}_j = \mathbf{E}_0 \cdot e^{j(\omega t + \alpha_j)} \tag{5.27}$$

where $\alpha_j = \mathbf{k} \cdot \mathbf{r}_j + \phi_j$ and especially $\mathbf{r}_j$ represents the directed distance from the reference plane at which the phase is ϕ_j at $t = 0$.

Summation of two waves will lead to interference. For example, if there are $\mathbf{E}_{01} = \mathbf{E}_{01}e^{i(\omega t + \alpha_1)}$ and $\mathbf{E}_{02} = \mathbf{E}_{02}e^{i(\omega t + \alpha_2)}$, the sum of two waves is:

$$\mathbf{E}_{1+2} = \mathbf{E}_0 e^{i(\omega t + \alpha)} \tag{5.28}$$

where the resulting amplitude is:

$$\mathbf{E}_0 = \sqrt{\mathbf{E}_{01}^2 + \mathbf{E}_{02}^2 + \mathbf{E}_{01}\mathbf{E}_{02}\cos(\alpha_2 - \alpha_1)} \tag{5.29}$$

Also the resulting phase is:

$$\tan(\alpha) = \frac{E_{01}\sin(\alpha_1) + E_{02}\sin(\alpha_2)}{E_{01}\sin(\alpha_1) + E_{02}\sin(\alpha_2)} \tag{5.30}$$

Both equations above are derived from the cosine law, as shown in Fig. 5.3.

There are two extreme cases of the interference which depend on the phase difference $\delta = a_1 - a_2$:

Destructive interference: δ equals a integer multiple of 2π
Constructive interference: δ equals π plus a integer multiple of 2π

5.2.5 Polarization, Incidence and Reflection of light

For reflection of light, the surface and the incoming beam define the plane of incidence (Fig. 5.4). The incoming beam and the surface normal are in the plane of incidence. The angle of incidence α is defined as the angle between the incoming beam and the surface

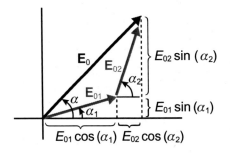

Figure 5.3: Summation of Two Vectors.

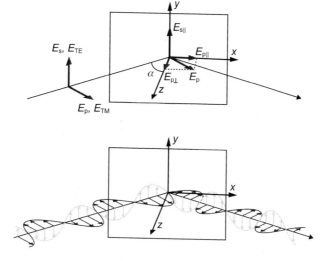

Figure 5.4: Incident and Reflected Light.

normal. The reflected beam is also in the plane of incidence. As we will explain later, the angle of reflection γ equals the angle of incidence α.

TM and TE Modes

The electric field can be separated into two projections: one is parallel to the plane of incidence and defined as E_p or E_{TM}, and the other is perpendicular to the plane of incidence and defined as E_s or E_{TE}. The electric field at the surface can also be separated into two projections: one parallel to the surface plane $E_{\parallel}$, and one perpendicular to the surface plane $E_{\perp}$. Only the p-component has a field component perpendicular to the surface plane.

With the introduction of an interface, a specific coordinate system is created (Note that the coordinate system used in Fig. 5.4 is different from the coordinate system used in Fig. 5.2):

- The plane of incidence is defined by the incoming beam and the surface normal
- The plane of incidence defines the x-z plane
- The z-axis is the surface normal
- The x-axis is in the surface
- The y-axis is in the surface and perpendicular to the plane of incidence.

The electric field components are separated into components parallel (p, TM) and perpendicular (s, TE) *to the plane of incidence*:

- The s, TE component is completely *in the plane of the surface* (parallel to the y-axis)
- The p, TM component has a component perpendicular to the surface plane (i.e. parallel to z-axis) and a component parallel to the surface (i.e. parallel to x-axis).

We define wave vectors $\mathbf{k}_i$, $\mathbf{k}_t$, and $\mathbf{k}_r$ for each of incident, transmitted, and reflected light, respectively. The magnitudes of the vectors are:

$$|\mathbf{k}_i| = \frac{\omega}{c} n_0 = \frac{2\pi}{\lambda} n_0 \tag{5.31}$$

$$|\mathbf{k}_t| = \frac{\omega}{c} n_1 = \frac{2\pi}{\lambda} n_1 \tag{5.32}$$

$$|\mathbf{k}_r| = \frac{\omega}{c} n_0 = \frac{2\pi}{\lambda} n_0 \tag{5.33}$$

We need to consider the boundary conditions at the surface between two media (Fig. 5.5).

Boundary condition for the fields

$$E_{0\parallel} = E_{1\parallel} \tag{5.34}$$

when there is no surface charge, Boundary condition for the electric flux density becomes:

$$\varepsilon_0 E_{0\perp} = \varepsilon_1 E_{1\perp} \tag{5.35}$$

Boundary conditions for magnetic field and magnetic flux density are:

$$H_{0\parallel} = H_{1\parallel} \tag{5.36}$$

$$B_{0\perp} = B_{1\perp} \tag{5.37}$$

As consequences of the boundary conditions:

$$e^{j(\omega_i t - \mathbf{k}_i \cdot \mathbf{r})} = e^{j(\omega_r t - \mathbf{k}_r \cdot \mathbf{r})} = e^{j(\omega_t t - \mathbf{k}_t \cdot \mathbf{r})} \tag{5.38}$$

All the angular frequencies must be the same:

$$\omega_i = \omega_t = \omega_r \tag{5.39}$$

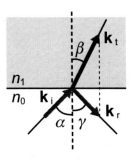

Figure 5.5: Boundary Conditions.

All the wave vectors must be in one plane:

$$(\mathbf{k}_i \cdot \mathbf{r})_{z=0} = (\mathbf{k}_r \cdot \mathbf{r})_{z=0} = (\mathbf{k}_t \cdot \mathbf{r})_{z=0} \tag{5.40}$$

Snell's Law, Total Internal Reflection

From the first equation of Eq. 5.40, we find the law of reflection:

$$\sin \alpha = \sin \gamma \tag{5.41}$$

or

$$\alpha = \gamma \tag{5.42}$$

From the other equation of Eq. 5.40, we find Snell's law, which is expressed as:

$$n_0 \sin \alpha = n_1 \sin \beta \tag{5.43}$$

where α is the incident angle and β is the transmission angle. The incident angle α is smaller than the transmission angle β if the wave is incident on medium 1 from medium 0, which has a larger refractive index than medium 1 ($n_0 > n_1$). In this situation, the transmission angle β increases with α until it reaches at $\pi/2$. When β is equal to $\pi/2$, the transmission wave propagates along the interface. The additional increase of α will result in no refracted transmission wave, and the incident light is totally reflected. The value of the total internal reflection angle, or critical angle α_c, can be found by substituting the transmitted angle $\beta = \pi/2$ in Snell's law.

Now the angle of total reflection is:

$$\sin \alpha_c = \frac{n_1}{n_0} \sin \beta, \tag{5.44}$$

where α_c is called the critical angle.

Fresnel Equations

Solving the boundary condition of Eqs 5.34 and 5.36, we find the Fresnel equations.

$$r_{\text{TE}} = \left(\frac{E_{0r}}{E_{0i}}\right)_{\text{TE}} = \frac{n_0 \cos(\alpha) - n_1 \cos(\beta)}{n_0 \cos(\alpha) + n_1 \cos(\beta)} \tag{5.45}$$

$$t_{\text{TE}} = \left(\frac{E_{0t}}{E_{0i}}\right)_{\text{TE}} = \frac{2n_0 \cos(\alpha)}{n_0 \cos(\alpha) + n_1 \cos(\beta)} \tag{5.46}$$

$$r_{\text{TM}} = \left(\frac{E_{0t}}{E_{0i}}\right)_{\text{TM}} = \frac{n_1 \cos(\alpha) - n_0 \cos(\beta)}{n_0 \cos(\beta) + n_1 \cos(\alpha)} \tag{5.47}$$

$$r_{TM} = \left(\frac{E_{0r}}{E_{0i}}\right)_{TM} = \frac{2n_0 \cos(\alpha)}{n_0 \cos(\beta) + n_1 \cos(\alpha)} \tag{5.48}$$

The other conditions Eqs 5.35 and 5.37 are automatically satisfied.

Brewster's Angle

There is an angle α_B of incidence at which only the component normal to the incident plane (i.e. parallel to the surface) will be reflected. The component polarized parallel to the incident plane (E_p or E_{TM}) will not be reflected at α_B. Such an angle is called the **Brewster's angle.** The condition for Brewster's angle is:

$$r_{TM}(=r_p) = 0. \tag{5.49}$$

When unpolarized light is incident at this angle, the light that is reflected from the surface is therefore perfectly polarized.

$$\tan \alpha_B = \frac{n_1}{n_0} \tag{5.50}$$

5.2.6 Evanescent Field

Consider a case of total internal reflection, with the interface between two different media with refractive indexes of n_0 and n_1 ($n_0 > n_1$). When the incident angle α is larger than the critical angle α_c, there will be no refracted waves transmitted through the horizontal interface for the impinging light. However, the mathematical description of such total internal reflection implies the fact that there will be evanescent waves formed at its interface. An evanescent wave is a confined electromagnetic wave exhibiting exponential decay with distance from the interface (Fig. 5.6). The properties of an evanescent wave can be derived from Snell's law.

With the total internal reflection condition ($n_0 > n_1$ and $\alpha > \alpha_c$), Snell's law can be rewritten as:

$$\sin \beta = \frac{n_0}{n_1} \sin \alpha > 1 \tag{5.51}$$

There is no solution for real-valued β satisfying Eq. 5.51, and $\cos \beta$ becomes imaginary under the given total internal reflection condition.

$$\cos \beta = \sqrt{1 - \sin^2 \beta} = \pm i\sqrt{\frac{n_0^2}{n_1^2} \sin^2 \alpha - 1}, (\sin \beta > 1), \tag{5.52}$$

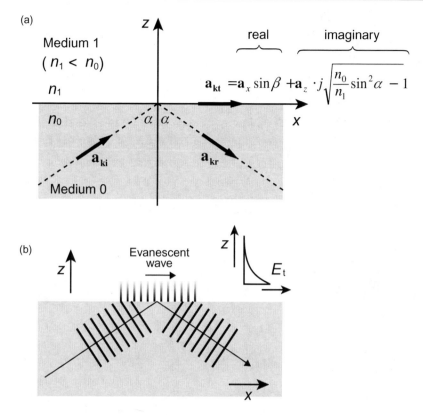

Figure 5.6: Evanescent Field Induced by Total Internal Reflection. (a) Wave Vector with Imaginary Component (Original). (b) Exponential Decay in z Direction.

Let us assume the unit vector $\mathbf{a_{kt}}$ which indicates the direction of propagation of a refracted wave:

$$\mathbf{a_{kt}} = \mathbf{a}_x \sin\beta + \mathbf{a}_z \cos\beta \tag{5.53}$$

then $\mathbf{k} \cdot \mathbf{r}$ becomes:

$$\mathbf{k} \cdot \mathbf{r} = k_t(\mathbf{a}_x \sin\beta + \mathbf{a}_z \cos\beta) \cdot (\mathbf{a}_x x + \mathbf{a}_y y + \mathbf{a}_z z) = k_t(x \sin\beta + z \cos\beta) \tag{5.54}$$

Now we can substitute it into the phasor expression of the electric field:

$$\mathbf{E}_t = \mathbf{E}_{t0}e^{j\mathbf{k}\cdot\mathbf{r}} = \mathbf{E}_{t0}e^{j(k_t x \sin\beta + k_t z \cos\beta)} \tag{5.55}$$

The incident angle term can be substituted for the transmission angle term:

$$\mathbf{E}_t = \mathbf{E}_{t0} \cdot \underbrace{e^{j\left(k_t x \frac{n_0}{n_1}\sin\alpha\right)}}_{\text{traveling wave}} \cdot \underbrace{e^{-k_t z \sqrt{\frac{n_0^2}{n_1^2}\sin^2\alpha - 1}}}_{\text{exponential decay}} \tag{5.56}$$

The positive sign is neglected in order to hold physical causality since it results in the electric field becoming infinitely large as z increases. Equation 5.56 demonstrates the evanescent wave along the interface. The first exponential term shows the wave propagates along the interface, which is the x-direction in this case, while the second exponential term indicates the exponential attenuation normal to the interface in medium 1.

Note that on the average, no electromagnetic energy can be transmitted to the medium 1, since the reflectivity on the interface becomes unity for the total internal reflection condition.

$$r_{TE} = \left(\frac{E_{0r}}{E_{0i}}\right)_{TE} = \frac{n_0 \cos(\alpha) - n_1 \cos(\beta)}{n_0 \cos(\alpha) + n_1 \cos(\beta)} \tag{5.57}$$

$$r_{TM} = \left(\frac{E_{0t}}{E_{0i}}\right)_{TM} = \frac{n_1 \cos(\alpha) - n_0 \cos(\beta)}{n_0 \cos(\beta) + n_1 \cos(\alpha)} \tag{5.58}$$

$$R_{TE} = |r_{TE}|^2 = r_{TE} \cdot r_{TE}^* = 1 \tag{5.59}$$

$$R_{TM} = |r_{TM}|^2 = r_{TM} \cdot r_{TM}^* = 1 \tag{5.60}$$

The decay length is the reciprocal of the exponential attenuation term. Using $k_t = 2\pi n_1/\lambda$,

$$L = \frac{1}{k_t \sqrt{\frac{n_0^2}{n_1^2} \sin^2\alpha - 1}} = \frac{\lambda}{2\pi n_1 \sqrt{\frac{n_0^2}{n_1^2} \sin^2\alpha - 1}} \tag{5.61}$$

5.3 Waveguide-Based Molecular Sensors

Optical principles and measurement techniques are very well established and are valuable for designing molecular sensors. In particular, essential techniques such as transmitting light over multiscale distances and various media, establishing secure communications, and fabricating miniaturized optical and optoelectronic devices are all based on an understanding of guided-wave optics.

Described in this section are a class of optical molecular sensors based on light waveguides. The main benefit of waveguides is fairly obvious; they are designed to guide or confine lights. With waveguides, a large scale optical experiments or measurements may be assembled into compact, usable configurations. Several components such as microfluidic channels, light sources, and detectors may be readily integrated with measurement systems.

We will first describe the fundamentals of waveguide design and theories of guided modes, followed by introduction of applications in molecular sensing.

5.3.1 Introduction

Structures of waveguides commonly used for sensing applications are summarized in Fig. 5.7. We categorize optical waveguides into the following three types:

1. Slab waveguide

    Slab waveguides are a waveguide with a planar geometry. They are also called planar waveguides. They are often fabricated with a thin transparent film on a substrate with an increased refractive index that can guide light waves by total internal reflection. The top surface is usually used as the sensing site. The advantages of planar structure include easier fabrication and a larger sensing site.

2. Rectangular waveguide

    Rectangular waveguides are waveguides with a rectangular cross section, and are also called strip waveguides, or channel waveguides. One type of waveguide is etched out of a substrate, other types are buried in an etched groove. They are often used for measurements that utilize interferometry or optical resonation, where the path lengths have to be strictly confined to control interference.

3. Circular waveguide (optical fibers)

    Circular waveguides, or commonly referred to as optical fibers are the most common form of light waveguide used for optical communication. The advantage of optical fibers for sensing applications is their capability to be used as a probe. Typically, one end of a fiber is used as a sensing site. The fiber tip can be brought into the sensing sites of in-situ or in-vivo measuring applications.

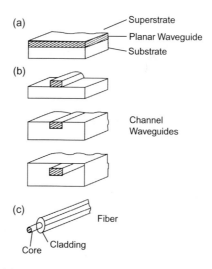

Figure 5.7: Basic Waveguide Structures: (a) Slab Waveguide. (b) Rectangular (Channel) Waveguide. (c) Circular Waveguide (Optical Fiber) [1].

5.3.2 Principles of Wave Propagation in Waveguide

The most fundamental principle behind guided light waves is total internal reflection, where light is confined in a medium with a higher refractive index.

As shown in Fig. 5.8, when the incident angle θ_1 is larger than the critical angle θ_c (see Section 5.2.5 for theoretical discussion), the light propagates inside the medium. Especially when the dimensions of a waveguide are relatively larger compared to the wavelength of light, this basic consideration based on geometrical optics provides sufficient information to describe the optical properties of several designs of waveguides.

Waveguide Modes

In order to design an efficient waveguide, we need to focus on the characteristics of propagating light as an electromagnetic wave, which is described by Maxwell's equations, or more specifically, electromagnetic wave equations.

Here we use slab waveguide as a simplified model of waveguide. It usually consists of a thin film with a high refractive index that works as the waveguiding layer (Fig. 5.9). For other waveguides, namely rectangular and circular waveguides, different coordinate systems or boundary conditions may need to be introduced. However, they still share most of the fundamental principles with slab waveguides.

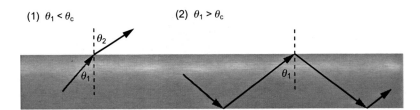

Figure 5.8: Total Internal Reflection.

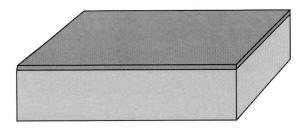

Figure 5.9: Basic Planar Waveguide Structure.

Depending on the propagation angle, wavelength, polarization, refractive indices, and the waveguide thickness, some waves are well confined in waveguide and allowed to propagate.

An important concept in guided-wave optics is "modes," which are solutions, or eigensolutions, to the wave equations for a specific wavelength and polarization with the boundary conditions given by the properties of the materials and interfaces. The modes are given as particular standing wave patterns of transverse distribution that are maintained at all the distances along the waveguide axis. In concept, they are very similar to finding a solution of vibrational modes of a mechanical cantilever (see Chapter 6, Section 6.2.2). Transverse modes are found for both TE and TM guided waves. The total electromagnetic field can be a combination of multiple modes. A simple case of finding propagation modes is described in Problem 5.9.

Figure 5.10 illustrates modes in a slab waveguide. The air, film, and substrate have refractive indices of n_1, n_2, and n_3, respectively. Index n_2 has to be larger than n_1 and n_3 for total internal reflection to happen. A guided wave is represented by a zigzag wave as shown in Fig. 5.10(a). The mode number is related to the number of waves as shown in Fig. 5.10(b). For $m = 0$ mode, it has the largest incident angle and as m increases, the incident angle gets close to the critical angle.

Light Coupling

The wavenumber of a guided wave along the direction of propagation is called the propagation constant. The propagation constant of a waveguide is typically larger than the wavenumber of light that propagates in vacuum or air. In order to couple light from free

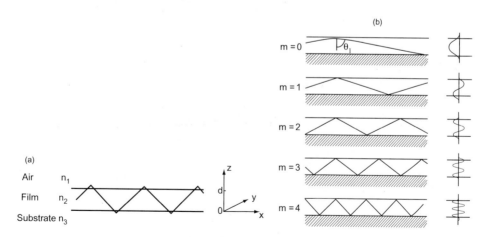

Figure 5.10: Waveguide Modes. (a) Simplified Structure of a Slab Waveguide (b) Mode Number Is Related to the Number of Waves.
From [97].

space into a waveguide, techniques to match wavenumbers needs to be introduced. Light coupling essentially is a process of matching wave numbers of two different media.

There are mainly three ways to couple light to a planar waveguide: gratings, prism, and end-fire coupling, as shown in Fig. 5.11.

1. Gratings fabricated on the top surface will add a grating vector to the light wave, and can therefore match the wavevectors in the waveguide.
2. Optical power may be coupled into or out of a planar waveguide by use of a prism. A prism with a higher refractive index comparing to air is placed close to the waveguide surface with a thin air gap (500−1000 nm) in-between. An optical wave is incident into the prism such that it undergoes total internal reflection within the prism at an angle larger than the critical angle. The transverse field distribution extends outside the prism and decays exponentially in the space separating the prism and the slab. If an appropriate interaction distance is selected, the wave is coupled to a mode of the slab waveguide. The operation can be reversed to make an output coupler, which extracts light from the slab waveguide into free space.
3. Finally, light can be coupled into a waveguide by directly focusing it at one end. For such "end-fire coupling," we have to collimate the beam into the planar surface of the waveguide. For optical fibers, this is the most commonly used method for light coupling. Because of the small dimensions of the waveguide slab, focusing and alignment are usually difficult and the coupling is not always efficient.

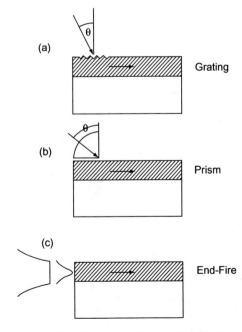

Figure 5.11: The Three Methods of Light Coupling.
From [1].

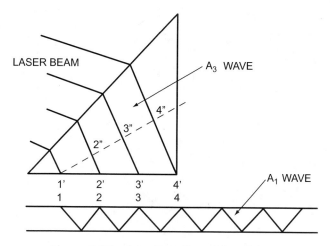

Figure 5.12: Light Coupling with a Prism.
The coupling becomes effective if the fields intensities at these points are in phase with a zigzag wave in the waveguide. *From [97]*

Let us take a closer look at light coupling with a prism. The energy from light in the prism excites the surface of the waveguide at the points 1, 2, 3, 4..., as shown Fig. 5.12. The coupling becomes effective if the fields intensities at these points are in phase with a zigzag wave in the waveguide. When this condition is satisfied, the amplitude of the zigzag wave in the coupling region increases roughly as the number of zigzags. An overall coupling efficiency of over 90% can be achieved with a well designed prism.

5.3.3 Slab Waveguide

In the following three sections, we will describe designs and fabrications of different types of waveguide structures, namely slab waveguide, rectangular waveguide, and circular waveguide, each of which is followed by introduction of applications in molecular sensing. We will start from the most basic structure of waveguide, slab waveguide.

Implementation

Slab waveguides take advantage of the large sensing site easily obtained on the top surface of the wave guiding layer. They mostly utilize fluorescence excitation by an evanescent field induced on the surface. When light propagates within a slab waveguide, an evanescent field is generated at the surface (see Section 5.2.6 for theories). The evanescent field decays exponentially in the direction perpendicular to the waveguide surface. This characteristic can be used to specifically probe the surrounding medium by exciting only molecules near the waveguide surface. Fluorescence from the excited molecules are guided back and measured by a detector. Figure 5.13 shows different configurations for (1) Coupling excitation light into the waveguide, (2) coupling fluorescence back into the waveguide, and (3) detection of the fluorescence signals.

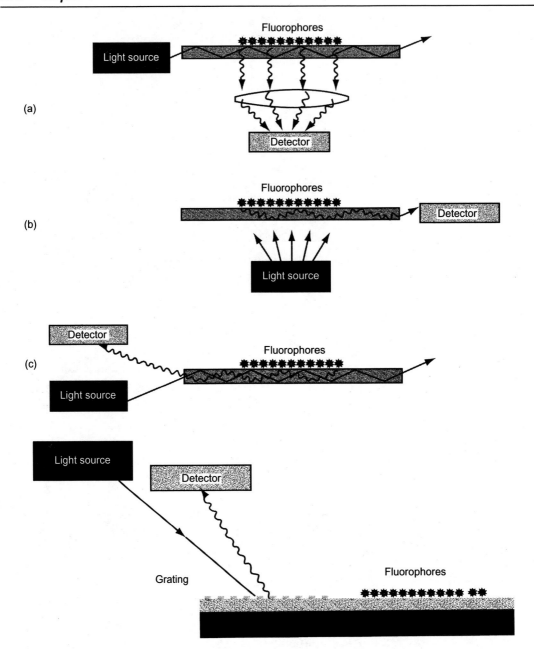

Figure 5.13: Different Configurations for Slab Waveguide Molecular Sensors.
Redrawn from Gizeli and Lowe, Biomolecular sensors, Taylor & Francis 2012, p.161,165.

Applications

Slab waveguide sensors can be specific to an analyte by functionalization of the waveguide surface. Some slab waveguide sensors have been utilized in bacterial sensing applications [2,3]. Other types of slab waveguide sensors have also been designed as an immunosensor [4]. This device uses antibody−antigen binding to detect the presence of an antigen on the waveguide surface, which offers a highly specific detection method for many different pathogens. It has applications in a clinical setting for diagnosis of pathologies.

Clerc et al utilized an output grating coupler for direct immunosensing [4]. The sensor uses a He−Ne laser light with a wavelength of 633 nm, which is end fired into a waveguide from the output of the waveguide. The outcoupling angle is measured to find the change in the refractive index induced by the sensing event. The waveguide consists of a 150−180 nm dip coated $SiO_2−TiO_2$ core guiding layer on top of a glass substrate. The core layer has a higher refractive index than the surrounding layers. A grating is embossed on the outer surface of the core. Light coupled out of the waveguide is detected by a position sensitive sensor which detects the change of the outcoupling angle. The outcoupling angle change can be related to the change in effective index of the film surface [4]. In this sensor, the binding of an antigen to bound antibodies at the surface causes a change in effective index at the surface. This can be used to detect antigen concentrations of less than 1 nanomolar. Antibodies are bound to the sensor surface by the intermediate avidin. Avidin is first adsorbed to the surface. Biotinyated IgG antibodies are then immobilized to avidin through avidin−biotin affinity binding. This is a very strong binding and ensures strong contact with the surface. The sensor can therefore be used to sense very small concentrations of antigens. One disadvantage of this sensor is the drift effect. The refractive index of the surface gradually increases while the sensor is contacted with a solution. This effect can be reduced by certain drift correction methods described by Clerc and Lukosz [4].

Plowman et al developed a multiple analyte immunoassay using a silicon oxynitride integrated optical waveguide [5]. Figure 5.14 shows a schematic of the multi-analyte assay. Three capture antibodies were adsorbed to the surface in channels separated by areas of blocking protein. A rubber gasket with holes clamped on the waveguide was used to pattern antibody capture layers. They used a sandwich immunoassay format, where tracer antibodies, specific for one channel each, are introduced for fluorescence signals. Fluorescent labels on the tracer antibodies are excited by the evanescent field. Multi-analyte assay responses, for clinically significant ranges of creatine kinase MB (0.5−100 ng/mL), cardiac troponin I (0.5−100 ng/mL), and myoglobin (5−500 ng/mL) were compared with responses from single analyte assay and good agreement ($R^2 = 0.97−0.99$) was observed.

An interesting material used as the waveguiding layer is a nanoporous silicon substrate [6]. It was used to increase the sensing properties of the waveguide as well as to allow for the

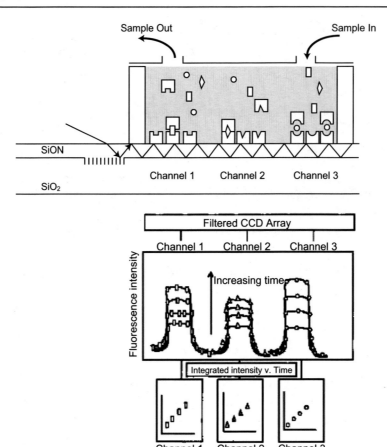

Figure 5.14: Multi-analyte Assay Integrated on a Silicon Oxynitride Waveguide.
From [5].

label free detection of analytes. The waveguide consists of a low porosity (high index) layer, a high porosity (low index) layer, and air gap (Fig. 5.15). Utilization of two different porosities enabled creation of waveguides from the same material. In addition, effective surface area would be increased and allow for greater interaction with certain molecules. The use of silicon in the waveguide is also advantageous to the already established semiconductor market.

5.3.4 Rectangular Waveguide (Channel Waveguides)

The behavior of rectangular waveguides is described in a similar way to that of slab wave guides. Although the confinement becomes two-dimensional and mode shapes are different, the essential ideas are the same as for the planar waveguides.

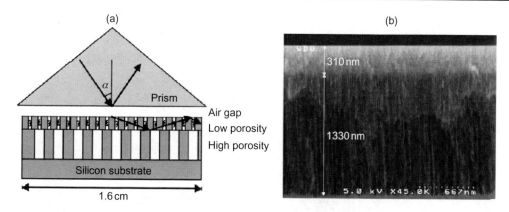

Figure 5.15: (a) Porous Silicon Waveguide Consists of Two Layers Porous Silicon with Different Porosities (b) Cross-sectional SEM of the Waveguide [6].

Interferometers

One of the advantages of rectangular waveguides is that it is possible to precisely define the light path length so it can be easily utilized for application based on optical interference. In a typical experimental configuration, a waveguide is split into two channels. One is used as a reference, and the other contains a sensing site. The effective path of the sensing site is changed by molecular adsorption. The signal is measured as an interference of the signals from the two waveguides. For interferometric applications, waveguides usually needs to be "single mode" waveguide, which allows only a single mode of light propagation. Cross-sectional dimensions of the single mode fiber is very close to the wavelength of the transmitted light, in order not to allow other modes to exist.

Figure 5.16 shows the typical design of a waveguide based interferometer for biosensing, where the injection of a streptavidin solution was measured [7]. Multiple waveguides for interferometry were integrated in the following way. First, a buffer layer of 2.5 μm thick SiO_2 was formed by thermal oxidation. A 350 nm thick siliconoxynitride layer with a refractive index of 1.55 was then deposited by PECVD (plasma enhanced chemical vapour deposition). A 55 nm thick rib was formed by standard lithography and RIE (reactive ion etching) to define a single mode waveguide. The width w. of the waveguide is 2.5 μm, which was limited by the photolithographic process [8].

Schmitt et al utilized a commercial planar Ta_2O_5 waveguide to construct a Young interferometer configuration (Fig. 5.17), where interference between the reference and the sensing signal was measured [9]. They discussed the sensing capability of the sensor for molecular sensing applications. The interferometer demonstrated an effective refractive index resolution of 9×10^{-9}, which corresponds to a surface coverage of ≈ 13 fg mm^{-2}.

(a)

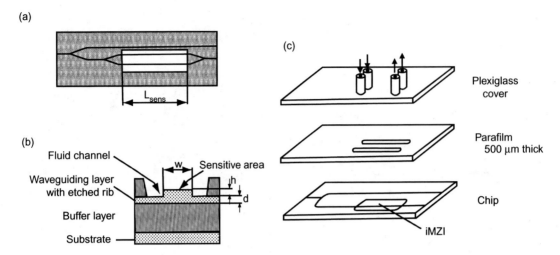

(b)

Fluid channel

Waveguiding layer
with etched rib

Buffer layer

Substrate

w

Sensitive area

h

d

(c)

Plexiglass
cover

Parafilm
500 μm thick

Chip

iMZI

Figure 5.16: Mach–Zehnder Interferometer-based Biosensor [7].

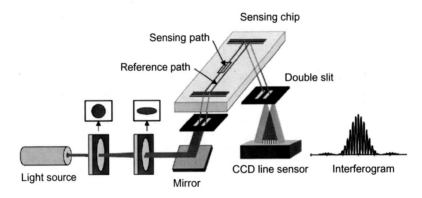

Figure 5.17: Young Interferometer Constructed with Planar Waveguides [9].

Silicon Ring Resonators

Another important method for channel waveguides is the use of a silicon ring oscillator. It typically consists of a straight waveguide coupled to a ring shaped waveguide that functions as an optical resonator. Many researchers study this structure in the hope of using it for optical logic devices that can be integrated onto a silicon chip [10]. Another important application area is molecular sensing [11].

Figure 5.18 shows a photograph of a ring oscillator along with the schematic of the experimental setup studied by De Vos et al. [11]. When the wavelength of the incident light

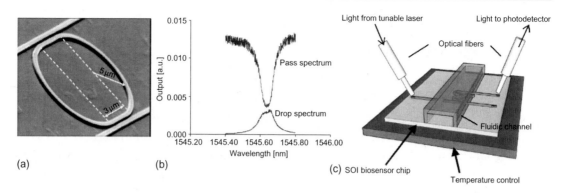

(a) (b) (c) SOI biosensor chip

Figure 5.18: Molecular Sensing Based on a Silicon Ring Resonator. (a) Photograph of a Ring Resonator. (b) Dip in the Transmission at the Resonant Frequency. (c) Experimental Setup. [11].

matches the condition for resonance, a sharp dip in the transmission is observed. The condition for optical resonance is simply given in the following equation:

$$\lambda_{\text{resonance}} = \frac{L}{m} \cdot n_{\text{eff}}, \tag{5.62}$$

where L is the round trip length of the ring, m is the cavity mode order ($=1,2\ldots$), and n_{eff} is the effective index of the ring waveguide.

We can measure changes in the refractive index in the sensing site by measuring the shift in the resonance spectrum. A similar idea is applied to the SPR sensors described in the next section. It is important to note that silicon is not transparent for visible light and the wavelengths is usually around 1500 nm, which is the wavelength commonly used for fiber-optics communication.

De Vos et al. theoretically estimate the sensitivity for bulk refractive index changes as 10^{-5} which corresponds to theoretical sensitivity of 1 fg molecular mass. They experimentally demonstrated detection of avidin−biotin binding with concentrations down to 10 ng/ml. Another group devised a spiral shaped cavity resonator and demonstrated streptavidin protein binding with a detection limit of $\sim$3 pg/mm^2, or a total mass of $\sim$5 fg [12].

One obvious advantage of this method is that silicon waveguides are suitable for mass production. Previously, this technique has been commercialized. The Maverick™ MT-ADA™ (multitier antidrug antibody assay) from Genalyte, Inc. is a commercially available multi-analyte assay based on silicon ring oscillators. The company claims that a multi-analyte assay requires very a small sample volume of 2 μL and it takes 15 minutes to complete the assay.

5.3.5 Circular Waveguide (Optical Fibers)

Principles of Wave Propagation in Waveguide

Optical fibers were first discovered in the mid-19th century when scientists had discovered that light could propagate through materials with higher refractive index, relative to a surrounding lower refractive index. Since this time, several technological advancements have propelled this phenomenon to a wide array of applications in many industries including telecommunications, healthcare, automotive, and information management. Just as rectangular waveguides, these fibers utilize the principle of total internal reflection to propagate light certain variable distances.

Basic Structure of a Waveguide

Optical fibers possess a cylindrical geometry and therefore have a surrounding material (cladding) of uniform refractive index, as opposed to potentially multiple refractive indices as seen in planar waveguides [13,14]. A visual representation of this optical fiber technology can be seen in Fig. 5.19.

The core is the component material with the higher index of refraction. In step-index fibers, this region is uniform. Interestingly enough, this region can be of variable permittivity. Graded-index fibers are optical fibers whereby the core is not homogeneous and various indices exist. The index is what ultimately determines the mode of the wave that is guided.

Waveguide Modes in a Circular Waveguide

Although a different coordinate system (i.e., cylindrical coordinate system), has to be employed for mathematical analysis, the basic idea behind the waveguide mode is the same as that of slab waveguide.

Single-mode fibers are generally constructed for the purposes of sending energy or information through long distances. This gives rise to a small diameter core which only lets a single mode to exist. Multi-mode fibers, however, allow light with a range of modes to propagate through a large aperture. The practical purposes for multi-mode are generally meant for information transfer over shorter distances.

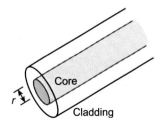

Figure 5.19: Circular Waveguide.

Applications

With the growing expansion of fiber optics to various forms of communication, optical fiber are also a natural fit for biomedical applications. Their accessibility to the inside of the body may enable in vivo measurements. The small diameter of the fiber allows it to relay important information on the microscopic level.

A hollow fiber sensor (Fig. 5.20) is a type of sensor that utilizes optical fiber as a miniature probe microscope. Schulz et al used competitive reversible binding of molecules and measured changes in molecular concentrations by means of fluorescence intensities [15]. They immobilized a glucose receptor, concanavalin A (ConA) to the Inner wall of a dialysis fiber (diameter = 0.3 mm).

$$\text{Con A} + \text{glucose} \rightleftharpoons \text{Con A} - \text{glucose} \quad (1)$$
$$\text{Con A} + \text{FITC-dextran} \rightleftharpoons \text{Con A} - \text{FITC-dextran} \quad (2)$$

When the concentration of glucose increases, reaction (1) will be displaced to the right and (2) to the left, resulting in releasing of dextran from Con A. The concentration of the released dextran is found by measuring the intensity of fluorescence from FITC (a type of fluorescence dye. See Section 5.6) that labels the dextran molecules.

There are many variations in hollow fiber sensors. Figure 5.21 shows an improved design of a fiber sensor from the Shultz group [16]. In this case, ConA instead of dextran is fluorescently labeled and is bound to dextran matrix immobilized within porous beads. In the absence of glucose, fluorescent-labeled ConA is bound to dextran immobilized inside.

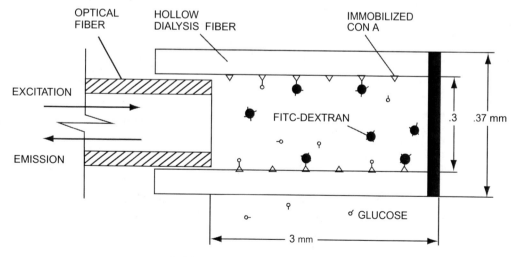

Figure 5.20: Hollow Fiber Sensor.
From [15].

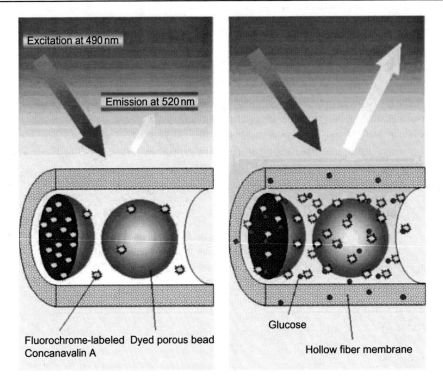

Excitation at 490 nm

Emission at 520 nm

Fluorochrome-labeled Dyed porous bead
Concanavalin A

Glucose

Hollow fiber membrane

Figure 5.21: Optical Fiber Chemical Sensors.
From [16].

The dextran matrix is colored to prevent fluorescence excitation of ConA bound within the beads. After glucose diffuses through the hollow fiber membrane, ConA is displaced and excited by the excitation from the fiber. The fluorescence goes back to the fiber to be measured. The new sensor demonstrated a glucose detection range extending from 0.15 to 100 mM (see Chapter 4, Section 4.2.2 for a discussion on glucose concentration in human blood).

5.3.6 Discussion on Waveguide Sensors

Optical fibers have been utilized in a variety of application areas. The number of commercially available products based on optical fibers is astounding. Optical fibers have been integrated into virtually every industry where communicating information is critical to success. In optical fibers, information is transferred in the form of light, and nothing can travel faster than at the speed of light.

In this chapter we have reviewed several types of waveguide-based molecular sensors. Waveguide-based chemical sensors designed for oxygen, carbon dioxide, and ammonia are a few sensors that may be applied to industries such as healthcare, environmental, and manufacturing [17]. The usability of waveguides gives them a great potential for in-vivo localized sensing in biomedical applications. Glucose sensing is a good example, and similar techniques may be introduced for detection of biomarkers related to tumors or other diseases.

5.4 Surface Plasmon Resonance (SPR) Sensors

5.4.1 Introduction

The focus of recent studies on nanophotonics is to better understand and utilize the properties of **surface plasmons** [18], which are oscillations coupled at the metal and dielectric interfaces. Utilization of surface plasmons have enabled the use of distinguishable optical phenomena not only for molecular sensors but also for emerging nanodevices such as nanoantenna [19–21], nano grating coupler [22] and plasmonic near-field scanning optical microscopy (NSOM) [23]. Taking advantage of the highly confined near-field properties, it has been proven that efficient optical devices featuring low physical profile can be realized.

In this section we discuss sensors based on measurement of **surface plasmon resonance** (SPR). We start from general theories followed by introduction of commercially available products. Finally, we discuss recent advancement in molecular sensing.

5.4.2 Principles of Surface Plasmons

Surface electromagnetic waves that propagate in a direction parallel to the metal/dielectric interface are called **surface plasmon polaritons** (SPPs). Since they propagate along the interface as a wave, they are sometimes called **surface plasmon waves** (SPWs).

SPR occurs when the frequency of light matches the natural frequency of oscillating surface electrons. These oscillations are very sensitive to changes on the sensing event on the surface, such as adsorption and binding of molecules, since the waves propagate exactly on the interface of the metal and the dielectric external medium (e.g. air or water). The terms SPP, SPW, and SPR are often used interchangeably or even confusingly, since the light coupling into surface plasmon is associated with resonance of surface plasmons. However, when a resonant coupling of light arises in nanometer-sized structures such as metal particles or nanoantennas, it should be better called a **localized surface plasmon resonance** (LSPR), because "waves" or "polaritons" may not apply to cases where surface plasmon does not propagate as a regular wave.

Mathematical Description

In order to describe the properties of surface plasmons, we use an intuitive approach and treat each material as a homogeneous continuum, described by a frequency-dependent

dielectric constant. The response of materials to external fields usually depends on the frequency of the applied field. This frequency dependence is based on the fact that a material's polarization does not respond instantaneously but arises after the field is applied, which is represented by a phase difference. We treat permittivity as a complex function of the angular frequency of the applied field $\varepsilon(\omega)$. For the generalized refractive index and dielectric constant:

$$\varepsilon(\omega) = \varepsilon'(\omega) + i\varepsilon''(\omega)$$

By using complex numbers for the dielectric constant, we can specify the magnitude and phase. The values of the real part ε' and imaginary part ε'' are related to the stored energy and the dissipation of energy within the medium, respectively. When surface plasmons exist, the real part of the metal dielectric constant must be negative and its magnitude must be greater than that of the dielectric. This condition is satisfied in the wavelengths range of IR-visible lights for air/metal and water/metal interfaces, where the real part of dielectric constant of a metal is negative and those of air and water are positive.

Now let us look at the case of surface plasmon on thick metal film. Consider a waveguide composed of an infinite metal and an infinite dielectric material with a boundary interface in-between two media (Fig. 5.22).

The permittivity of the metal and that of the dielectric material are:

$$\varepsilon_m = \varepsilon'_m + j\varepsilon''_m \tag{5.63}$$

$$\varepsilon_d = \varepsilon'_d + j\varepsilon''_d \tag{5.64}$$

Since the launched SPPs at the interface decay exponentially along the transverse direction (z-direction) to that of wave propagation, most of the electromagnetic energy is confined in proximity to the interface and easily affected by changes in dielectric properties of

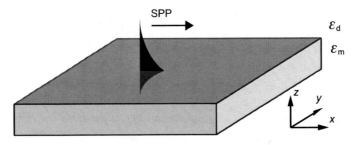

Figure 5.22: Surface Plasmon Polarization, Excited at the Interface Between Spectrally-thick Metal Film and Semi-infinitely Thick Dielectric Medium.

surroundings. With such squeezed field distribution, one may tailor surface plasmons by presenting surface corrugations and abnormalities to out-couple the part of surface optical energy.

Surface plasmons, which are waves in strict transverse magnetic (TM) mode, in the given configuration (see Fig. 5.23) can be described and understood by solving the boundary condition of the Maxwell's equation. The magnetic field distribution of excited SPP with the wavenumber of is given by:

$$H_{yd}(x, z) = H_0 \exp[-jk_{spp}x - \alpha_d z] \quad (z > 0) \tag{5.65}$$

$$H_{ym}(x, z) = H_0 \exp[-jk_{spp}x + \alpha_m z] \quad (z < 0) \tag{5.66}$$

where

$$\alpha_d = [-\varepsilon_d k_0{}^2 + k_{spp}{}^2]^{1/2}$$

and

$$\alpha_m = [-\varepsilon_m k_0{}^2 + k_{spp}{}^2]^{1/2}$$

It should be noted that there is only a single mode of surface plasmons that can exist on thick metal film by following the transverse resonant mode of operation (Fig. 5.23). Due to the high quality factor of transverse resonance and confined field distribution, the energy of surface wave decays quickly due to the ohmic losses within the metal as the wave propagates along the interface. For instance, the expected propagation length $L = |2 \, \text{Re}[jk_{spp}]|^{-1}$ of SPP is in general contained about $20-30 \, \mu\text{m}$ in the visible range. In this respect, it is essential to fabricate devices within a compact physical dimension to allow for utilizing SPP before the most energy dissipates.

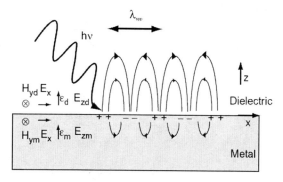

Figure 5.23: Surface Plasmon on Metal-Dielectric Waveguides.

Now we try to find a **dispersion relation**, which is a relationship between ω and k. For waveguides and other complex optical devices, the dispersion relation contains most of the important information about wave propagation in the structure.

By solving the Eqs 5.65 and 5.66 with the Maxwell equation, we obtain the dispersive wavenumber of SPP:

$$k_{spp} = \frac{\omega}{c}\sqrt{\frac{\varepsilon_m \varepsilon_d}{\varepsilon_m + \varepsilon_d}} \tag{5.67}$$

It can be rewritten as:

$$\omega = ck_{spp}\sqrt{\frac{\varepsilon_m + \varepsilon_d}{\varepsilon_m \varepsilon_d}} \tag{5.68}$$

This is the dispersion relation for surface waves.

Excitation of Surface Plasmons by Light

Here we consider excitation of surface plasmons by light. Similar to our previous discussions on light coupling, we need to match the boundary conditions such that the wave number k_{ph} and frequency ω of the incoming photon match to the wavenumber k_{spp} and frequency ω of the surface plasmon. We find matching by using a dispersion curve, which is a a plot of ω versus k. Figure 5.24 shows dispersion curves for surface plasmons, air, and glass.

The dispersion curve for air is simply given by $\omega = ck_{ph}$. Dispersion curves for air and surface plasmons do not intersect, which indicates that it is not possible to match the light propagating in air and surface plasmons. For this reason, surface plasmons are nonradiative. However, if light propagates in a medium whose index n is larger than 1, the dispersion curve becomes $\omega = ck_{ph}/n$ and there can be intersection as shown in the figure. In practice, this can be achieved by passing the incident light through a medium such as glass, on

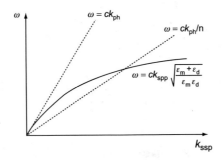

Figure 5.24: Dispersion Curves for SPR Matching.

which a metal film is deposited. Details of coupling between the evanescent wave and the surface plasmon is described in Appendix 5B.

5.4.3 Experimental Configuration

We will examine how light radiation is coupled to the surface plasmon resonance. There are two popular ways (Fig. 5.25):

- Kretschmann configuration (Fig. 5.25a): the metal film is evaporated onto the glass block. The light illuminates the glass block, and an evanescent wave penetrates through the metal film. The plasmons are excited at the outer side of the film.
- Otto configuration (Fig. 5.25b): a thin metal film is positioned close to the prism wall so that an evanescent wave can interact with the plasma waves on the surface and hence excite the plasmons.

The wavenumber along the direction of propagation can be tuned to match SPR simply by changing the angle of incidence of the p-polarized radiation at the prism/dielectric interface. A triangular or cylindrical prism is usually used to couple evanescent wave with SPR. Mechanical rotational stages are often employed to achieve measurements of angular responses.

Figure 5.26(a) shows a simplified diagram of SPR measurement. Figure 5.26(b) shows the reflectivity curves for p-polarized and s-polarized (flat dotted line) radiation with gold and silver films deposited on a sapphire prism ($n = 1.766$). The wavelength was $\lambda = 632.8$ nm.

Strong absorptions at the resonance angle were observed in angular scans. The angles of the minimum reflectance correspond to the matching of the laser excitation and SPRs on the gold and silver surface. SPR was not induced by s-polarized light as predicted. The dispersion of SPR is no longer dictated simply by the dielectric/metal boundary because it is perturbed by the presence of the additional coupling prism (see Appendix 5B).

Figure 5.27(a) shows a change in the plasmon dispersion curve as a result of a change in the refractive index at the interface of a metal film in a Kretschmann geometry.

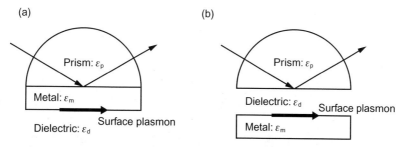

Figure 5.25: Kretschmann (a) and Otto (b) Configurations.

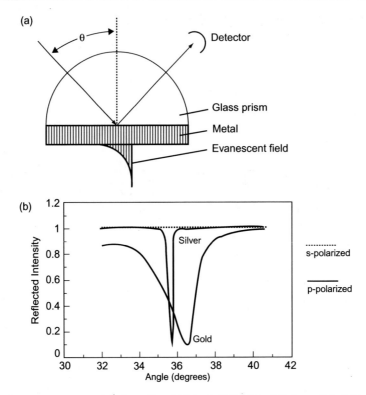

Figure 5.26: SPR Measurement. (a) Coupling With A Cylindrical Prism. (b) SPR Responses with Au and Ag films.
(a) from [1], (b) adapted from [24].

The solid lines show fit curves obtained using Fresnel theory. The change is observed as a peak shift in the angular scan as shown in Fig. 5.27(b).

Since SPR is confined exactly on the metal−dielectric interface, it constitutes a sensor which is sensitive to index changes specifically on the metal surface, which is the main principle of the SPR based molecular sensor. Molecular absorption at the metal interface can be measured as a peak shift in the angle scan.

5.4.4 SPR Sensor for Molecular Sensing

For molecular sensing, the surface of the dielectric layer is functionalized with recognizing elements such as antibodies or DNAs. Metals commonly used for SPR systems are gold or silver, which is suitable for biomolecule functionalization, as we discussed in Chapter 2. Once analyte molecules bind to the surface of the SPR sensor, changes in dielectric constant at the metal interface is utilized as the sensing effect. Figure 5.28 shows the

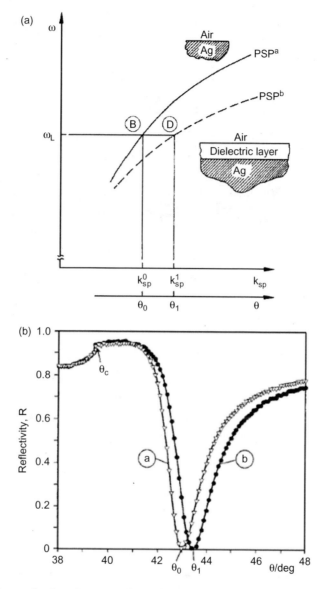

Figure 5.27: (a) Change in the Plasmon Dispersion Curve. (b) Peak Shift Induced in a SPR Curve.
From [1].

conceptual sketch of the molecular detection system in a SPR sensor. The change is measured as a shift of the absorption peak in a similar way as we illustrated in Fig. 5.26.

Major companies that provide commercially available SPR systems include Biacore AB, Affinity Sensors, Windsor Scientific Limited, BioTul AG, Nippon Laser and Electronics Lab [26]. Among products from those companies, the Biacore biosensor is most popular

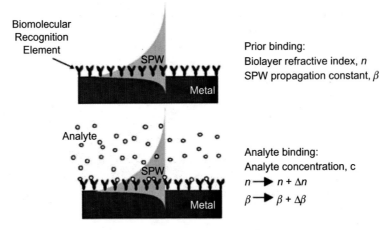

Figure 5.28: Optical Transduction.
From [25].

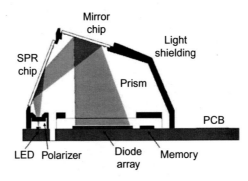

Figure 5.29: Composition of the Texas Instruments SPR Sensor.
From [98].

and considered to be one of the most accurate, precise and sensitive systems. Measureable concentration range for analytes with high molecular weights is $10^{-5}-10^{-9}$ M for direct assay, and $10^{-3}-10^{-11}$ M for sandwich assays. For analytes with low molecular weight, minimum concentration ranges $10^{-3}-10^{-9}$ M for inhibition assays.

Figure 5.29 shows the composition the SPR module developed by Texas Instruments. One characteristic feature of the device is the use of a photodetector array which eliminates the need of mechanical rotation in the measurement of angular scans. The angular absorption curve is directly mapped onto the photodetector array, reducing the experimental time and complexity involved in mechanical rotation.

The SPR module of the Biacore system is coupled with a microfluidic chip that supplies the anayte solution at a control flow rate. The schematic of the microfluidic channel design is

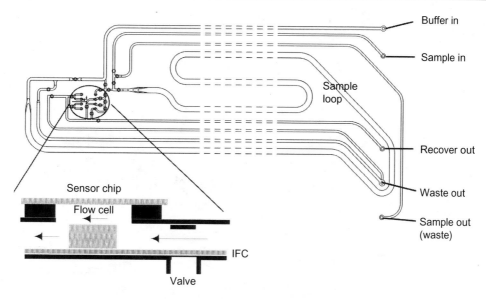

Figure 5.30: Microfluidic Channels Used in the Biacore System.
From Gizeli and Lowe, Biomolecular sensors, Taylor & Francis 2012, p 250.

shown in Fig. 5.30. For an example of the Biacore 3000 system, the flow rate can be controlled between $1-5000$ μl/min. It also allows a precise temperature control better than $\pm 3 \times 10^{-3}$°C, which corresponds to a measurement error of approximately 0.3 pg/mm^2.

Figure 5.31 shows example measurement results obtained from the Biacore system. Interleukin-2 (IL-2), which is a type of cytokine signaling molecule in the immune system, is injected at a flow rate of 100 μL/min with different concentrations of 233, 78, 26, 8.6, 2.9, and 0 nM in 10 mM sodium phosphate. The system uses SPR responses to monitor the refractive index change as molecules interact at the sensor surface. Solid curves represent the best fit of the binding responses to a simple one-to-one bimolecular reaction model of:

$$A + B \underset{k_d}{\overset{k_a}{\rightleftharpoons}} AB$$

which gives the relationship

$$\frac{d[RL]}{dt} = k_a[R_{tot}][L] = (k_a[L] + k_d)[RL] \tag{5.69}$$

The association (k_a) and dissociation (k_d) rate constants were $(4.66 \pm 0.04) \times 10^6$ M^{-1} s^{-1} and 0.0420 ± 0.0002 s^{-1}, respectively. One RU approximately corresponds 10^{-6} in refractive index, 10^{-4}° angle change, and 1 pg/mm^2 for protein.

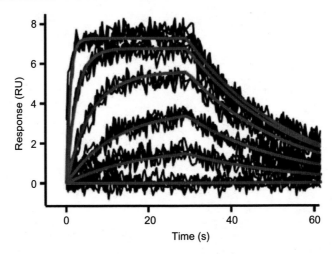

Figure 5.31: Global Analysis of Biosensor Data.
From [26].

Recent Advances in SPR Sensors

Several interesting studies have been recently reported to improve SPR sensors. One approach is integration of microfluidic systems that enables parallel high throughput screening.

Luo et al reported on a polydimethylsiloxane (PDMS) microfluidic device containing an array of micro chambers, each of which contains a gold spot designed for SPR measurements [27]. The flow cell is coupled with a triangular prism as shown in Fig. 5.32. After the surface of the gold spots is coated with biotin BSA (bovine serum albumin), anti-biotin antibody is injected into the flow cell at different concentrations. Immunoreaction was detected and characterized in about 10 min at subnanomolar sensitivity. When additional gold nanoparticles are selectively coupled to the immunocomplex to amplify SPR signals, the sensitivity was improved to the 10−100 pM level but the time needed for measurement increased to about 60 min.

Another approach is utilization of nanostructures. Localized surface plasmon resonance (LSPR) in nanoscale metal patterns has been a topic of interest in the study of SPR.

Specifically, the peak extinction wavelength, λ_{max} is known to be very sensitive to nanoparticle size, shape, and local ($\sim 10-30$ nm) external dielectric characteristics (see Section 5.7.2 for details of metal nanostructure and extinction and scattering measurements). Nanoscale metal pattern can be utilized to a substrate SPR measurement with a locally amplified sensitivity. Haes et al prepared triangular silver nanoparticles (~ 100 nm wide and 50 nm high) on a glass substrate by nanosphere lithography (NSL),

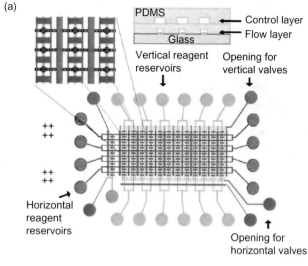

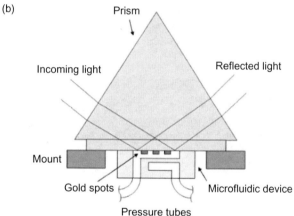

Figure 5.32: SPR Microfluidic Device for Immunoassays Based on Surface Plasmon Resonance. (a) Top View Schematic of the Array of Microflow Channel. (b) Prism Coupled to the Cell.
From [27].

where a self-assembled array of nanospheres serves as a shadow mask during the process of metal deposition with an evaporator [28].

Exposure of biotin-functionalized Ag nanotriangles to 100 nM streptavidin caused a red shift of 27.0 nm in the peak extinction wavelength. They demonstrated a strategy to amplify the optical signal by introducing biotinylated Au colloids. They concluded the limit of detection (LOD) for their LSPR nanobiosensor is in the low-picomolar to high-femtomolar region. A conceptual sketch of the experiments is shown in Fig. 5.33(c). An atomic force microscopy (AFM) image of the Ag nanoparticles is shown in Fig. 5.33(a).

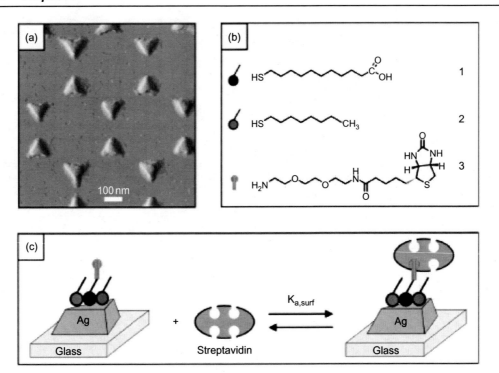

Figure 5.33: Localized Surface Plasmon Resonance Spectroscopy of Triangular Silver Nanoparticles. (a) AFM Image of the Triangular Silver Nanoparticles. (b) Surface Chemistry of the Ag Nanobiosensor. (c) Conceptual Sketch of the Biosensor [28].

Several interesting studies related to utilization of LSPR have been reviewed in [29]. Recent techniques include the use of nanohole arrays, where LSPR is directly excited by perpendicularly incident light, which minimizes the alignment requirements for light coupling. The signal induced by detection of live viruses was measured with a simple experimental setup with a CCD [30].

LSPR used in light scattering techniques are described in Section 5.7. LSPR found in metal nanoparticles will also be described in Section 5.7.

5.4.5 Discussion on SPR Sensors

In this section, we first introduced the theoretical background and practical applications of SPR-based molecular sensors. Surface plasmons are electron density waves that travel along the surface. These density waves depend on an electric field that collects electrons from the bulk. Only p(TM)- polarized light can excite surface plasmons. A dispersion relation, which indicates the condition for matching energy and momentum at the interface, determines when surface plasmons can be excited.

We then described how SPR sensors are implemented in the experimental setup. Because the **k**-vector of the surface plasmon wave is always larger than the **k**-vector of light in a medium, a prism or grating has to be used in the experiments. SPR is detected by measuring the angle under which the reflected light has minimal intensity. Binding is measured under constant flow and high flow rates and low receptor.

Finally we introduced a popular commercial system and experimental results of measurements of biomolecules. The technique does not use labels and can detect binding kinetics (k_{on}, k_{off}, K_A). Binding constants are in good agreement with other methods, but share a small underestimation. Mechanical and temperature stability determine the sensitivity, which is about 1 pg/mm^2. We also introduced recent advancement in the study of SPR sensing.

In conclusion, SPR based molecular sensors are well studied and are the most practically used molecular sensing systems.

5.5 Absorption Spectroscopy

5.5.1 Optical Density Measurement

The **optical density** or **absorbance** of a material is a logarithmic intensity ratio of the light falling upon the material, to the light transmitted through the material.

$$A(\lambda) = -\log_{10}\frac{I_0}{I_1} \tag{5.70}$$

where I_0 and I_1 are the intensities of incident and transmitted lights, respectively. Although sometimes confusing, the **transmittance** I_1/I_0 or the **percent transmittance** $100 \cdot I_1/I_0[\%]$ is also commonly used. Absorbance is dependent on the wavelength (a filter designed to absorb lights of all visible wavelengths equally is specifically called a **neutral density filter** or **ND filter**), and is often measured as a function of the radiation wavelength, which is called the **absorption spectrum**.

The absorption spectrum is a very important quantitative measure to evaluate optical properties of solutions. It provides several types of information such as the concentration of a solute, shapes or sizes of suspended particles, and biological activities. Absorbance of a solution is usually measured by preparing a sample in a cuvette or a microwell plate.

5.5.2 Beer–Lambert Law

Let us consider a case where the incident light I_0 with the cross sectional area of S passes through a liquid with the thickness of d. The liquid contains particles with the **absorption cross section** σ at the volume concentration of n.

Absorption of light for a small path increment dx is written as

$$dI = -I \cdot \frac{\sigma n S\, dx}{S} = -\sigma n I \cdot dx \tag{5.71}$$

or

$$\frac{dI}{dx} = -\sigma n I \tag{5.72}$$

Integration of Eq. 5.72 gives

$$I(x) = I_0 \cdot e^{-\sigma n x} \tag{5.73}$$

For the thickness $x = d$,

$$\frac{I}{I_0} = e^{-\sigma n d} \tag{5.74}$$

From the definition earlier, the absorbance of this solution becomes

$$A = \log_{10} \frac{I}{I_0} = \log_{10} e^{-\sigma n d} = -\frac{1}{\ln 10} \cdot \sigma n d \tag{5.75}$$

This relationship states that **the absorbance is proportional to the concentration and the thickness**, which is known as the **Beer–Lambert law** (Fig. 5.34), and is more commonly expressed with the **molar extinction coefficient** ε and molar concentration c in the following way:

$$A = \log_{10} \frac{I}{I_0} = -\varepsilon c d. \tag{5.76}$$

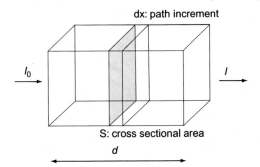

dx: path increment

I_0

I

S: cross sectional area

d

Figure 5.34: Beer–Lambert Law.

Comparing Eqs 5.75 and 5.76 and using $n = N_A c$ (N_A: Avogadro constant), one can find the relationship between absorption cross-section and the molar extinction coefficient.

$$\varepsilon = \frac{\sigma n}{\ln 10 \cdot c} = \frac{\sigma N_A}{\ln 10} \tag{5.77}$$

The term ln 10 is needed because base 10 (as in Eqs 5.70 and 5.75) rather than base e (as in Eqs 5.73 and 5.74) is more commonly used in practice. We also have to be careful with the units for conversion. In optical measurements, [cm] and [L] are commonly used for distances and volumes, respectively. An absorption cross section σ, a molar extinction coefficient ε and molar concentration are usually given in [cm^2], [L/mol/cm] and [mol/L], respectively.

5.5.3 Sensors Based on Absorption Spectroscopy

Optical Oximetry

One interesting application of absorption measurement for biomedical sensing is optical oximetry. It is a method that monitor hemoglobin saturation in a patient.

Figure 5.35 shows absorption spectra of hemoglobin and oxyhemoglobin (hemoglobin that contains bound O_2), which are present in arterial blood. The absorbances are dependent on the content of oxygen. A simple implementation of finding the ratio between the

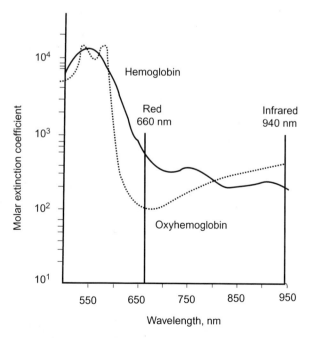

Figure 5.35: Absorption Spectrum of Blood.
From Harsanyi G 2000 Sensors in Biomedical Applications, CRC Press, Boca Raton, chapter 6.2.4, p 205.

hemoglobin and oxyhemoglobin contents are comparing the absorbances at two wavelengths, namely red and infrared. Blood saturation (oxygen content) SaO_2 can be correlated with the ratio of absorbances at red and IR, as shown in Fig. 5.36.

Measurement of arterial oxygenation is important in clinical management of ill or anesthetized patients. The practical technique used in optical oximetry is pulse oximetry [31,32], where pulsing absorbance change due to the heartbeat is measured at wavelengths for red and IR. The method allows exclusion of changes from venous blood, tissues of skin, bone, muscle, fat, and other artificial elements such as nail polish [33] by finding the change that occurs as a result of arterial pulsation.

A typical form of pulse oximetry is with a small device that clips a fingertip (Fig. 5.37a). Each side of the clip has either an LED or a detector. This is a type of measurement called transmission pulse oximetry, where a thin part of body section has to be chosen for measurement. An alternative method is reflectance pulse oximetry [34−36] as shown in

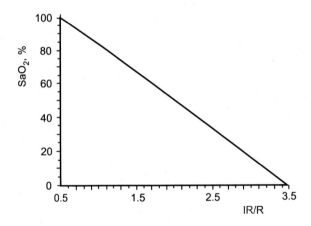

Figure 5.36: Blood Saturation as a Function of IR/R Intensity Ratio.
From Harsanyi G 2000 Sensors in Biomedical Applications, CRC Press, Boca Raton, chapter 6.2.4, p 207.

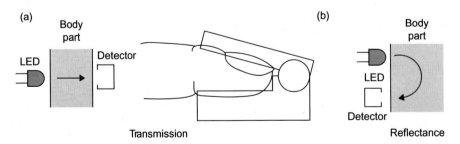

Figure 5.37: Two Types of Pulse Oximetry. (a) Transmission Pulse Oximetry. (b) Reflectance Pulse Oximetry.

Fig. 5.37(b). It is suitable for application to more general body parts including limbs, forehead and chest.

Glucose Sensing

As we described in Chapter 4, the majority of glucose sensors are electrochemical sensors based on the measurement of reducing properties of glucose (see Section 4.2.2). However, noninvasive spectroscopic measurements [37] of transmitted or reflected light may be more advantageous since they can potentially eliminate several problems related to blood drawing.

Mid-infrared (MIR: $3-50$ μm) transmission spectroscopy has been used for in vitro blood glucose measurement by Shen et al [38]. The spectral range of $950-1200$ cm^{-1} ($8.3-10.5$ μm in wavelengths) was used to find stretching modes of $C-C$ and $C-O$ in glucose molecules. A 4-vector partial least-squares calibration model was used for a calibration set consisting of samples from 14 patients. A standard-error-of-prediction of 0.59 mM for an independent test set of 14 samples has been demonstrated. They also suggested use of two specific wavenumbers to determine the glucose concentration with a prediction error of 0.95 mM. The difficulty in MIR spectroscopy is significant background absorption by other molecules in blood including water.

Near-infrared (NIR: $0.78-3$ μm) region, in contrast, is more suitable for in vivo sensing, because $90-95\%$ of NIR light passes through the epidermis. Burmeister et al evaluated six measurement sites of human body for transmission NIR spectroscopy. The wavenumber region used for the blood glucose sensing extends from $6500-5500$ cm^{-1} ($1.54-1.82$ μm), which corresponds to first overtones of $C-H$ and $O-H$ stretching modes [39]. They demonstrated a standard error prediction of 3.4 mM with measurements across the tongue [40]. Malin et al demonstrated the use of NIR diffuse reflectance spectroscopy over the $1050-2450$ nm wavelength range for blood glucose monitoring. Validation with an independent test set showed a mean standard error of 1.03 mM [41].

Techniques based on **Raman Spectroscopy** and **light scattering** will be discussed in Section 5.7.2. Examples of implantable glucose sensors will be introduced in Chapter 7, Section 7.3.6.

5.6 Fluorescence Spectroscopy

5.6.1 Basics of Fluorescence

Fluorescence is emission of a photon by a substance as a result of an electron transition from an excited state to the ground state. Fluorescence is utilized for several types of applications such as lighting, labeling, optical microscopy, chemical sensing, and analysis of material structures.

Energy Levels

Fluorescence is usually studied with polycyclic aromatic molecules. The electronic state of a molecule is defined by the distribution of negative charges and molecular geometry. Each electronic state is further subdivided into vibrational energy levels associated with periodic motion of atoms in the molecule. A **Jablonski diagram** is often used to illustrate the energy levels of a molecule. An example of a Jablonski diagram is shown in Fig. 5.38. The states are arranged vertically by energy.

In Fig. 5.38, electronic states are illustrated as S(0), S(1), S(2), ... The vibrational mode v are denoted as S(i) = v_i. For example, the lowest vibrational energy level of the first excited state is denoted as S(1) = 0.

Transitions between energy states are indicated by arrows. A transition associated with fluorescence excitation or fluorescence emission (radiative transition) is indicated by a straight arrow. A nonradiative transition is indicated by a wiggly arrow.

Horizontally arranged states are grouped by spin multiplicity. There are cases where the electron undergoes an intersystem crossing into a triplet state T(1), which has the higher spin multiplicity. In this case, transition from T(1) to the lower energy state is "forbidden," and only occurs at a slower time scale. Luminescence observed in this process is called **phosphorescence**. Phosphorescence emission is a very slow process (it could take up to several hours) and is often utilized for special painting purposes such as paints for exit signs and clock dials.

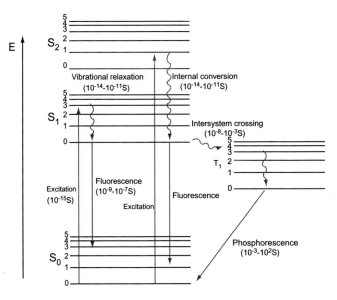

Figure 5.38: Jablonski Diagram.

Important Features of Fluorescence

Several important features are found for absorption and fluorescence spectrum. The energy of the absorbed photon is always larger than the energy of the emitted photon. As a result, the wavelength of emission is always longer than that of excitation (exceptions are cases of multi-photon excitation). The difference in wavelength between the band maxima of absorption and fluorescence spectra is called the **Stokes shift**. The Franck−Condon energy diagram in Fig. 5.39 illustrates the vibrational wave functions of energy levels for the internuclear distance. Since absorption and emission are very fast processes, the internuclear distance does not change, and thus the transitions are represented as vertical lines. The **Franck−Condon principle** states that an electronic transition occurs more likely when the vibrational wave functions of the two levels overlap more significantly.

According to the Franck−Condon principle, the probability of an excited electron transition is related to the degree of similarity between wave functions of the energy states. As a result, some transitions are more probable for both absorption and emission than others. For example in Fig. 5.38, excitation from $S(0) = 0$ to $S(1) = 2$, will be most probably followed by emission from $S(1) = 0$ to $S(0) = 2$. This principle leads to the **mirror image rule**, where fluorophores often have a visible band substructure in the fluorescence spectrum which resembles the mirror image of that of the absorption spectrum (see Fig. 5.40).

The average time the electron stays in an excited state before photon emission is referred to as the **lifetime**. Fluorescence lifetime typically ranges in the order of 10^{-9} to 10^{-7} [sec].

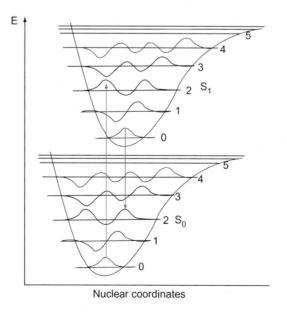

Figure 5.39: The Franck−Condon Energy Diagram.

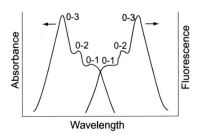

Figure 5.40: The Mirror Image Rule.

The lifetime τ is equal to the time after which the intensity I drops to $1/e$ of its initial value I_0. Fluorescence intensity I typically follows a simple exponential decay:

$$I(t) = I_0 \cdot \exp(-t/\tau) \tag{5.78}$$

Equation 5.78 can be rewritten with the **decay rate** $\Gamma = 1/\tau$ as

$$I(t) = I_0 \cdot \exp(-\Gamma t) \tag{5.79}$$

Both radiative and nonradiative transitions are included in the decaying process as:

$$\Gamma_{\text{total}} = \Gamma_R + \Gamma_{NR} \tag{5.80}$$

where Γ_R and Γ_{NR} are the contributions of radiative and nonradiative processes, respectively. The fluorescence **quantum yield** is a measure which indicates the efficiency of the fluorescence process. Quantum yield Q is defined as:

$$Q = \frac{\Gamma_R}{\Gamma_R + \Gamma_{NR}}$$

Quantum yield Q also means the ratio of the number of photons emitted to the number of photons absorbed.

5.6.2 Fluorescence Spectroscopy and Imaging

Fluorescence microscopy is a technique where samples stained with fluorescent dyes are observed with a **fluorescent microscope**. Figure 5.41 (left) shows the basic composition of a fluorescence microscope. Essential components for fluorescence microscopes are the light source, the excitation filter, the dichroic mirror, and the emission filter. The light source is usually a xenon lamp, mercury lamp, or a tungsten halogen lamp, which has a wide band of emission. The excitation light passes through the excitation filter, is reflected by the dichroic mirror and illuminates the sample. When a laser or a monochromatic light source

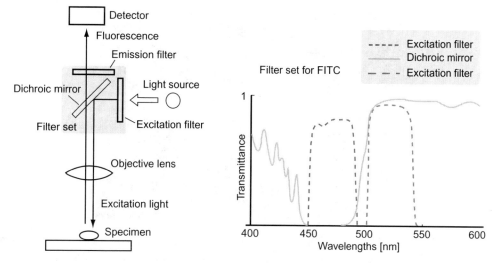

Figure 5.41: Schematic of a Fluorescence Microscope (Left) and the Spectra of a Fluorescence Filter Cube (Right).

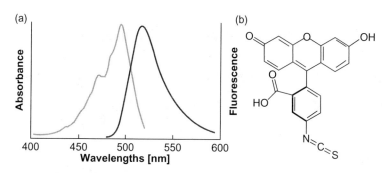

Figure 5.42: (a) Excitation and Emission Spectra of Fluorescein Isothiocyanate (FITC). (b) Structural Formula of FITC.

Plotted based on data from Jackson Immuno Research.

is used for excitation, the excitation filter is not necessary. Fluorescence from the sample passes through the dichroic mirror and the emission filter. A set of an excitation filter, a dichroic mirror, and an emission filter is usually packed in a 2.5 cm (~1 inch) sized box and called a filter cube or a filter set. Sets of filters have to be chosen for different types of fluorescent dyes. Figure 5.41 (right) shows an example of transmission spectra of a filter cube, which consists of a band-pass excitation filter, a long-pass dichroic mirror, and a band-pass emission filter designed for fluorescein isothiocyanate (FITC).

FITC is one of the most commonly used fluorescent dyes. Figure 5.42 shows the excitation and emission spectra of FITC. As we discussed in the previous section, the emission

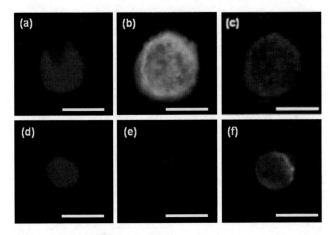

Figure 5.43: Immunofluorescent Imaging of a Cancer Cell and a White Blood Cell. (a) DAPI (Emission ∼460 nm, Blue), (b) CK (Emission ∼520 nm, Green) and (c) CD45 (Emission ∼610 nm, Red) Fluorescence Images of a Cancer Cell. (d) DAPI, (e) CK and (f) CD45 Images of a White Blood Cell. Bars = 10 μm.
From [42].

spectrum of FITC is similar to the mirror image of the absorption spectrum. The structural formula is also shown in Figure 5.42(b). Most fluorescent dyes are polycyclic aromatic hydrocarbons like FITC. The isothiocyanate group ($-N=C=S$) reacts with amino terminal and primary amines in proteins. FITC has been used for the labeling of antibodies.

Immunofluorescence is the most important technique for molecular sensing and imaging. The specificity of an antibody chemically linked to a fluorophore is used to stain specific biomolecule targets. Multiple fluorescent dyes are functionalized with specific antibodies to measure the concentration or visualize the spatial distribution of target molecules. Immunofluorescence is often used with other non-antibody stains such as DAPI, which binds to adenine- and thymine-rich (AT-rich) regions in DNA.

Figure 5.43 is an example of cell identification based on immunofluorescence. Photographs shown in fluorescence images are a cancer cell (COLO205: colon cancer cell line) and a leukocyte (white blood cell) using a color CCD camera. The cells are fluorescently stained with DAPI for DNA staining, FITC functionalized with an antibody against cytokeratin (protein found in epithelial tissue) and AlexaFluor 568 dye functionalized with an antibody against CD45 (protein found on leukocytes). Cubes designed for DAPI, FITC and AlexaFluor 568 are switched for taking each photograph. The cancer cells are (a) DAPI positive, (b) CK positive, and (c) CD45 negative, while the leukocytes are (d) DAPI positive, (e) CK negative, and (f) CD45 positive.

Most organic dyes are subject to photochemical destruction called **photobleaching**. After a certain number of photons are emitted from a dye, the dye loses its fluorescence functionality due to photo bleaching. Alexa 488 and DyLight 488 are commercial products that have been tailored for improved photostability with similar absorption and emission spectra as those of FITC. Most of filter sets designed for FITC can be used for them.

5.6.3 Quantum Dots

A **quantum dot** (QD) is a nanometer scale semiconductor crystal where the electrons are subject to three-dimensional quantum confinement [43]. The confinement becomes significant when the size of the crystal is smaller than the characteristic length called the **exciton Bohr radius**, which is approximately the most probable distance between a hole and an electron (an **exciton** is a hydrogen-atom-like quasiparticle formed when an electron and a hole interact by Coulomb attraction.). QDs can be fabricated in various forms. Some are integrated on bulk semiconductor devices by lithographic approaches [44,45]. Here, we mainly discuss **colloidal QDs**, which are chemically synthesized semiconductor nanoparticles that have found applications similar to those of organic fluorescent dyes. Colloidal QDs are typically between 5 and 50 nm, and have been made of ZnS, CdS, ZnSe, CdTe, and PbSe [46]. The most important feature of colloidal QDs is their size dependent fluorescence emission. The emission wavelength can be tailored simply by choosing the particles of a specific diameter. The emission wavelengths of QDs reported ranged from the UV to the infrared [46].

The mechanism of fluorescence emission is explained with an energy diagram as we described for silicon-based FETs in Chapter 4 and basics of fluorescence in Chapter 5, Section 5.6.1. A QD is a semiconductor, and has a conduction band and a valence band with the band gap as we discussed for silicon in Chapter 4. Due to the quantum confinement, the energy levels of a QD is quantized. The molecular orbitals in the valence band are bonding orbitals, which are full. The orbitals in the conduction band are vacant antibonding orbitals. The highest occupied molecular orbital (the top of valence band) and the lowest unoccupied molecular orbital (bottom of conduction band) are abbreviated as HOMO and LUMO, respectively (Fig. 5.44).

When excited, an electron in HOMO moves to an orbital in the conduction band, leaving a hole in HOMO (Fig. 5.45). Fluorescence emission is a result of a transition of the electron from LUMO back to HOMO, which can be also considered as a hole-electron recombination. The emission wavelength is defined by the energy difference between LUMO and HOMO (Fig. 5.45). When considering the emission energy, we have to take the quantum confinement into account as well as the band gap found in a bulk semiconductor. The emission energies are dependent on the size of QDs, because the quantum confinement effect is stronger for smaller quantum dots. If made with the same material, the emission wavelength is shorter (i.e., the emission energy is larger) for a smaller QD.

Because of this simple structure of the energy levels, the emission wavelength of a QD is defined mainly by the band gap, which can be tuned by the size of the quantum dot. Unlike

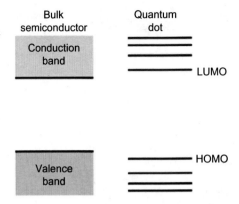

Figure 5.44: Energy Levels of a Bulk Semiconductor and a Quantum Dot.

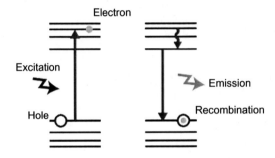

Figure 5.45: Fluorescence Excitation and Emission of a Quantum Dot.

organic fluorescent materials, mirror images of absorption and emission spectra are not observed. Figure 5.46 shows emission spectra of six different QD dispersions. Absorption of the 510-nm-emitting QDs is also shown in the figure.

Most available QDs are made of CdSe cores capped with a one to two monolayers of ZnS. This form is called a core-shell structure. The shell passivates the core surface and protects it from surrounding solutions. Substantial improvement in the PL yield was observed with the core-shell structure, because it prevents crystal defects which lead to unwanted electron or hole traps.

QDs are typically prepared in organic solvent. Surface capping of QDs with hydrophilic ligands allows the solubility in water and further chemical attachment of functional biomolecules.

QDs functionalized with biomolecules are alternatives to organic fluorescent dyes as labels for biological imaging and sensing. Comparison of organic dyes and QDs are summarized

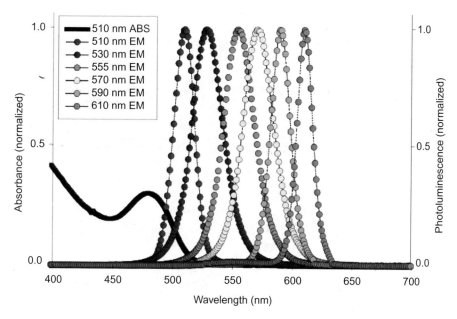

Figure 5.46: Emission Spectra of Different QDs.
From [46].

in [47]. The most significant advantage of QDs may be their photostability. Figure 5.47 shows photostability comparison between QDs and Alexa 488 [48].

The photographs are lapse fluorescence images of a mouse 3T3 fibroblast cell. For the photographs in the upper row, nuclear antigens are labeled with red QDs, and Microtubules are labeled with ALEX 488 dye conjugated to an antibody. For the lower row another cell was stained with opposite labeling. The measured changes in fluorescence intensities are shown in Fig. 5.47(b). The "antifade medium" in the figure is a commercially available medium that delays the process of photobleaching.

The largest drawback of QDs as a fluorescent marker is the size. Typically fluorescent dyes are smaller than a nanometer (see the structural formula in Fig. 5.42), while QD sizes are 5–50 nm. Although extracellular targeting with QDs is frequently reported, cellular delivery of QDs is challenging. Compared to organic dyes, internal labeling strategies with QDs are still behind [47].

5.7 Light Scattering

Light scattering is a general category for several different physical phenomena. It can be considered as the deflection of a ray from a straight path due to irregularities such as small particles or defects. Here, we first introduce types of light scattering that are important in

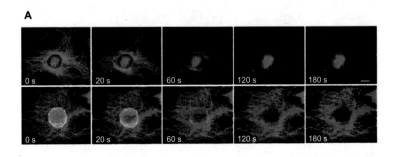

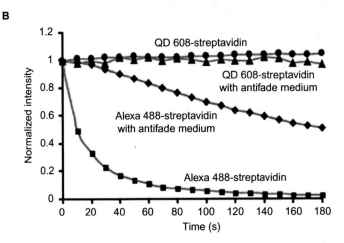

Figure 5.47: Comparison Between Fluorescent Dye (Alexa 488) and Quantum Dots. (A) Mouse 3T3 Fibroblast Cell Stained with QDs and Alexa 488 Dye. QDs Demonstrate Higher Photo Stability. (B) Fluorescence Intensity Plot for QDs and Alexa 488.
From [48].

optical sensing and imaging applications, followed by discussions on the principles of optical measurement methods that are essential in molecular sensing and imaging.

5.7.1 Forms of Light Scattering

Light scattering is divided into two types, namely elastic and inelastic scattering. Rayleigh scattering and Mie scattering are forms of elastic scattering where the energy of the incident light is conserved. Raman scattering is a form of inelastic scattering, where transfer of energy is involved. Raman scattering will be discussed later.

Rayleigh Scattering

Rayleigh scattering is a type of scattering by molecules and particulate matter that is much smaller than the wavelength. The intensity of scattered light has a very strong dependence on the size of the particles and the wavelengths. The wavelength dependence explains the reason why the sky appears blue: A light of shorter wavelength (violet, blue, and green) will scatter more than that of longer wavelengths (yellow and red). This scattering makes the sky look blue during daytime regardless of the directions and the position of the sun. When the sun becomes closer to the horizon, the thickness of the air that the sunlight passes through becomes larger, and lights of shorter wavelengths are scattered away before reaching the surface of the earth. The sky appears yellow—red. Rayleigh scattering is a main cause of signal loss in optical fibers.

Mie Scattering

Mie scattering is the scattering described as the Mie solution [99]. It is the solution to Maxwell's equations for the scattering by an isotropic, homogeneous, dielectric sphere. Although the Mie solution applies to a particle with any diameter, it is usually used for particles whose size is similar to the wavelength. The Mie solution allows theoretical calculation of the electric and magnetic fields inside and outside a spherical particle. It is useful to discuss scattering and resonances in particles with different diameters and materials. Examples of applications of Mie solution to find resonances in metal nanoparticles will be described in Section 5.7.2.

5.7.2 Methods Based on Light Scattering Measurements

Dark Field Microscopy

A dark field microscope is a type of microscope that enables microscopic observation of scattered lights from the sample. It is constructed with a simple design of optics and is very useful to study scattering characteristics of nanomaterials.

Figure 5.48 illustrates a schematic of a dark field microscope. The condenser lens focuses the light onto the sample from the other side of the objective lens. An opaque disc is placed before the condenser lens to blocks the central part of the light from the source, leaving illumination only from an outer ring. The diameter of the disc is designed so the ring illumination light does not goes directly into the objective lens, and only scattered lights are collected and observed. In order for this to happen, the NA of the condenser lens has to be larger than that of the objective. In the dark-field microscopy, the specimen appears to be bright with dark background in that object scatters the illuminated light. Table 5.1 shows the comparison between bright-field, dark-field, and fluorescence microscopies.

Static Light Scattering

Static light scattering is a method that measures the intensity of the scattered light with different measuring angles. It is called "static" because time-averaged intensity is measured in contrast to dynamic light scattering, where fluctuations in light intensity are measured.

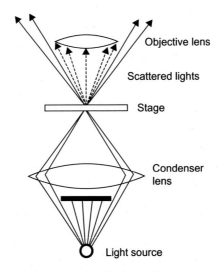

Figure 5.48: Dark Field Microscope.

Table 5.1: Comparisons of Bright-Field, Dark-Field, and Fluorescence Microscopies

	Bright-Field	**Dark-Field**	**Fluorescence**
Key components	Two lenses and a condenser	Two lenses, a condenser, and opaque disk	Two lenses, filters, and dichroic mirror
Image	Bright background Dark specimen	Dark background Bright specimen	Dark background Fluorescent specimen
Specimen	Absorbs illumination light	Scatters illumination light	Absorbs light of shorter wavelengths

Static light scattering is used to determine the average molecular weight of particles such as polymers or proteins.

Scattering is dependent on polarizability of a solution, which is related with the size and the shape of the solute molecules. There are several methods [49,50] that correlate angular intensity measurements and the characteristics of molecules. Measurement at multiple observation angles allows one to calculate characteristics related to the molecular weight of the solute. Figure 5.49 shows a typical experimental setup for static light measurement. A laser beam is focused onto a sample prepared in a cylindrical cell. The scattered light can be observed from different angles.

Dynamic Light Scattering

Dynamic light scattering is based on the measurement of time-domain fluctuation of scattered light intensity [51]. The technique is used to find diameters of suspended particles.

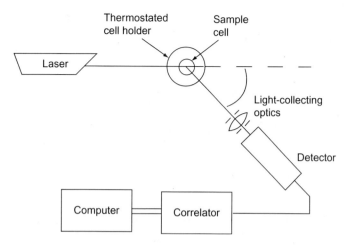

Figure 5.49: Static Light Scattering.
Redrawn from Turner et al [100] Biosensors.

It can be used for particles sized about nanometer to micrometer with concentrations of 10^4 to 10^{10} particles/mL. Many instruments allow both dynamic light scattering and static light scattering with the same experimental setup.

The principle is that the time scale of particle random motion is dependent on the diameter of particles. More specifically, the diffusion constant D is used to discuss the particle motion. In order to find characteristic time scales from the measured signal, the autocorrelation function of the recorded intensity fluctuation is used.

The diffusion constant of spherical particles with the radius r in liquid is given as:

$$D = \frac{k_B T}{6\pi\eta r},$$

where k_B is the Boltzmann constant, T is the absolute temperature, η is viscosity.

The autocorrelation function of light scattering in a monodisperse suspension (i.e. suspension of particles with a uniform diameter) is found to be:

$$g(t) = \exp\frac{-t}{\tau_q} \tag{5.81}$$

where τ_q is the time delay given with the diffusion constant D, viewing angle θ and the wavelength λ in the following way:

$$\tau_q = \frac{1}{DK^2} \qquad \left(K = \frac{4\pi}{\lambda}\sin\frac{\theta}{2}\right) \tag{5.82}$$

For polydisperse solution (i.e. suspension of particles with different diameters), the autocorrelation function is given as integral of exponential functions along time delays for all the particle diameters.

$$g(t) = \int_0^\infty G(\tau) \exp\frac{-t}{\tau} \cdot d\tau \tag{5.83}$$

where $G(\tau)$ is the normalized distribution of time delays. Figure 5.50(a) shows examples of time delay distributions for particle suspensions with narrower and broad diameter distributions. Figure 5.50(b) shows a case with two peaks in the diameter distribution.

Laser Doppler Velocimetry

Laser Doppler velocimetry (LDV) is a technique to measure the velocity of a flow based on the measurement of light scattering caused by particles in the flow [53,54].

According to the Doppler effect, the frequency of reflected radiation from a flowing particle is shifted from that of the incident light. The change in the frequency is a function of the velocity of the particle. Figure 5.51 illustrates the principle of LDV. The two coherent laser beams beam1 and beam2, with the directions given as unit vectors $\mathbf{k}_1$ and $\mathbf{k}_2$, respectively, intersect each other in the flow. Particles with the velocity $\mathbf{V_P}$ is flowing in the intersection and the scattered lights are monitored. The direction from the particle to the observer is given with a unit vector $\mathbf{k}_s$. Frequencies of the scattered light from beam1 and beam2 are

$$f_1 = \frac{c - \mathbf{V_P} \cdot \mathbf{k}_1}{c - \mathbf{V_P} \cdot \mathbf{k}_1} f_0 \tag{5.84}$$

and

$$f_2 = \frac{c - \mathbf{V_P} \cdot \mathbf{k}_2}{c - \mathbf{V_P} \cdot \mathbf{k}_2} f_0 \tag{5.85}$$

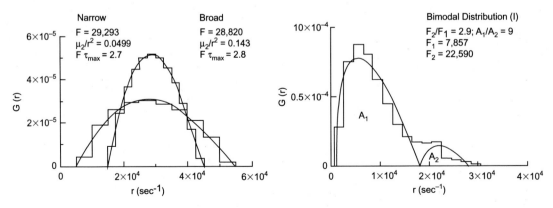

Figure 5.50: Dynamic Light Scattering.
From [52].

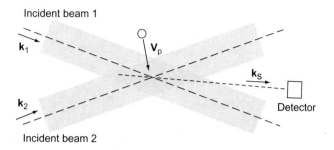

Figure 5.51: Laser Doppler Velocimetry.

respectively, where is the unit vector for, and $\mathbf{V_P}$ is the velocity of the particle. The frequency difference, or Doppler shift, between f_1 and f_2 can be found by measuring heterodyned scattered signals. The basic idea is that the summation of two slightly different frequencies creates a new frequency as:

$$\sin(\theta + \Delta\theta) + \sin(\theta - \Delta\theta) = 2\sin\theta\cos\Delta\theta \qquad (5.86)$$

LDV is a very important and useful technique because it is a noncontact, noninvasive method that can be used for a wide variety of flows. For molecular sensing, measurement of blood flow [55–57] is one of the important application areas studied earlier [55]. Other measurements include electrophoretic light-scattering studies of bacterial cells [58], metal, and semiconductor nanoparticles [59,60].

Raman Spectroscopy

Raman Scattering

In contrast to elastic scattering, where the energy is conserved before and after the process, **Raman scattering** is a process where the incident light interacts with molecules and some of the energy is lost or increased. **Raman spectroscopy** is a technique based on measurement of Raman scattering, which provides information about vibrational and rotational modes in molecules.

In Raman scattering, an incident photon allows a molecule to transit from one vibrational or rotational state to another. The energy difference between the two states results in a shift in the frequency of emitted photon.

When the new energy state is higher than the initial state, the scattering is called **Stokes Raman scattering**. The frequency of the emitted photon will be shifted to lower. The shift is called a **Stokes shift**. When the new state is lower than the initial state, it is called anti-Stokes Raman scattering. The shift is called anti-Stokes shift.

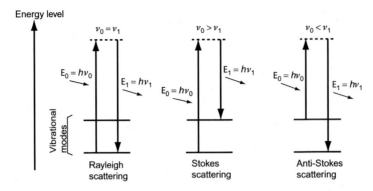

Figure 5.52: Stokes Scattering and Anti-Stokes Scattering.

Spectroscopic techniques are crucial in separating the weak Stokes and anti-Stokes scattering from the intense Rayleigh scattering of the excitation laser light. Figure 5.52 shows energy diagrams for types of scattering found in Raman spectroscopy.

Although Raman scattering has some similarities with fluorescence excitation, they are different phenomena.

In the process of fluorescence excitation, the incident light is completely absorbed by the molecule, which is transferred to an excited state. Fluorescence emission occurs after a certain time characterized as fluorescence lifetime ($\sim 10^{-9} - 10^{-7}$ s). Raman scattering, on the other hand, is a spontaneous effect.

The wavelengths of fluorescence emission are defined by the energy levels of the molecule. Fluorescence emission is observed only when it is excited by a photon within a certain wavelength range. Raman scattering can take place by any frequency of excitation. Peaks of scattering maintain a constant separation from the excitation frequency. Unlike the case with fluorescence spectroscopy, the absolute frequency or wavelength of a scattered light does not have a significant meaning. In Raman spectroscopy, the energy shift from the excitation is discussed as the Raman shift

$$\Delta w = \frac{1}{\lambda_0} - \frac{1}{\lambda_1}, \qquad (5.87)$$

where λ_0 and λ_1 are the wavelengths of excitation and scattering, respectively. Raman shift is usually measured in wavenumbers [cm^{-1}] (kayser). The transferred energy ΔE is found from the following relationship:

$$\Delta E = h\nu_0 - h\nu_1 = hc\left(\frac{1}{\lambda_0} - \frac{1}{\lambda_1}\right), \qquad (5.88)$$

where h is Planck constant, c is speed of light, ν_0 and ν_1 are the frequencies of excitation and scattering lights.

Raman Spectroscopy

Raman spectroscopy is a powerful tool in chemistry and solid-state physics. It is used to find vibrational information specific to chemical bonds and structure of molecules. Knight et al reported characterization of diamond films based on Raman spectroscopy in 1998 [25]. Although diamond films are all composed of carbon atoms, the characteristics of C-C bonds are different in different materials. Each band in the Raman spectrum corresponds to a specific vibrational mode within the molecule. The band at 1580 cm^{-1} found in graphitic carbon (Fig. 5.53a) is known as the G-band. The band corresponds to vibration of sp^2 carbon atoms, and is not significant in amorphous carbon (Fig. 5.53b). The band at 1357 cm^{-1} in amorphous carbon is known as the D-band. The D band corresponds to vibrational mode associated with graphene edges and found in disordered polycrystalline and noncrystalline graphitic carbons. The D-band is often called the disorder band. The D/G intensity ratio is often used as a measure of the quality of carbon materials.

Raman spectroscopy also plays a critical role in characterizing newly discovered carbon nanomaterials such as graphene or carbon nanotubes.

The important feature of monolayer graphene is the so-called G′ band at about 2700 cm^{-1}. The G′ band is also found in graphite (see Fig. 5.53a) but not as significant as in graphene. Figure 5.54 is an example of the Raman spectrum of a graphene edge, which clearly shows the D, G, and G′ bands

Carbon nanotubes are formed by graphene sheets rolled at specific angles (chiral angles, see Chapter 4, Section 4.4.1). In Fig. 5.55, we can observe the D, G, and G′ bands as expected.

The important modes specifically found in nanotubes are radial breathing mode or RBM bands, which are observed in $100-300 \text{ cm}^{-1}$. The bands correspond to the radial expansion and contraction. The RBM bands are characteristic resonant peaks of single-walled carbon nanotubes (SWCNTs). They are not observed in MWNT, because the multi-walls restrict radial mode vibrations. The frequency of RBM bands are correlated to the diameter of nanotubes.

Surface Enhanced Raman Spectroscopy

Surface enhanced Raman spectroscopy (SERS) utilizes enhanced Raman scattering of molecules adsorbed on surfaces of metals such as silver or gold. One theory that explains the enhancement is based on localized surface plasmons on the metal surface [64]. Experimentally measured enhancement factors can be as much as 10^{14} to 10^{15} [65]. Many utilize silver or gold nanoparticles prepared in a solution [66] or immobilized on a substrate [65,67]. Others utilize metal thin films deposited on substrates [68,69]. Metal surfaces are often roughened to have similar enhancement effect as with nanoparticles. Local enhancement with a metal tip of scanning probe microscopy has been also reported [70].

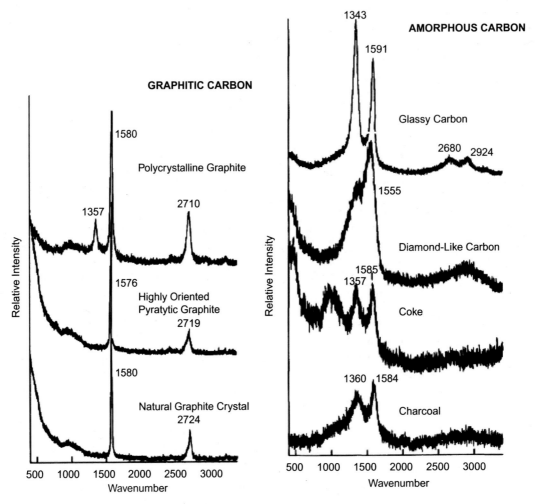

Figure 5.53: Raman Spectra of Carbon Materials.
From [61].

Metal nanoparticles can be functionalized to detect specific target molecules. Figure 5.56(a) shows an example of highly sensitive detection of prostate specific antigen (PSA), which is commonly used as a prostate cancer marker [68]. Gold nanoparticles with a diameter of 30 nm are functionalized with PSA-specific antibodies and molecular labels which show intense Raman scattering (Raman scatterer). The gold nanoparticles enhance the signal from the Raman scatterer. The detection scheme was based on sandwich enzyme-linked immuno-sorbent assay (ELISA), where antigens are first captured by primary antibodies immobilized on a gold coated glass substrate, and then labeled by Raman scattering gold nanoparticles functionalized with secondary antibodies. Figure 5.56(b) shows Raman spectra for different PSA concentrations and a dose–response curve. Detection limit of 1 pg/mL in human

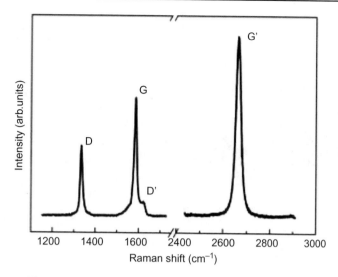

Figure 5.54: Raman Spectrum of a Graphene Edge.
From [62].

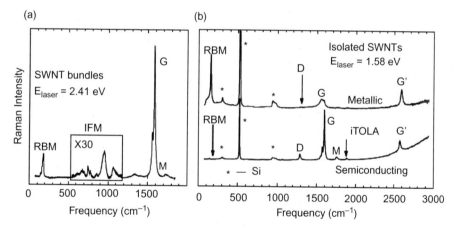

Figure 5.55: Raman Spectra of SWNT Bundles. (a) Raman Spectra from HiPco SWNT Bundles. (b) Raman Spectra from a Metallic (Top) and a Semiconducting (Bottom) SWNT at the Single Nanotube Level. The Spectra Show the RBM D-band, G-band and G'-band Features.
From [63].

serum has been achieved (PSA levels of more than 4 ng/mL may be considered suspicious in clinical cancer screening).

In [71], surface enhanced Raman spectroscopy was used for in-vivo targeting and detection. Gold nanoparticles (60 nm diameter) are functionalized with the epidermal growth factor receptor (EGFR) and Raman scattering molecules and injected to a mouse bearing a tumor. Raman spectra from the tumor and the lever locations were measured 5 hours after

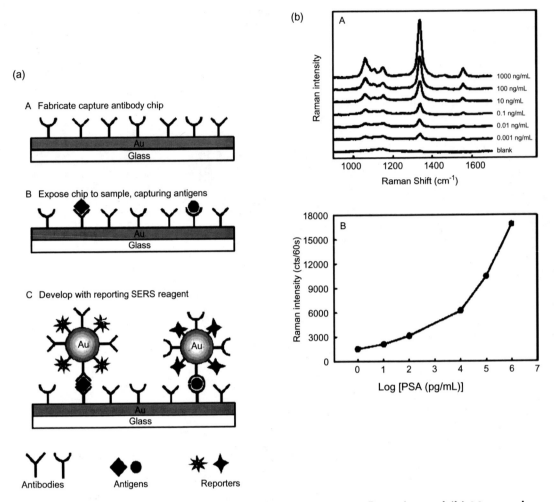

Figure 5.56: (a) Surface Enhanced Raman Spectroscopy Configuration and (b) Measured Signals for Various PSA Concentrations.
From [68].

injection. Figure 5.57(a) and (b) were measured with targeted and untargeted nanoparticles, respectively. Raman scattering of the gold nanoparticles are clearly observed from the lever site in the targeted measurement.

Raman Spectroscopy for Glucose sensing

SERS has been utilized for in vivo and in vitro glucose sensing (Fig. 5.58). Duyne et al. used the metal film over nanospheres (FON) by depositing a thin Ag flim onto self assembled close-packed 2D array of nanospheres. This process creates periodic arrays of nanoscale metal nanoparitcles. The AgFON substrates were then modified with a mixed self assembled monolayer (SAM) consisting of decanethiol (DT) and mercaptohexanol (MH)

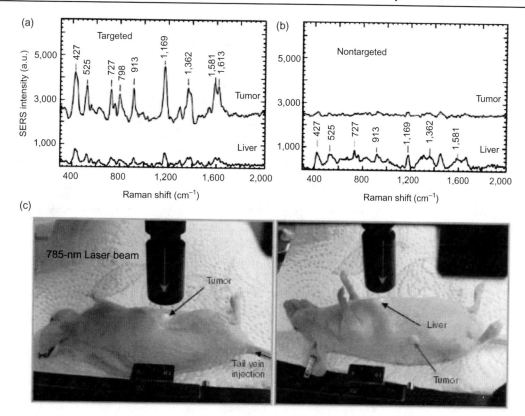

Figure 5.57: In vivo Raman Spectroscopy for Tumor Detection.
(a, b) SERS spectra measured at the tumor and liver locations with targeted (a) and nontargeted (b) nanoparticles. (c) Photographs of a laser beam focusing on the tumor site or on the anatomical location of liver. *From [71].*

and used for the SERS measurements of glucose sensing [64]. The substrate is implanted in a rat and in vivo measurements were also performed [72].

Non invasive glucose measuments were also reported by another group [73]. During the test, 461 Raman spectra of human skin from 17 subjects were measured along with glucose reference measurements provided by standard capillary blood analysis. Predicted glucose concentrations versus the reference values showed correlations with the mean absolute errors of $7.8 \pm 1.8\%$ (mean $\pm$ std dev) with mean R^2 values of 0.83 ± 0.10.

5.7.3 Metal Nanoparticles

Gold nanoparticles have long been used to color stained glass [74]. Different colors of stained glass are due to size- and shape-dependent absorption spectra defined by localized SPR.

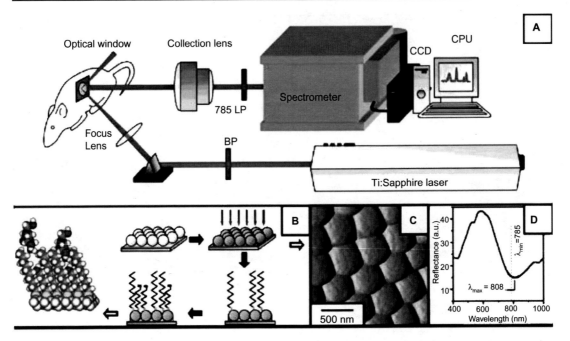

Figure 5.58: In vivo Raman Spectroscopy for Glucose Sensing.
From [72].

Localized SPR in noble metal nanoparticles has been intensively studied as the key phenomenon in highly sensitive and spatially resolved optical sensing of molecules. It has similarities to propagating surface plasmon as we discussed in Section 5.4. It is electronic excitation at metal–air interface that is excited by light. One of unique characteristics found in localized surface plasmon is strong size and shape dependence [75–77]. Incident light generates both propagating plasmon waves and localized standing resonance, which are highly dependent on the excitation wavelength, the size and shape of the particle, and dielectric constant of the surrounding medium. Figure 5.59 illustrates optical excitation of metal nanoparticles. The resonance characteristics are determined by scattering spectra. Figure 5.60 shows dark field images and spectra of nanoparticles (a) before, (b) during, and (c) after immersion in an oil with a refractive index of 1.44 [78].

Nanospheres

The Mie solution provides accurate evaluation of spheres with an arbitrary diameter [79]. Figure 5.61 shows Mie extinction properties of water suspended gold nanoparticles with different diameters.

Figure 5.61 shows extinction spectra of 30- and 100-nm spheres along with contributions from absorption and scattering. Most energy is absorbed for the smaller particle, while scattering is more important for the larger particles.

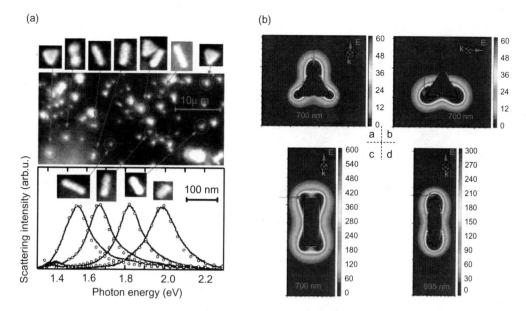

Figure 5.59: Localized Surface Plasmon Resonance (a) Shape Dependent Scattering Spectra [76] and (b) Electric Near-field Profile Calculated Using Discrete Dipole Approximation Formalism [77].

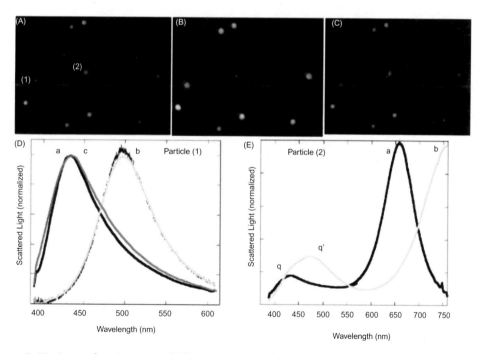

Figure 5.60: Scattering Spectra of Silver Nanoparticles (A) Before, (B) During, and (C) After Immersion of Index Oil (Each Showing Different Colors). (D) Spectra Measured for Blue Particles (a) Before, (b) During, and (c) After Immersion Oil. (E) Spectra Measured for Red Particles (a) Before, (b) During, and (c) After Immersion in Oil [78].

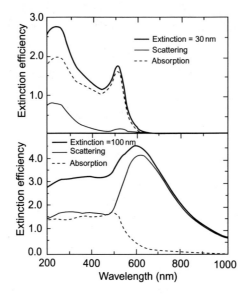

Figure 5.61: Mie Scattering Calculated for Gold Nanoparticles with Different Diameters.
From [79].

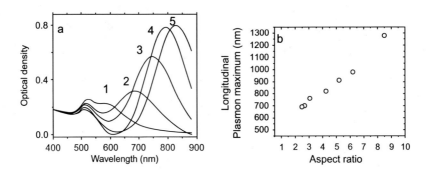

Figure 5.62: Optical Densities of Nanorods with Different Aspect Ratios.
From [80].

Nanorods

Generally, it is not possible to find analytical solutions for non-spherical particles with arbitrary shapes. Rod or cylindrical shaped nanoparticles, usually called **nanorods** (Fig. 5.62), are relatively simple, and several approaches have been made to find analytical solutions for scattering and absorption in metal nanorods [80].

Gold nanorods were synthesized using a seed-mediated reaction [81,82]. The aspect ratio of the gold nanorods can be tuned by controlling the amount of $AgNO_3$ added to the reaction mixture. Gold nanorods typically have two characteristic absorption peaks correlated to the size and shape [83]. The longer peak is mainly defined by the aspect ratio of the nanorods.

Other Shapes

Recent progress has allowed synthesis of various shapes of metal nanoparticles.

The finite-difference time-domain (FDTD) method is a very commonly utilized numerical analysis method to find solutions to general cases of nanoparticles with different shapes. It finds the solution to Maxwell's equations as a function of time by using discretized partial derivatives of time and space. Numerical analysis is very useful because it can also analyze interaction between multiple particles, which is crucial for study of nanoscale optical resonators or antennas.

Sensing and Imaging Applications

Because of the characteristic scattering spectra, metal nanoparticles have a potential as an optical marker for biomedical imaging and sensing. They have been used for labeling of cancer cells [84]. SPR scattering images and absorption spectra were measured from cultured cells labeled with gold nanoparticles conjugated with anti-EGFR antibodies. Figure 5.63 shows nonmalignant epithelial cells (HaCaT) and malignant oral epithelial cells (HOC 313 and HSC 3) incubated with functionalized gold nanoparticles. The nanoparticles used have an average size of 35 nm.

The antibody-conjugated nanoparticles specifically and homogeneously bind to the cancer cells with 600% greater affinity than to the noncancerous cells. Three images of each kind of cells are shown for reproducibility.

Nath et al reported a method to measure biomolecular interactions in real time at the surface of a self-assembled monolayer of colloidal gold prepared on glass [85]. They used the change in the absorbance spectrum of the gold particles as a function of molecular binding to the surface of the immobilized colloids. The spectrophotometric sensor demonstrated a detection limit of 16 nM for streptavidin. They described this method as a type of calorimetric sensing (see Chapter 6, Section 6.4.2 on calorimetry).

One problem of metal nanoparticles as an optical marker is that the individual optical characteristics of particles can be easily affected by neighboring particles. Aggregations of nanoparticles show totally different scattering spectra compared to that of individual particles. Nanoparticles should be separated from each other for labeling purposes. Coating particles with a layer of dielectric material is a solution to this problem. A method introduced by Stober [86] was used to grow a silica shell around gold nanorods [87].

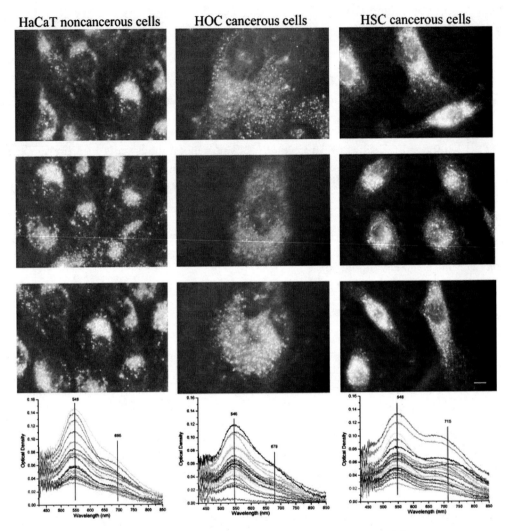

Figure 5.63: Cancer Cell Labeling with Gold Nanoparticles.
Light scattering images and microabsorption spectra of HaCaT noncancerous cells
(left column), HOC cancerous cells (middle column), and HSC cancerous cells (right column)
after incubation with anti-EGFR antibody-conjugated gold nanoparticles. Scale bar 10 μm for all
images. *From [84].*

5.8 Near-Field Scanning Optical Microscopy

The resolution of a conventional microscope is limited by Fraunhofer diffraction of light
(see Appendix 5.A). **Near-field scanning optical microscopy** (NSOM), also known as
scanning near-field optical microscopy (SNOM), is a type of **scanning probe microscopy**
(SPM) that overcomes the resolution limit of conventional microscopes by utilizing a

sub-wavelength sized light source [88–90]. The light source created at the scanning tip measures maps of sub-wavelength optical properties as well as nanoscale topographic profiles. The optical resolution of NSOM is directly related to the size of the light source at the tip, while the capability of topographic measurement relies on the same force sensing scheme as used in AFM. Principles and implementation of force sensing with a silicon based scanning probe is described in Chapter 6, Section 6.2. A quartz tuning fork based sensing described in Chapter 6, Section 6.3.7 is also commonly used for NSOM.

Figure 5.64(a) shows a typical implementation of NSOM. As the light source at the probe tip scans the sample surface, the nanoscale optical interaction between the light source and the sample is recorded by a photo detector to form optical images. In many practical applications the probe tip is fixed in the focal point of the detector optics and the sample stage moves to allow the light source to scan. The photo detector is typically located under the sample with the inverted microscope configuration. Side-viewing microscopes are also commonly used to detect scattered light.

Figure 5.64(b) summarizes types of NSOM tips. The most common type shown in (1) utilizes a nanometer-sized aperture, which defines the size of the light source. Usually a metal coated tapered optical fiber is used in this scheme [89]. Type (2) utilizes plasmonic enhancement that occurs at the metal tip. Localized surface plasmon resonance [91] (see Section 5.4) arises when excited by an external illumination. An example of this scheme is the tip enhanced Raman spectroscopy introduced in Section 5.8.1. Some of type (2) probes

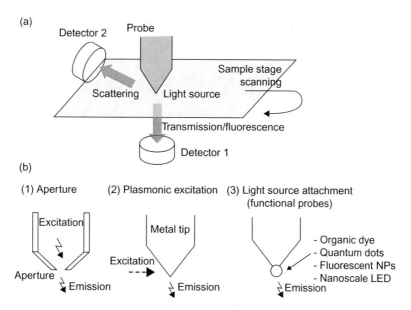

Figure 5.64: Overview of the Near-field Scanning Nanophotonic Microscopy. (a) Schematic of a Standard Near-field Scanning Optical Microscopy (NSOM). (b) Types of NSOM Tips.

are integrated with a nano resonator (i.e., a nanoscale metal pattern [23]), which improves the plasmonic enhancement at the tip. Probes of type (3) are attached with a small-sized light emitting element at the tip. Such fluorescent materials include organic dyes [92], QDs [93], and nanodiamonds [94]. Probe tips attached with a metal nanoparticle [21,95] may be categorized as type (2), since it relies on the localized surface plasmon resonance (LSPR) on the particle.

5.8.1 Aperture-Based NSOM Tip

Figure 5.65 illustrates mode propagation in a type (1) tapered metal-coated fiber tip. Tapered structures are made by pulling heated optical fibers or chemical etching. Metal is deposited by thermal evaporation. In an aluminum-coated tapered waveguide ($\varepsilon_m = -34.5 + j8.5$ and $\varepsilon_d = 2.16$) with a propagating wavelength of 488 nm, only HE_{11} mode is allowed for inner diameters between ~ 250 and ~ 160 nm. Beyond ~ 160 nm, the HE_{11} mode goes into cutoff, and the mode field decays exponentially. Lower energy efficiency is one of the drawbacks of tapered fiber based system.

5.8.2 Metal Plasmonic Tip

Figure 5.66 shows an example of a type (2) NSOM probe with a metal tip. In this example, the tip is integrated with a grating resonator that enhances the intensity of surface plasmons at the metal tip.

One advantage of this method is that a mechanically very sharp tip compared with an aperture probe is possible. The resolution is approximately the size of the tip apex (20–30 nm). The disadvantage is the strong background illumination which makes fluorescent measurement difficult. Metal tip based probes are usually used for scattered light based imaging. Figure 5.67 shows an example of tip-enhanced Raman spectroscopy (TERS), where a silver coated tip employed in a NSOM setup is used to enhance the

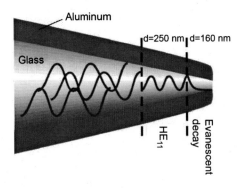

Figure 5.65: Mode Propagation in a Fiber-Based Aperture Tip [89].

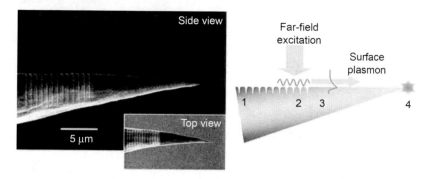

Figure 5.66: Metal Tip Probe With an Integrated Plasmonic Resonator.
From [23].

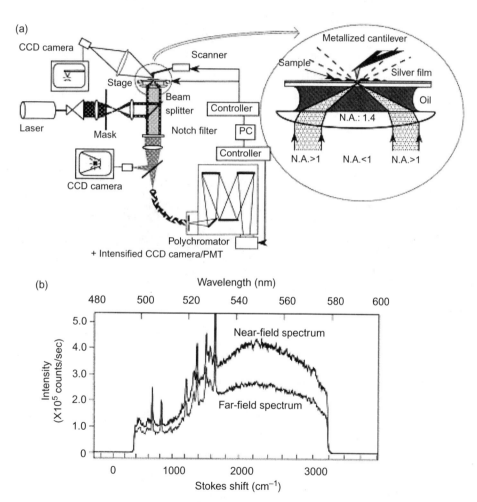

Figure 5.67: Tip Enhanced Raman Microscopy (a) Experimental Setup and (b) Amplified Near-field Signal Obtained with A Metallic Tip.
From [70].

Raman signals [70]. Figure 5.67(a) shows the experimental setup. Total internal reflection is used to create an evanescent field that locally excites organic dye, Rhodamine 6G molecules (RH-6G) on the surface. Figure 5.67(b) shows the amplification of the near-field signal due to the field enhancement of the metallic tip. This amplification effect is not observed with a bare silicon tip which was not coated with silver.

5.8.3 Functional NSOM Tip

A novel approach for type (3) probe we introduce here is integration of nano-scale QD-based light emitting diodes [96] (QDLEDs). One of the advantages of using QDLEDs is that the emission wavelengths can be easily tailored by choosing QDs with proper diameters. Figure 5.68 shows multicolor emission from the QDLEDs integrated at the tip. The emission spectra of the LEDs are shown in Fig. 5.68. The narrow bandwidth of QDLEDs provides considerable potential as excitation sources for fluorescence imaging. NSOM fluorescence imaging with the QDLED probe is performed to demonstrate a 50 nm-order resolution.

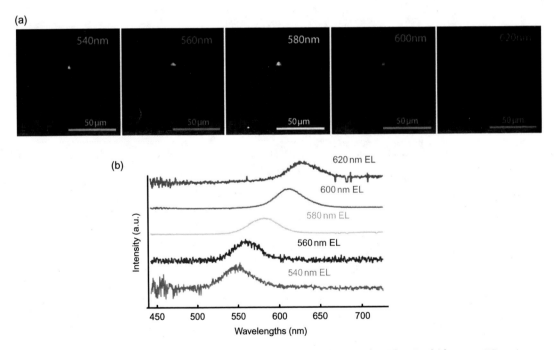

Figure 5.68: (a) Electroluminescence (EL) from QDLED Probe Tip. Scale bars = 50 μm. (b) Electroluminescence Spectra of QDLEDs at Probe Tips.
From [96].

Problems

P5.1 Reflection of light

1. Find the critical angle for indexes $n_1 = 1.43$, $n_2 = 1.00$ (Fig. P5.1).
2. Find the Brewster's angle for the for indexes $n_1 = 1.00$, $n_2 = 1.52$.

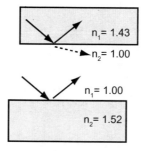

Figure P5.1

P5.2 Reflection and Refraction

A laser beam is reflected on an air–liquid interface. The angle between the incident and reflected beams is 140°. The refracted light is transmitted at 45°. The refractive index of air is $n_1 = 1.0$. What is the refractive index of the liquid?

P5.3 Multi-Stack Thin Film Filters

Calculate the output angle α for a light beam with an incident angle of α_0 after passing through the 5 layers of the structure in Fig. P5.3.

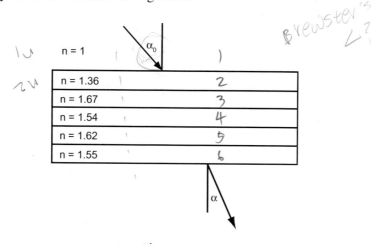

Figure P5.3

P5.4 Evanescent Field

The evanescent field induced by total internal reflection is expressed with Eq. 5.56.

$$\mathbf{E}_t = \mathbf{E}_{t0} \cdot \underbrace{e^{j\left(k_t x \frac{n_0}{n_1} \sin\alpha\right)}}_{\text{traveling wave}} \cdot \underbrace{e^{-k_t z \sqrt{\frac{n_0^2}{n_1^2} \sin^2\alpha - 1}}}_{\text{exponential decay}}$$

Prove that the energy of the incident wave is not transferred in z direction. When the field is a time-periodic sinusoidal electromagnetic field, we can use time-averaged Poynting vector defined in the following way:

$$\mathbf{S} = \frac{1}{2}\text{Re}(\mathbf{E} \times \mathbf{H}^*)$$

The Poyinting vector indicates the energy flux density in a certain direction in average time.

1. Find $\mathbf{H}$ from Faraday's Law, assuming there is no magnetization component and μ_0.
2. Calculate the Poynting vector $\mathbf{S}$ and explain the result.

P5.5 Polarization

1. Consider two vectors

$$\mathbf{E}_1 = \begin{bmatrix} \sqrt{3} \\ \sqrt{3} \end{bmatrix} \cdot e^{j(\omega \cdot t + 60°)}, \quad \mathbf{E}_2 = \begin{bmatrix} -1 \\ \sqrt{3} \end{bmatrix} \cdot e^{j(\omega \cdot t - 30°)}$$

We define a new vector

$$\mathbf{E}_3 = \mathbf{E}_1 + \mathbf{E}_2$$

Simplify $\mathbf{E}_3$ in the form of

$$\mathbf{E}_3 = \begin{bmatrix} E_x \\ E_y \end{bmatrix} = \begin{bmatrix} E_{0x} \cdot e^{j(\omega \cdot t + \alpha)} \\ E_{0y} \cdot e^{j(\omega \cdot t + \beta)} \end{bmatrix}$$

2. Illustrate a pattern drawn by the real part of $\mathbf{E}_3$, namely $\begin{bmatrix} \text{Re}(E_x) \\ \text{Re}(E_y) \end{bmatrix}$.
3. Repeat (1) and (2) for

$$\mathbf{E}_1 = \begin{bmatrix} \sqrt{3} \\ 3 \end{bmatrix} \cdot e^{j(\omega \cdot t + 60°)}, \mathbf{E}_2 = \begin{bmatrix} 1 \\ \sqrt{3} \end{bmatrix} \cdot e^{j(\omega \cdot t - 30°)}$$

P5.6 Polarization

Let us consider a plane wave propagating along z axis as shown in the following equation and Fig. P5.6.

$$\mathbf{E}_1 = \begin{bmatrix} E_x \\ E_y \end{bmatrix} = \begin{bmatrix} 2e^{j(\omega \cdot t + 30° - z)} \\ 2e^{j(\omega \cdot t - 60° - z)} \end{bmatrix}$$

1. Describe the pattern drawn by the real part of $\mathbf{E}_1$ at $z = 0$.
2. At $z = -1$, we put filter 1 that transmits 100% of x component and 5% of y component. Describe the pattern drawn by the real part of $\mathbf{E}_2$ at $z = 0$. This type of filter is called a polarizer.
3. We remove filter 1 and put filter 2 that delays the phase of x component by 90°. The y component is not affected by the plate. Describe the pattern drawn by the real part of $\mathbf{E}_3$ at $z = 0$. This type of filter is called a quarter wave plate.
4. Filter 1 and filter 2 can be expressed by simple 2×2 matrixes in the way shown below. Find such matrixes.

$$\mathbf{E}_2 = \begin{bmatrix} a_1 & b_1 \\ c_1 & d_1 \end{bmatrix} \mathbf{E}_1, \quad \mathbf{E}_3 = \begin{bmatrix} a_2 & b_2 \\ c_2 & d_2 \end{bmatrix} \mathbf{E}_1$$

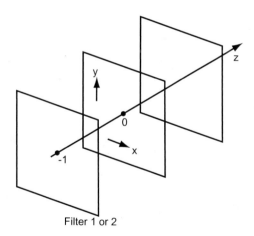

Filter 1 or 2

Figure P5.6

P5.7 Interferometry

Figure P5.7 shows the concept of interferometry-based distance measurement. A plane light wave $E = E_0\, e^{j(\omega t - kz)}$ propagates in the z direction. The wavenumber k is given as $k = 2\pi/\lambda$,

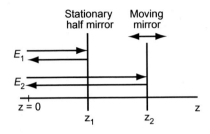

Stationary   Moving
half mirror   mirror

E_1

E_2

$z = 0$

z_1

z_2

z

Figure P5.7

where λ is the wavelength. The half of the propagating light, E_1, is reflected by a stationary half mirror at $z = z_1$. The other half E_2 is reflected by a moving mirror $z = z_2$. Show that the field intensity of the reflected light measured at $z = 0$ can be used to measure the position of the moving mirror.

P5.8 Interference

When a thin film of silicon dioxide is grown on a silicon wafer, the wafer shows different colors depending on the thickness of the oxide. Calculate the colors generated by oxide films with thickness 220 nm, 440 nm and 500 nm. Use the refractive index of thermal oxide $n = 1.46$ (Fig. P5.8).

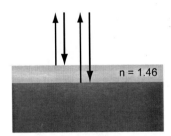

$n = 1.46$

Figure P5.8

P5.9 Propagation Modes

Let us consider a waveguide consisting of two mirrors and air (Fig. P5.9). A light with a wavelength of λ is propagating with an angle φ to the mirrors. The wave is polarized in the x direction. In a waveguide mode, a standing wave arises in the y direction. For this to happen, the field just before reflection at point A and the field just before reflection at C must be in phase. Find the condition for waveguide modes.

(a)

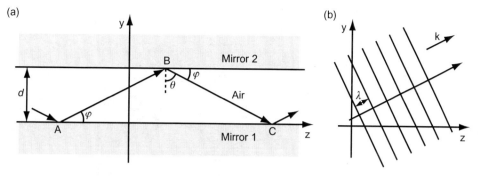

Figure P5.9

P5.10 Photometric Waveguide Transducers

1. Describe the basic elements of a planar waveguide. Explain how light can be transmitted in the waveguide.
2. Describe three ways to couple light into a waveguide.
3. Figure P5.10 shows a fiber-based glucose sensor design. Describe the glucose sensing mechanism.

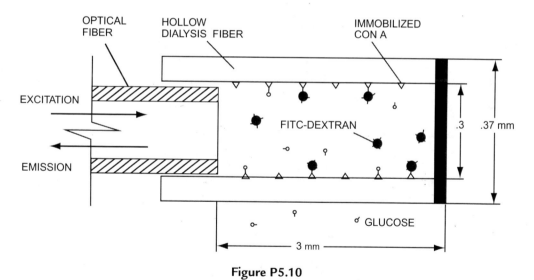

Figure P5.10

P5.11 Waveguide Based Sensors

Figure P5.11 shows a Mach Zehnder waveguide molecular sensor. We use an input light wave given as

$$E_{in} = E_0\, e^{j(wt-kx)},$$

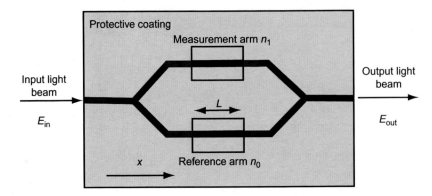

Figure P5.11: Mach Zehnder Waveguide Biosensor.

where $k = 2\pi/\lambda$, and λ is the wavelength in air. A part of the waveguide with a length of 1 μm is surface-functionalized as the measurement arm. The effective refractive index is n_1. On the other side is a reference arm with the same dimensions and the effective refractive index of n_0. When target analytes bind to the measurement arm, effective refractive index of the sensing part changes by Δn.

1. Describe the operation principle of the Mach Zehnder waveguide biosensor. Write down the mathematical of expression for the resulting output light wave E_{out}.
2. How can the device be used for rapid detection of bacteria DNA? How we can use the reference arm in order to efficiently measure the changes induced by molecular adsorption?

P5.12 Surface Plasmon Resonance Based Sensors

Figure P5.12 shows a prism coupler-based SPR measurement using the Kretschmann configuration. The relative dielectric constants of the cylindrical prism, metal, and dielectric material are 2.28, $-25 + j1.44$, and 1.76 respectively. The wavelength of the incident beam

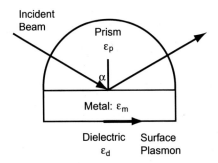

Figure P5.12

is 836 nm and the thickness of the metal layer is 200 nm. Find the optimum incident angle of the given structure (see Appendix 5.B).

P5.13 Texas Instruments SPR Biosensors

The schematic of a popular SPR sensor from Texas Instruments is shown in Fig. P5.13.

1. What does SPR stand for? Circle the sensing element in the figure.
2. Explain the functions of the sensor components showing in the figure.

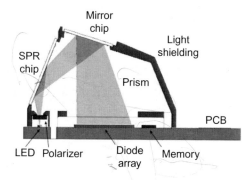

Figure P5.13

P5.14 Absorbance

1. A laser beam with the initial intensity of I_0 passes through a neutral density filter with the optical density of OD (=absorbance) = 1 (Fig. P5.14). What is the intensity of transmitted light?
2. When we have two filters with the optical density of OD = 1, What is the total optical density?
3. A suspension of colloidal Quantum dot QDot525 from Life Technologies has the molar extinction coefficient of 360000 cm^{-1} M^{-1} at 405 nm. If we prepare a sample with the concentration of 0.1 μM in a cuvette with the inside width of 1 cm, what is the optical density at 405 nm?

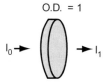

Figure P5.14

4. Calculate absorption cross section σ of QDot525. Use Avogadro constant 6.02×10^{23} mol^{-1}.

P5.15 Molecular Oximetry Sensor Design

You are requested to design an optical absorption-based sensor to measure the oxygen content in blood. The absorption spectrum of hemoglobin and oxyhemoglobin are given in Fig. P5.15.

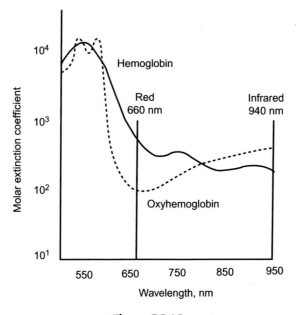

Figure P5.15

1. Based on the information giving in Fig. P5.15, describe the operation principle for such measurement. List any relevant mathematical equation.
2. Draw a sensor device design schematic and label key components.

P5.16 Fluorescence, Raman Scattering

1. Explain the difference between fluorescence emission and Raman scattering. Use keywords "vibrational modes," "lifetime," "anti-Stokes shift."
2. Explain what the mirror image rule is and why it happens.
3. Explain why the mirror image rule is not observed with quantum dots.

P5.17 Fluorescence and Waveguides

As shown in Fig. P5.17, the top surface of a flat waveguide ($n = 1.67$) is covered with buffer ($n_1 = 1.22$). The bottom surface of the waveguide is a cladding material. A fluorescent protein is bound to the surface of a waveguide, and its fluorescence is measured by capturing the light from the waveguide. Assuming you have a multimode fiber, estimate the maximum fluorescence — as a percentage of the total fluorescence emitted — that can be coupled back into the waveguide for the setup shown below. Assume that there are no losses once the light is coupled into the waveguide and on its way to the detector.

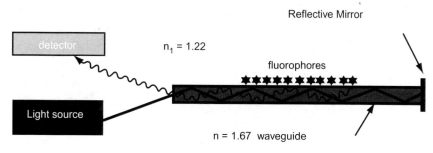

Figure P5.17

P5.18 Quantum Dot, Nanoparticles

A suspension of gold nanoparticles of diameter $10-15$ nm shows red color. This has been long used to stain glasses. CdSe quantum dots with the diameter of 10 nm also show red color when illuminated with a UV light. Briefly describe the mechanisms of how these colors are produced and what the differences are.

P5.19 Laser Doppler Velocimetry

A Laboratory bought a new laser Doppler velocimeter, where two coherent laser beams of a wavelength of 650 nm intersect each other at the measuring point. However, the intersection angle 2θ is unknown since the manual is lost. A study which used the identical device reports a result that the measured Doppler frequency with a sample which moves 150 mm/s is 20 kHz. Find the value of 2θ.

P5.20 Bright Field Microscopy and Dark Field Microscopy

We will perform microscopic observation of a material whose absorption and scattering properties are given in Fig. P5.20.

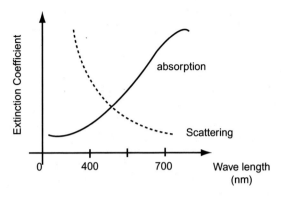

Figure P5.20

1. If we use a bright field microscope, what would be the color observed? Explain why.
2. If we use a dark field microscope, what would be the color observed? Explain why.

References

[1] Knoll W. Interfaces and thin films as seen by bound electromagnetic waves. Annual Review of Physical Chemistry 1998;49:569−638.
[2] Barrios CA, Gylfason KB, Sánchez B, Griol A, Sohlström H, Holgado M, et al. Slot-waveguide biochemical sensor. Optics Letters 2007;32:3080−2.
[3] Kim N, Park I-S, Kim W-Y. *Salmonella* detection with a direct-binding optical grating coupler immunosensor. Sensors and Actuators B: Chemical 2007;121:606−15.
[4] Clerc D, Lukosz W. Direct immunosensing with an integrated-optical output grating coupler. Sensors and Actuators B: Chemical 1997;40:53−8.
[5] Plowman T, Durstchi J, Wang H, Christensen D, Herron J, Reichert W. Multiple-analyte fluoroimmunoassay using an integrated optical waveguide sensor. Analytical Chemistry 1999;71:4344−52.
[6] Rong G, Najmaie A, Sipe JE, Weiss SM. Nanoscale porous silicon waveguide for label-free DNA sensing. Biosensors and Bioelectronics 2008;23:1572−6.
[7] Weisser M, Tovar G, Mittler-Neher S, Knoll W, Brosinger F, Freimuth H, et al. Specific bio-recognition reactions observed with an integrated Mach−Zehnder interferometer. Biosensors and Bioelectronics 1999;14:405−11.
[8] Brosinger F, Freimuth H, Lacher M, Ehrfeld W, Gedig E, Katerkamp A, et al. A label-free affinity sensor with compensation of unspecific protein interaction by a highly sensitive integrated optical Mach−Zehnder interferometer on silicon. Sensors and Actuators B: Chemical 1997;44:350−5.
[9] Schmitt K, Schirmer B, Hoffmann C, Brandenburg A, Meyrueis P. Interferometric biosensor based on planar optical waveguide sensor chips for label-free detection of surface bound bioreactions. Biosensors and Bioelectronics 2007;22:2591−7.
[10] Almeida VR, Barrios CA, Panepucci RR, Lipson M. All-optical control of light on a silicon chip. Nature 2004;431:1081−4.
[11] De Vos K, Bartolozzi I, Schacht E, Bienstman P, Baets R. Silicon-on-Insulator microring resonator for sensitive and label-free biosensing. Opt. Express 2007;15:7610−5.

[12] Xu D, Densmore A, Delâge A, Waldron P, McKinnon R, Janz S, et al. Folded cavity SOI microring sensors for high sensitivity and real time measurement of biomolecular binding. Optics Express 2008;16:15137−48.

[13] Adams M. An Introduction to Optical Waveguides, vol. 198. Chichester: JohnWiley & Sons; 1984.

[14] Okamoto K. Fundamentals of Optical Waveguides. New York: Academic Press; 2010.

[15] Schultz JS, Mansouri S, Goldstein IJ. Affinity sensor: a new technique for developing implantable sensors for glucose and other metabolites. Diabetes Care 1982;5:245−53.

[16] Ballerstadt R, Schultz JS. A fluorescence affinity hollow fiber sensor for continuous transdermal glucose monitoring. Analytical Chemistry 2000;72:4185−92.

[17] Brandenburg A, Edelhaäuser R, Hutter F. Integrated optical gas sensors using organically modified silicates as sensitive films. Sensors and Actuators B: Chemical 1993;11:361−74.

[18] Raether H. Surface plasmons on smooth and rough surfaces and on gratings. Springer Tracts in Modern Physics 1988;111:.

[19] Curto AG, Volpe G, Taminiau TH, Kreuzer MP, Quidant R, van Hulst NF. Unidirectional emission of a quantum dot coupled to a nanoantenna. Science 2010;329:930−3.

[20] Hanke T, Krauss G, Träutlein D, Wild B, Bratschitsch R, Leitenstorfer A. Efficient nonlinear light emission of single gold optical antennas driven by few-cycle near-infrared pulses. Physical Review Letters 2009;103:257404.

[21] Kühn S, Håkanson U, Rogobete L, Sandoghdar V. Enhancement of single-molecule fluorescence using a gold nanoparticle as an optical nanoantenna. Physical Review Letters 2006;97:017402.

[22] Akimov A, Mukherjee A, Yu C, Chang D, Zibrov A, Hemmer P, et al. Generation of single optical plasmons in metallic nanowires coupled to quantum dots. Nature 2007;450:402−6.

[23] Ropers C, Neacsu C, Elsaesser T, Albrecht M, Raschke M, Lienau C. Grating-coupling of surface plasmons onto metallic tips: a nanoconfined light source. Nano Letters 2007;7:2784−8.

[24] Sambles J, Bradbery G, Yang F. Optical excitation of surface plasmons: an introduction. Contemporary Physics 1991;32:173−83.

[25] Homola J. Present and future of surface plasmon resonance biosensors. Analytical and Bioanalytical Chemistry 2003;377:528−39.

[26] Rich RL, Myszka DG. Advances in surface plasmon resonance biosensor analysis. Current Opinion in Biotechnology 2000;11:54−61.

[27] Luo Y, Yu F, Zare RN. Microfluidic device for immunoassays based on surface plasmon resonance imaging. Lab on a Chip 2008;8:694−700.

[28] Haes AJ, Van Duyne RP. A nanoscale optical biosensor: sensitivity and selectivity of an approach based on the localized surface plasmon resonance spectroscopy of triangular silver nanoparticles. Journal of the American Chemical Society 2002;124:10596−604.

[29] Anker JN, Hall WP, Lyandres O, Shah NC, Zhao J, Van Duyne RP. Biosensing with plasmonic nanosensors. Nature Materials 2008;7:442−53.

[30] Yanik AA, Huang M, Kamohara O, Artar A, Geisbert TW, Connor JH, et al. An optofluidic nanoplasmonic biosensor for direct detection of live viruses from biological media. Nano Letters 2010;10:4962−9.

[31] Yelderman M, New Jr W. Evaluation of pulse oximetry. Anesthesiology 1983;59:349−51.

[32] Yoshiya I, Shimada Y, Tanaka K. Spectrophotometric monitoring of arterial oxygen saturation in the fingertip. Medical and Biological Engineering and Computing 1980;18:27−32.

[33] Brand TM, Brand ME, Jay GD. Enamel nail polish does not interfere with pulse oximetry among normoxic volunteers. Journal of Clinical Monitoring and Computing 2002;17:93−6.

[34] König V, Huch R, Huch A. Reflectance pulse oximetry−principles and obstetric application in the Zurich system. Journal of Clinical Monitoring and Computing 1998;14:403−12.

[35] Mendelson Y, Ochs BD. Noninvasive pulse oximetry utilizing skin reflectance photoplethysmography. Biomedical Engineering, IEEE Transactions on 1988;35:798−805.

[36] M. Nogawa, T. Kaiwa, S. Takatani, A novel hybrid reflectance pulse oximeter sensor with improved linearity and general applicability to various portions of the body, in: Engineering in Medicine and Biology Society, 1998. Proceedings of the 20th Annual International Conference of the IEEE, 1998, pp. 1858−1861.

[37] Oliver N, Toumazou C, Cass A, Johnston D. Glucose sensors: a review of current and emerging technology. Diabetic Medicine 2009;26:197−210.

[38] Shen Y, Davies A, Linfield E, Elsey T, Taday P, Arnone D. The use of Fourier-transform infrared spectroscopy for the quantitative determination of glucose concentration in whole blood. Physics in Medicine and Biology 2003;48:2023.

[39] Burmeister JJ, Arnold MA. Evaluation of measurement sites for noninvasive blood glucose sensing with near-infrared transmission spectroscopy. Clinical Chemistry 1999;45:1621−7.

[40] Burmeister JJ, Arnold MA, Small GW. Noninvasive blood glucose measurements by near-infrared transmission spectroscopy across human tongues. Diabetes Technology & Therapeutics 2000; 2:5−16.

[41] Malin SF, Ruchti TL, Blank TB, Thennadil SN, Monfre SL. Noninvasive prediction of glucose by near-infrared diffuse reflectance spectroscopy. Clinical Chemistry 1999;45:1651−8.

[42] Hoshino K, Huang Y-Y, Lane N, Huebschman M, Uhr JW, Frenkel EP, et al. Microchip-based immunomagnetic detection of circulating tumor cells. Lab on a Chip 2011;11:3449−57.

[43] Reed M, Randall J, Aggarwal R, Matyi R, Moore T, Wetsel A. Observation of discrete electronic states in a zero-dimensional semiconductor nanostructure. Physical Review Letters 1988;60:535−7.

[44] Leobandung E, Guo L, Wang Y, Chou SY. Observation of quantum effects and Coulomb blockade in silicon quantum-dot transistors at temperatures over 100 K. Applied Physics Letters 1995;67:938−40.

[45] Moreau E, Robert I, Gérard J, Abram I, Manin L, Thierry-Mieg V. Single-mode solid-state single photon source based on isolated quantum dots in pillar microcavities. Applied Physics Letters 2001;79:2865−7.

[46] Medintz IL, Uyeda HT, Goldman ER, Mattoussi H. Quantum dot bioconjugates for imaging, labelling and sensing. Nature Materials 2005;4:435−46.

[47] Resch-Genger U, Grabolle M, Cavaliere-Jaricot S, Nitschke R, Nann T. Quantum dots versus organic dyes as fluorescent labels. Nature Methods 2008;5:763−75.

[48] Wu X, Liu H, Liu J, Haley KN, Treadway JA, Larson JP, et al. Immunofluorescent labeling of cancer marker Her2 and other cellular targets with semiconductor quantum dots. Nature Biotechnology 2002;21:41−6.

[49] Debye P. Molecular-weight determination by light scattering. The Journal of Physical Chemistry 1947;51:18−32.

[50] Zimm BH. Molecular theory of the scattering of light in fluids. The Journal of Chemical Physics 1945;13:141.

[51] Pecora R. Dynamic Light Scattering: Applications of Photon Correlation Spectroscopy. Springer; 1985.

[52] Gulari E, Gulari E, Tsunashima Y, Chu B. Photon correlation spectroscopy of particle distributions. The Journal of Chemical Physics 1979;70:3965.

[53] Yeh Y, Cummins H. Localized fluid flow measurements with an He−Ne laser spectrometer. Applied Physics Letters 1964;4:176−8.

[54] Drain LE. *250 p*. The Laser Doppler Techniques, vol. 1. *Chichester, and New York*: *Wiley-Interscience*; 1980.

[55] Holloway GA, Watkins DW. Laser Doppler measurement of cutaneous blood flow. Journal of Investigative Dermatology 1977;69:306−9.

[56] Shepherd A, Riedel G. Continuous measurement of intestinal mucosal blood flow by laser-Doppler velocimetry. American Journal of Physiology-Gastrointestinal and Liver Physiology 1982;242:G668−72.

[57] Riva C, Grunwald J, Sinclair S. Laser Doppler Velocimetry study of the effect of pure oxygen breathing on retinal blood flow. Investigative Ophthalmology & Visual Science 1983;24:47−51.

[58] Van Loosdrecht M, Lyklema J, Norde W, Schraa G, Zehnder A. Electrophoretic mobility and hydrophobicity as a measured to predict the initial steps of bacterial adhesion. Applied and Environmental Microbiology 1987;53:1898−901.

[59] Wagner AJ, Bleckmann CA, Murdock RC, Schrand AM, Schlager JJ, Hussain SM. Cellular interaction of different forms of aluminum nanoparticles in rat alveolar macrophages. The Journal of Physical Chemistry B 2007;111:7353–9.

[60] Shevchenko EV, Talapin DV, Kotov NA, O'Brien S, Murray CB. Structural diversity in binary nanoparticle superlattices. Nature 2006;439:55–9.

[61] Knight DS, White WB. Characterization of diamond films by Raman spectroscopy. Journal of Materials Research 1989;4:385–93.

[62] Malard L, Pimenta M, Dresselhaus G, Dresselhaus M. Raman spectroscopy in graphene. Physics Reports 2009;473:51–87.

[63] Dresselhaus MS, Dresselhaus G, Saito R, Jorio A. Raman spectroscopy of carbon nanotubes. Physics Reports 2005;409:47–99.

[64] Stiles PL, Dieringer JA, Shah NC, Van Duyne RP. Surface-enhanced Raman spectroscopy. Annual Review of Analytical Chemistry 2008;1:601–26.

[65] Nie S, Emory SR. Probing single molecules and single nanoparticles by surface-enhanced Raman scattering. Science 1997;275:1102–6.

[66] Kneipp K, Wang Y, Kneipp H, Perelman LT, Itzkan I, Dasari RR, et al. Single molecule detection using surface-enhanced Raman scattering (SERS). Physical Review Letters 1997;78:1667–70.

[67] Xu H, Bjerneld EJ, Käll M, Börjesson L. Spectroscopy of single hemoglobin molecules by surface enhanced Raman scattering. Physical Review Letters 1999;83:4357–60.

[68] Grubisha DS, Lipert RJ, Park H-Y, Driskell J, Porter MD. Femtomolar detection of prostate-specific antigen: an immunoassay based on surface-enhanced Raman scattering and immunogold labels. Analytical Chemistry 2003;75:5936–43.

[69] Otto A, Mrozek I, Grabhorn H, Akemann W. Surface-enhanced Raman scattering. Journal of Physics: Condensed Matter 1992;4:1143.

[70] Hayazawa N, Inouye Y, Sekkat Z, Kawata S. Metallized tip amplification of near-field Raman scattering. Optics Communications 2000;183:333–6.

[71] Qian X, Peng X-H, Ansari DO, Yin-Goen Q, Chen GZ, Shin DM, et al. In vivo tumor targeting and spectroscopic detection with surface-enhanced Raman nanoparticle tags. Nature Biotechnology 2007;26:83–90.

[72] Stuart DA, Yuen JM, Shah N, Lyandres O, Yonzon CR, Glucksberg MR, et al. In vivo glucose measurement by surface-enhanced Raman spectroscopy. Analytical Chemistry 2006;78:7211–5.

[73] Enejder AM, Scecina TG, Oh J, Hunter M, Shih W-C, Sasic S, et al. Raman spectroscopy for noninvasive glucose measurements. Journal of Biomedical Optics 2005;10:031114–0311149.

[74] Murphy CJ. Nanocubes and nanoboxes. Science 2002;298:2139–41.

[75] Maier SA, Atwater HA. Plasmonics: localization and guiding of electromagnetic energy in metal/dielectric structures. Journal of Applied Physics 2005;98: 011101-10

[76] Kuwata H, Tamaru H, Esumi K, Miyano K. Resonant light scattering from metal nanoparticles: practical analysis beyond Rayleigh approximation. Applied Physics Letters 2003;83:4625–7.

[77] Hao E, Schatz GC. Electromagnetic fields around silver nanoparticles and dimers. The Journal of Chemical Physics 2004;120:357.

[78] Mock JJ, Smith DR, Schultz S. Local refractive index dependence of plasmon resonance spectra from individual nanoparticles. Nano Letters 2003;3:485–91.

[79] Kelly KL, Jensen TR, Lazarides AA, Schatz GC. Modeling metal nanoparticle optical properties. Metal Nanoparticles: Synthesis, Characterization, and Applications 2002;:89–118.

[80] Lee K-S, El-Sayed MA. Dependence of the enhanced optical scattering efficiency relative to that of absorption for gold metal nanorods on aspect ratio, size, end-cap shape, and medium refractive index. The Journal of Physical Chemistry B 2005;109:20331–8.

[81] Jana NR, Gearheart L, Murphy C. Seed-mediated growth approach for shape-controlled synthesis of spheroidal and rod-like gold nanoparticles using a surfactant template. Advanced Materials 2001;13:1389–93.

[82] Nikoobakht B, El-Sayed MA. Preparation and growth mechanism of gold nanorods (NRs) using seed-mediated growth method. Chemistry of Materials 2003;15:1957−62.

[83] Gou L, Murphy CJ. Fine-tuning the shape of gold nanorods. Chemistry of Materials 2005;17:3668−72.

[84] El-Sayed IH, Huang X, El-Sayed MA. Surface plasmon resonance scattering and absorption of anti-EGFR antibody conjugated gold nanoparticles in cancer diagnostics: applications in oral cancer. Nano Letters 2005;5:829−34.

[85] Nath N, Chilkoti A. A colorimetric gold nanoparticle sensor to interrogate biomolecular interactions in real time on a surface. Analytical Chemistry 2002;74:504−9.

[86] Stöber W, Fink A, Bohn E. Controlled growth of monodisperse silica spheres in the micron size range. Journal of Colloid and Interface Science 1968;26:62−9.

[87] Chen Y-S, Frey W, Kim S, Homan K, Kruizinga P, Sokolov K, et al. Enhanced thermal stability of silica-coated gold nanorods for photoacoustic imaging and image-guided therapy. Optics Express 2010;18:8867−78.

[88] Betzig E, Trautman J, Harris T, Weiner J, Kostelak R. Breaking the diffraction barrier: optical microscopy on a nanometric scale. Science 1991;251:1468−70.

[89] Hecht B, Sick B, Wild UP, Deckert V, Zenobi R, Martin OJ, et al. Scanning near-field optical microscopy with aperture probes: fundamentals and applications. The Journal of Chemical Physics 2000;112:7761.

[90] Dunn RC. Near-field scanning optical microscopy. Chemical Reviews 1999;99:2891−928.

[91] Inouye Y, Kawata S. Near-field scanning optical microscope with a metallic probe tip. Optics Letters 1994;19:159−61.

[92] Michaelis J, Hettich C, Mlynek J, Sandoghdar V. Optical microscopy using a single-molecule light source. Nature 2000;405:325−8.

[93] Shubeita G, Sekatskii S, Dietler G, Potapova I, Mews A, Basché T. Scanning near-field optical microscopy using semiconductor nanocrystals as a local fluorescence and fluorescence resonance energy transfer source. Journal of Microscopy 2003;210:274−8.

[94] Cuche A, Drezet A, Sonnefraud Y, Faklaris O, Treussart F, Roch J-F, et al. Near-field optical microscopy with a nanodiamond-based single-photon tip. Optics Express 2009;17:19969−80.

[95] Kalkbrenner T, Ramstein M, Mlynek J, Sandoghdar V. A single gold particle as a probe for apertureless scanning near-field optical microscopy. Journal of Microscopy 2001;202:72−6.

[96] Hoshino K, Gopal A, Glaz MS, Vanden Bout DA, Zhang X. Nanoscale fluorescence imaging with quantum dot near-field electroluminescence. Applied Physics Letters 2012;101:. pp. 043118-043118-4

[97] Tien P. Integrated optics and new wave phenomena in optical waveguides. Reviews of Modern Physics 1977;49(2):361.

[98] Chinowsky T, Quinn J, Bartholomew D, Kaiser R, Elkind J. Performance of the Spreeta 2000 integrated surface plasmon resonance affinity sensor. Sensors and Actuators B: Chemical 2003;91(1):266−74.

[99] Mie G. Beiträge zur Optik trüber Medien, speziell kolloidaler Metallösungen. Annalen der physik 1908;330(3):377−445.

[100] Turner PF, Karube I, Wilson GS, Biosensors: fundamentals and applications, Oxford science publications; 1987.

Further Reading

Optics, electromagnetic waves
Born M, Wolf E. Principles of Optics. Cambridge: Cambridge University Press; 1999.
Hecht E. Optics. fourth ed. Boston: Addison-Wesley; 2001.
Cheng DK. Fundamentals of Engineering Electromagnetics. Upper Saddle River, NJ: Prentice Hall; 1992.
Balanis CA. Advanced Engineering Electromagnetics. New York: Wiley; 1989.
de Fornel F. Evanescent Waves : From Newtonian Optics to Atomic Optics. New York: Springer; 2001.

CHAPTER 6

Mechanical Transducers

Cantilevers, Acoustic Wave Sensors, and Thermal Sensors

Chapter Outline

6.1 Introduction

Mechanical sensors form a class of sensors which are sensitive to changes in mechanical properties. In combination with the micromachining technology, mechanical sensors such as cantilevers and acoustic sensors have been playing an important role in molecular detection. In most mechanical sensors for molecular sensing, binding events are detected using two general strategies: by detecting stresses induced on the cantilever surface or by detecting the change in resonant frequency. With a long history, piezoelectric and piezoresistive phenomena have been used in transduction schemes allowing for extremely sensitive response to deflection or mass change. Optical or other electrical measurements are also important in specific applications. In the following sections, we will introduce a few major classes of mechanical molecular sensors.

In Section 6.2, we describe a mechanical cantilever-based molecular sensor. In Section 6.2.2, we will discuss analytical characterization of cantilevers. Analysis of a cantilever is the fundamental subject that introduces the basic ideas behind structural analysis such as stress, strain, deflection, and resonant modes. Cantilever sensitivity scales with physical dimensions and varies with mechanical parameters including Young's modulus and the Poisson's ratio. In Section 6.2.3, we will introduce designs, implementation and applications of cantilever-based molecular sensors. Micromachined cantilevers are fabricated using photolithography techniques. In Section 6.2.4, the two primary methods of molecular detection, namely the deflection based method and the resonant frequency based method, will be discussed. Techniques for readout will be explored in Section 6.2.5. We will introduce four main readout methods: laser reflection, interferometry, piezoresistive, and capacitive methods. Microcantilever devices, although currently fast and inexpensive to use, have variable sensitivities that require frequent calibration and limit detection capabilities. In Section 6.2.6, we introduce applications of cantilever sensors for molecular sensing, followed by discussions on the limitation of cantilever sensors and techniques needed to push microcantilever technology past the current state in Section 6.2.7. Finally, we will describe recent advancement in the development of nanomaterial-based resonators in Section 6.2.8.

In Section 6.3 we describe acoustic wave-based biosensors. An acoustic wave sensor is a sensor that utilizes acoustic waves to derive information about its environment. Many sensors are operated at a resonant mode. The acoustic waveform can be altered by environmental properties which translate into a measurable electrical signal. The electrical transduction is based on the piezoelectric effect, which is a phenomenon by which a material produces a charge in response to an applied force and vice versa. We also describe mechanical resonators based on quartz tuning forks, the working principle of which also relies on piezoelectric effect.

In Section 6.4, we introduce several types of thermal sensors related to molecular sensing or biomedical applications. Thermal sensing is a measure to find the activities of molecules in chemical or biological systems. In many examples, they are not only based on mechanical transduction but utilizing combinations of electrical and optical techniques. Related sections in Chapter 4 and Chapter 5 will be indicated in each description.

6.2 Cantilever-Based Molecular Sensors

6.2.1 Introduction

Micromachined cantilevers have enabled measurements of forces at the single molecular scale. The unbinding force of a single antibody-antigen recognition event was measured to be 244 +/− 22 pN [125]. Interaction forces corresponding to 20, 16, and 12 base pairs of a DNA were measured with three distinct distributions centers at 1.52, 1.11, and 0.83 nN, respectively [126]. Unbinding force measurements with only a few avidin-biotin bonds were quantized in integer multiples of 160 +/− 20 pN [127]. Ionic and van der Waals interaction, are also theoretically estimated to be within the measureable force range [128]. The following is a summary of molecular interaction forces:

antibody-antigen [125]	244 pN
DNA (20bp) [126]	1.52 nN
avidin-biotin [127]	160 pN
ionic (NaCl, theoretical) [128]	2.1 nN
van der Waals (Ar-Ar, theoretical) [128]	12 pN

Detection of biomolecules using microstructures is based on molecular adsorption that results in measurable mechanical changes. Due to their miniature size, rapid response, high sensitivity, and label-free capability, microcantilevers can be integrated into lab on a chip or point of care designs that require molecular sensing capabilities [1].

A broad range of surface functionalization techniques can be applied to the microcantilever surface (see Chapter 2, Section 2.5.3 for techniques for surface modification). In order to detect the presence of chemicals, the microcantilever must be loaded with a material that promotes the adsorption of a particular molecule to the surface [1]. Cantilever-based sensors can even detect ongoing processes such as DNA hybridization [2]. The surface coating can be tailored to respond to a very narrow range of molecules, resulting in a highly selective sensor that can ignore large concentrations of background molecules. Microcantilever sensors may be used to detect early stage cancers, the presence of viruses, and many other applications that require a response to minimal stimuli.

Molecular adsorption to the microcantilever surface increases the mass of the system, resulting in either a static mechanical deflection of the microcantilever or a shift in its resonance frequency. Both of these can be measured and used to determine the presence of a particular molecule or group of molecules. These mechanical changes are transduced by using techniques such as variation in piezoresistivity, piezoelectricity, optical beam deflection, interferometry, and capacitance [3,4]. As an extreme example, the sensitivity of carbon nanotube based mass sensor even reached as small as 1.7×10^{-24} g, allowing measurement of the binding energy of a xenon atom (see section 6.2.8 for details). Considering the much larger molecular weight of proteins (for example, IgG antibody has a molecular mass of ~160kDa, or 2.5×10^{-19} g per molecule) single molecular sensing may be potentially possible for most biomolecules. However, each analyte has different measuring requirements. For example, protein concentration is typically measured in liquid, and a nanotube sensor will not work due to the large viscosity of the liquid. The sensitivity limit is mainly defined by the configurations of the sensing schemes.

Cantilever based sensors can be mass-manufactured in a cost-effective fashion, allowing for easy commercialization. Microcantilevers can be produced in large arrays, allowing for rapid parallel processes to occur in a very small area [5]. It takes advantage of silicon micromachining techniques developed for integrated circuit process technology. This drives down both the manufacturing costs and the testing costs, making microcantilevers economically efficient.

Technological Background: Atomic Force Microscopy and Microfabrication

The field of cantilever-based molecular sensors followed the development of two independent technologies.

The first part is the invention of the atomic force microscopy (AFM) in 1986. Binning and Quate from Stanford University used a scanning tunneling microscope (STM) to measure the motion of a cantilever beam with an ultrasmall mass [6]. They measured the force acting between the cantilever tip and the sample surface as the tip approaches the sample surface. The force required to move the cantilever to the measurable distance can be as small as 10^{-18} N. They constructed the AFM in which images are obtained by raster scanning the surface with a probe that consists on the sharp tip mounted on a cantilever. Binning further discussed that microfabrication would be needed in order to make the cantilevers with a mass less than 10^{-10} kg and a resonant frequency greater than 2 kHz [6].

The other key technology is microfabrication. Micromachined cantilevers with integrated tips for measurements in AFM were realized in 1990 by Albrecht et al. at Stanford University [7]. They studied several methods of producing SiO_2- and Si_3N_4-based microcantilevers. Wolter et al from IBM also developed micromachined silicon sensors in 1991 [8]. They developed a batch process for the microfabrication of single crystal silicon force sensors. The process involved a combination of wet and dry etching techniques which are suitable for mass production [9].

Studies on microcantilevers for molecular sensing were initiated in the mid-1990s, based on the technologies developed for the AFM. Since then, the research area has been growing during the past years. Cantilevers offer a wide array of advantages in terms of sensing devices that has attracted a lot of attention. Potential applications to a wide variety of fields have been researched, especially in the field of medical diagnostics, homeland security and environmental monitoring [9].

6.2.2 Mechanics of Cantilevers

Here we describe basics of structural analysis of a cantilever. Theories of important mechanical characteristics such as force, stress, strain, moment, and radius of curvature will be discussed with a simple cantilever structure.

Young's Modulus

Let us consider a cuboid beam made of a uniform unidirectional material as shown in Fig. 6.1. The dimensions are width w, thickness t, and L in the length. When we apply a force F to the beam in the direction of the length, the beam elongates by a length ΔL. Within the range of the material's elasticity, the force F and the elongation ΔL follows **Hooke's law**, which states that the force needed to extend the material is proportional to the length of deformation.

The force F is also dependent on the dimensions of the beam. It is proportional to the cross sectional area $A = w \times t$ and is inversely proportional to the length L. Now the force F can be written as:

$$F = \frac{EA\Delta L}{L},\tag{6.1}$$

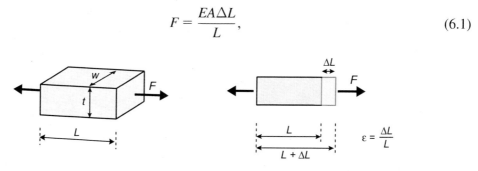

Figure 6.1: Young's Modulus.

where E is the constant of proportionality. In order to consider the material's mechanical characteristics independent of the dimensions of the structure, we can devide both sides of Eq 6.1 by the cross sectional are A:

$$\frac{F}{A} = E \cdot \frac{\Delta L}{L} \tag{6.2}$$

The intensity of the force per unit area acting on section A is called a **stress**:

$$\sigma = \frac{F}{A} \tag{6.3}$$

Change in length of the body over its original length, or change in length per unit length, is called a **strain**.

$$\varepsilon = \frac{\Delta L}{L} \tag{6.4}$$

Using Eqs 6.3 and 6.4, Eq 6.2 can be written as:

$$\sigma = E \cdot \varepsilon \tag{6.5}$$

$$\text{or} \quad E = \frac{\sigma}{\varepsilon} \tag{6.6}$$

Eq 6.5 or 6.6 describes the force-elongation relationship of a unit cubic volume. Now we define constant E as the **Young's modulus** of the material. Young's modulus is a measure of the material stiffness and is defined as the ratio of the stress to the strain. It is a material characteristic defined independent of the dimensions of the structure.

Basic Mechanical Characteristics

Stress σ: force per unit area

$$\sigma = \frac{F}{A}$$

Strain ε: change in length per unit length.

$$\varepsilon = \frac{\Delta L}{L}$$

Young's Modulus E: the ratio of the stress over the strain

$$\sigma = E \cdot \varepsilon$$

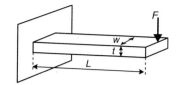

Figure 6.2: Model of Cantilever Bending.

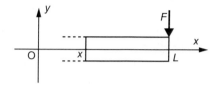

Figure 6.3: Part of the Cantilever Separated at Imaginary Cross-section.

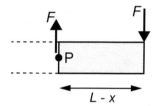

Figure 6.4: Force Equilibrium: Shear Force.

Analysis of Cantilever Bending

Figure 6.2 shows a simple model of bending of a mechanical cantilever. Force F is applied at the end of the cantilever beam. Dimensions are width w, thickness t, and length L.

Stress Distribution

Let us consider a part l of the cantilever separated at imaginary cross-section A at x (Fig. 6.3). Now all the forces to support the part come from the cross-section. Since part l is in mechanical equilibrium, the external forces and the moments of the external forces must satisfy equilibrium equations.

1. The sum of the external forces is zero.

    The force to counter force F is a shear force $-F$ at the cross-section. The intensity of the shear force (Fig. 6.4) is independent of the length of part l. If multiple external forces are applied to the part, F should be the summation of all the forces.
2. The sum of the moments of the external force is zero.

    The external force F produces moment M_F about the axis a in Fig. 6.4.

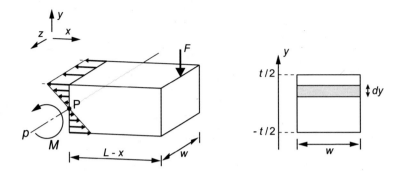

Figure 6.5: Moment Equilibrium: Moment Induced by Stress Distribution.

$$M_F = F \cdot (L - x) \tag{6.7}$$

The moment to counter M_F is created by a distribution of normal stresses $\sigma(y)$ at A as shown in Fig. 6.5. Moment M produced by $\sigma(y)$ about axis p must cancel the moment M_F induced by the external force

$$M_F + M = 0 \tag{6.8}$$

For the example shown in Fig. 6.3, the moment M_F is given by Eq 6.7. If multiple external forces and moments are applied to the part, M_F should be the summation of all the applied moments and the moments induced by all the applied forces.

Since the net force in x direction is zero, the integration of $\sigma(y)$ along the surface should be also zero. When we assume $\sigma(y)$ is linearly distributed along y, $\sigma(y)$ is expressed in the following way:

$$\sigma(y) = k \cdot y \tag{6.9}$$

M is the moment produced by $\sigma(y)$ about P:

$$M = \int_{-t/2}^{t/2} y \cdot w \cdot \sigma(y) \cdot dy = k \int_{-t/2}^{t/2} wy^2 \cdot dy, \tag{6.10}$$

In order to simplify the following discussion, we define term I as:

$$I = \int_{-t/2}^{t/2} wy^2 \cdot dy = \frac{1}{12} wt^3 \tag{6.11}$$

The term I is called the second momentum of area (Box 6.1).

Box 6.1 Second Momentum of Area

In Eq 6.11 term *I* is called the **second momentum of area** for cross-section *A* about axis *p*. The second momentum of area about *z* axis for an arbitrary shape *A* is defined as:

$$I = \iint_A y^2 \cdot dA = \iint_A y^2 \cdot dy \cdot dz,$$

where *y* is the distance from the axis *x* to area d*A*. By introducing second momentum of area, discussion in this section can be applied to beams with arbitrary shapes.

With I, the moment M is simply expressed as:

$$M = k \cdot I \tag{6.12}$$

We find the stress distribution function

$$\sigma(y) = \frac{M}{I} y \tag{6.13}$$

Comparing Eqs 6.9 and 6.13 and using Eq 6.8, we find

$$k = \frac{M}{I} = -\frac{M_F}{I} \tag{6.14}$$

Plugging Eq 6.7 into Eq 6.14, we obtain

$$\sigma(y) = \frac{F(L - x)}{I} y \tag{6.15}$$

As one can see in Eq 6.15 the stress becomes larger as the cross-section gets closer to the base, or $L - x$ becomes larger. The intensity is linearly correlated with the distance measured from the end $L - x$ (see Fig. 6.6).

Figure 6.6: Internal Stress Distribution.

Beam Bending

Equation 6.5 gives relationship between a normal stress and a strain. Using Young's modulus, the stress distribution in a beam can be correlated to the distribution of strain, which leads to bending of the beam. Let us consider a small slice of the beam with the length dx at $y = 0$.

Using Eqs 6.5 and 6.13,

$$\varepsilon(y) = \frac{\sigma(y)}{E} = \frac{M}{EI}y \tag{6.16}$$

As shown in Fig. 6.7, the slice is deformed following the linear stain distribution. The length at y is given as:

$$dl(y) = (1 + \varepsilon(y)) \cdot dx = \left(1 + \frac{M}{EI}y\right) \cdot dx, \tag{6.17}$$

which is longest at $y = t/2$ and shortest at $y = -t/2$. From the geometrical consideration, the center of curvature of the beam can be found as the position where $dl(y) = 0$.

The center of the curvature R becomes:

$$R = -\frac{EI}{M} \tag{6.18}$$

Note we use R in the signed form to consider direction of the bending.

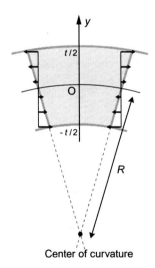

Center of curvature

Figure 6.7: Radius of Curvature.

The radius of curvature of a curve $y = y(x)$ is given in the following way

$$R = \frac{(1+y'^2)^{3/2}}{y''} \qquad (6.19)$$

In the case of our model, y' is negligibly smaller than 1.

$$R = \frac{1}{y''} \qquad (6.20)$$

This is a very important relationship to correlate curvature and the bending of a beam. From Eqs 6.18 and 6.20, we find

$$y'' = -\frac{M}{EI} \qquad (6.21)$$

Equation 6.21 can be used for general cases of beam bending (Box 6.2, and consider why?). For the cantilever case shown in Fig. 6.5, we can use Eqs 6.7 and 6.8 to substitute M in Eq 6.21 and correlate the beam bending with the external force:

$$y'' = -\frac{F(L-x)}{EI} \qquad (6.22)$$

Integrating Eq 6.22 twice and using the fixed-end boundary conditions of $y(0) = y'(0) = 0$, we find

$$y = \frac{Fx^2}{6EI}(3L - x) \qquad (6.23)$$

The maximum deflection is at $x = L$ and

$$y_{max} = \frac{L^3}{3EI} \cdot F \qquad (6.24)$$

Box 6.2 Analysis of General Cases of Beam Bending

The following equation applies to general cases of beam bending with different boundary conditions and applied forces:

$$y'' = -\frac{M}{EI}$$

Deflection of types of beams and applied forces are summarized in Appendix 6.A.

$$\text{or} \quad F = \frac{3EI}{L^3} \cdot y_{\text{max}} \tag{6.25}$$

The term $3EI/L^3$ defines the linear relationship between the applied force F and the deflection. It can be considered as the effective spring constant of the beam at $x = L$ in the y direction.

$$k_{\text{eff}} = \frac{3EI}{L^3} = \frac{Ewt^3}{4L^3} \tag{6.26}$$

Stoney's Formula

Bi-Layered Cantilever

Stoney described tension and delamination of thin metal films in 1909. Stoney's formula describes the bending of a bilayer cantilever (Fig. 6.8) due to the difference between the thermal expansion coefficients (see problem 6.13 and discussion on evaporation deposition in Chapter 2, Section 2.3.3). Now the equation is also used to discuss bending of cantilevers induced by molecular adsorption.

We consider a cantilever deposited with a thin film. The dimensions of the cantilever are the same as previously described in this section. The thickness of the film and Young's modulus are d and E_f respectively. Due to thermal expansion or molecular adsorption, there are mismatch forces F_f at the film/cantilever interface. Here we introduce effective forces F_f and M_f for the film and F and M for the cantilever body. There is no external force and the cantilever is in equilibrium.

The force equilibrium equation is

$$F_f = F \tag{6.28}$$

Assuming the film thickness d is negligible compared to the cantilever thickness t, the moment equilibrium equation is

$$F_f \cdot \frac{t}{2} = M + M_f \tag{6.29}$$

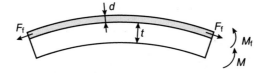

Figure 6.8: Bi-layered Cantilever.

Radius of curvature and the moment have been correlated in Eq 6.18

$$M = -\frac{EI}{R} \tag{6.18}$$

From this, we can find moments M and M_f by substituting term I using Eq 6.11:

$$M = \frac{EI}{R} = \frac{Et^3 w}{12R} \tag{6.30}$$

$$M_f = \frac{EI_f}{R} = \frac{E_f d^3 w}{12R} \tag{6.31}$$

We neglected the sign here. Plugging Eqs 6.30 and 6.31 into Eq 6.29,

$$F_f \cdot \frac{t}{2} = \frac{Et^3 w}{12R} + \frac{E_f d^3 w}{12R} \tag{6.32}$$

Since d is much smaller than t,

$$F_f = \frac{Et^2 w}{6R} \tag{6.33}$$

This is **Stoney's formula** for a cantilever beam. Stoney's formula can be expressed in several different ways.

Surface Stress

Surface stress σ_{surface} is the amount of work per unit area needed to stretch a surface. Figure 6.9 is the top view of the cantilever.

In this case of one-dimensional stretching, the new area $dA = w\Delta L$, the work $dW = F \cdot \Delta L$. Surface stress σ is given as:

$$\sigma_{\text{surface}} = dW/dA = F/w \tag{6.34}$$

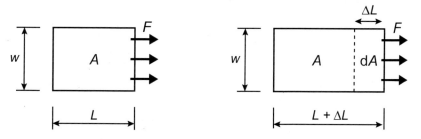

Figure 6.9: Surface Stress on a Cantilever.

the force F is given as

$$F = \sigma_{\text{surface}} \cdot w \tag{6.35}$$

Note that a surface stress has a different dimension (force per length) from that of a normal stress (force per area). Surface stress is often used when the film thickness is unknown or very thin. A surface-functionalized cantilever is a good example where surface stress rather than normal stress is easily applied.

Using surface stress (Eq 6.35 in 6.33), Stoney's formula is written as:

$$\sigma_{\text{surface}} = \frac{Et^2}{6R} \tag{6.36}$$

For two dimensional bending of a plate instead of a cantilever, Stoney's formula is modified as:

$$\sigma_{\text{surface}} = \frac{Et^2}{6(1 - \nu)R'} \tag{6.37}$$

where ν is Poisson's ratio.

Deflection and Stoney's Formula

The cantilever bending is found from R. Using Eqs 6.20 and 6.36,

$$y'' = \frac{1}{R} = \frac{6}{Et^2} \cdot \sigma_{\text{surface}} \tag{6.38}$$

Integrating this twice and using the fixed end boundary condition of $y(0) = y'(0) = 0$,

$$y = \frac{3}{Et^2} \cdot \sigma_{\text{surface}} \cdot x^2 \tag{6.39}$$

Deflection δ at the end is

$$\delta = \frac{3L^2}{Et^2} \cdot \sigma_{\text{surface}} \tag{6.40}$$

This is one of the most commonly used forms of Stoney's formula.

Stoney's Formula

The surface stress $\sigma_{surface}$ is found from the radius of curvature R

$$\sigma_{surface} = \frac{Et^2}{6R}$$

Deflection δ at the end is found from the surface stress $\sigma_{surface}$

$$\delta = \frac{3L^2}{Et^2} \cdot \sigma_{surface}$$

Resonant Frequency

Equation of Motion

Resonant frequency of a cantilever is found by solving the equations of motion for a small slice of a cantilever in free vibration, as shown in Fig. 6.10.

From force equilibrium:

$$dF = \rho A \cdot \frac{d^2 y}{dt^2} \cdot dx \tag{6.41}$$

$$\text{or} \quad \frac{dF}{dx} = \rho A \cdot \frac{d^2 y}{dt^2} \tag{6.42}$$

where ρ and A are the mass density and the cross sectional area of the beam, respectively.

We assume there is no rotational motion. From the moment equilibrium about point P:

$$dM = dx \cdot (F + dF) \tag{6.43}$$

or

$$F = \frac{dM}{dx} \tag{6.44}$$

From Eqs 6.42 and 6.44

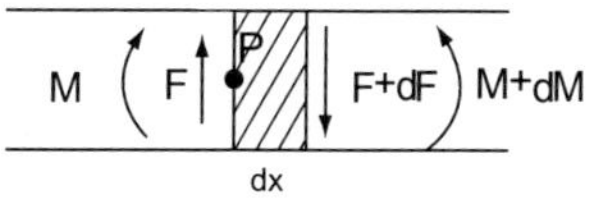

Figure 6.10: Equation of Motion.

$$\frac{d^2 M}{dx^2} = \rho A \cdot \frac{d^2 y}{dt^2} \tag{6.45}$$

Recalling Eq 6.21:

$$M = - EI \cdot y'' \tag{6.21}$$

Equation 6.45 can be rewritten as:

$$- EI \cdot \frac{d^4 y}{dx^4} = \rho A \cdot \frac{d^2 y}{dt^2} \tag{6.46}$$

This is the equation of motion to solve. Assuming $y(x, t)$ can be expressed in the form of $y = Y(x) \cdot T(t)$, Eq 6.46 is separated into two differential equations for $Y(x)$ and $T(t)$ in the following way:

$$\frac{d^4 Y}{dx^4} - C^4 Y = 0 \tag{6.47}$$

$$\frac{d^2 T}{dx^2} + (\alpha C)^4 T = 0, \tag{6.48}$$

where

$$\alpha^4 = \frac{EI}{\rho A} \tag{6.49}$$

Solutions to Eq 6.47 with the following boundary conditions give the shapes of cantilever bending.

$$Y = \frac{dY}{dx} = 0 \ (x = 0) \tag{6.50}$$

$$\frac{dY^2}{dx^2} = \frac{dY^3}{dx^3} = 0 \ (x = L) \tag{6.51}$$

where Eqs 6.50 and 6.51 are boundary conditions for the fixed end at $x = 0$ and the free end at $x = L$, respectively.

There are multiple solutions, or multiple vibration modes. From numerical analysis, the first three mode are given as:

$$C \approx \frac{1.875}{L}, \frac{4.694}{L}, \frac{7.855}{L} \tag{6.52}$$

Figure 6.11 shows $y = Y(x)$ plotted for the first three natural frequencies.

The general solution to Eq 6.48 gives the angular frequency.

$$T(t) = T_0\cos(\alpha C)^2(t - t_0), \tag{6.53}$$

In most cantilever sensors, the first natural frequency is used:

$$f = \frac{(\alpha C)^2}{2\pi} = \frac{1}{2\pi}\left(\frac{1.875}{L}\right)^2\sqrt{\frac{EI}{\rho A}} \tag{6.54}$$

Effective Spring Constant and Mass

The resonant frequency and the effective spring constant k_{eff} define the effective mass m_{eff} of the cantilever in the following way:

$$f = \frac{1}{2\pi}\sqrt{\frac{k_{\text{eff}}}{m_{\text{eff}}}} \tag{6.55}$$

Note that the effective mass is different from the actual mass m_0. For the case of a cantilever, we have already found k_{eff} and f in Eqs 6.26 and 6.54, respectively.

$$k_{\text{eff}} = \frac{3EI}{L^3} = \frac{Ewt^3}{4L^3} \tag{6.26}$$

$$f = \frac{1}{2\pi}\left(\frac{1.875}{L}\right)^2\sqrt{\frac{EI}{\rho A}} \tag{6.54}$$

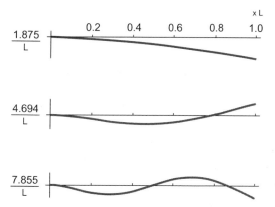

Figure 6.11: Mode Shapes of the First Three Natural Frequencies.

The effective mass m_{eff} can be found from Eqs 6.54, 6.26, and 6.55.

$$m_{\text{eff}} = \frac{3m_0}{(1.875)^4},$$ (6.56)

where m_0 is the actual mass. We can approximate m_{eff} as:

$$m_{\text{eff}} = 0.24m_0$$ (6.57)

Oscillating Cantilever as a Mass Sensor

The introduction of effective mass and stiffness allows us to use a simplified mass spring system to analyze the characteristics of mechanical cantilever.

When a mass Δm is attached to mechanically oscillating system, it induces a change Δf in the resonant frequency. This principle can be used for a mass sensor based on mechanical oscillation. The sensor is characterized by the following parameter called **responsivity** (see Chapter 1, Section 1.4 for general discussion):

$$Rsp = \frac{\Delta f}{\Delta m}$$ (6.58)

where Δf is the change in resonant frequency caused by the added mass Δm. It can be approximated by calculating a partial derivative of f with respect to m_{eff}:

$$\frac{\Delta f}{\Delta m_{\text{eff}}} \approx \frac{\partial f}{\partial m_{\text{eff}}} = -\frac{1}{2m_{\text{eff}}} \cdot \frac{1}{2\pi} \sqrt{\frac{k_{\text{eff}}}{m_{\text{eff}}}} = -\frac{f}{2m_{\text{eff}}}$$ (6.59)

Responsivity is the linear coefficient between the changes in mass and the resonant frequency.

$$\Delta f = Rsp \cdot \Delta m$$ (6.60)

Note that the definition in Eq 6.59 is based on an assumption that the deflection of the added mass Δm is the same as that of the effective mass m_{eff}, i.e. the added mass is attached at the end of the cantilever. In this case, $\Delta m_{\text{eff}} = \Delta m$. We introduce this definition because the main point of introducing effective mass and stiffness is that we model the sensor as a simple mass spring system. In reality, the responsivity depends on how the additional mass is attached to the cantilever. One can imagine that a cantilever is less sensitive to a change in mass near the base and more sensitive at the end. If a mass is uniformly added along the cantilever, $\Delta m_{\text{eff}} = 0.24\Delta m$, and the responsivity may be defined as $-f/(2m_0)$ instead of $-f/(2m_{\text{eff}})$ (consider why).

Responsivity

The linear coefficient between the changes in mass and the resonant frequency

$$Rsp = \frac{\Delta f}{\Delta m} = -\frac{f}{2m_{eff}}$$

Q factor and damping effect

With effective mass m and spring constant k, the cantilever system is considered as a damped harmonic oscillator with an externally applied sinusoidal force $f(w) = Fe^{j\omega t}$.

$$m\frac{d^2x}{dt^2} + c\frac{dx}{dt} + kx = Fe^{j\omega t}$$

Where c is the viscous damping coefficient which defines the viscous force proportional to the velocity of the mass.

The maximum displacement of x, or the oscillation amplitude, can be plotted as a function of the angular frequency ω of the applied force (by assuming $x = Xe^{j\omega t}$, we can find X/F as a function of ω from the differential equation). The plot has a significant peak at the resonant frequency. The resonant frequency is independent of the value of c, but the peak becomes less sharp as c becomes larger. Here we define the quality factor, or Q factor, which characterizes the system's bandwidth relative to the resonant frequency.

$$Q = \frac{\omega_r}{\Delta\omega}$$

Where $\Delta\omega$ is the half-power bandwidth, which is the width of the frequency band where the stored energy is larger than half of that at resonance. The value of Q is given with the coefficients m, k, and c in the following way:

$$Q = \frac{\omega}{\Delta\omega} = \frac{\sqrt{mk}}{c}$$

When the Q factor is low, the resonant peak becomes less significant and measurement of finding the change in the resonant frequency becomes more difficult. The Q factor is also defined as the ratio of the energy stored in the system to the energy dissipated per cycle. When the Q factor is higher, the rate of energy loss is low, and the peak at the resonant frequency becomes sharper.

6.2.3 Fabrication of Microcantilevers: Materials and Design Features

In this section, we describe fabrication techniques of microcantilevers. The majority of microcantilevers are made of silicon-based materials such as single crystal silicon, polysilicon, silicon nitride, and silicon oxide. We will first describe standard fabrication techniques for silicon-based cantilevers. We then describe the relatively new trend of fabrication techniques utilizing polymer materials.

Silicon-based Devices

Silicon microfabrication, initially developed for integrated circuit technology in the 1960s, has subsequently become the most well-established and used technique for fabricating microcantilevers [10]. Furthermore, techniques needed to integrate the wiring required for both actuation and readout of the cantilevers have extensively matured through much research and testing. Specific materials used in these devices include silicon, silicon nitride, and silicon oxide layers, which are often combined in different ways depending on the ultimate sensing goal.

The microfabrication of silicon cantilevers is very similar to the fabrication of the pressure sensor described in Chapter 2. It is based on techniques such as film deposition, photolithography, etching, and doping. The fabrication process typically begins with a monocrystalline wafer of silicon and the addition of multiple layers that will later define the detection scheme of the cantilever sensor.

Deposition involves the act of applying a thin film of material onto the surface of the wafer. This step can be performed by a variety of methods, including spin-coating, chemical or physical vapor deposition (CVD/PVD), or electroplating. CVD is useful for depositing silicon dioxide or silicon nitride layers that can serve as insulation, masks, and etch-stops during later fabrication processes. PVD and electroplating are used to deposit different metal films on the silicon wafer, which give additional functionalities to the cantilever. Such characteristics include producing reflective surfaces, electrode material, electronic interconnects, and chemically reactive binding sites.

Photolithography transfers specific patterns onto the wafer. After the pattern from the photomask has been transferred to the photoresist layer, the following etching step carves the shape defined by the photoresist onto the wafer. Etching is the process used to selectively remove specific parts of film or the wafer itself. After this step, the cantilever is "released" from the main wafer substrate, as the etching process defines the physical shape of the sensor. Two of the most commonly used techniques are anisotropic wet etching and deep reactive ion etching (DRIE). These techniques represent the two fundamental etching types: wet and dry. Anisotropic wet etching is named so because it is orientation dependent, and thus will produce sharp, well-defined features based on the crystal direction. DRIE, also an anisotropic process, relies on plasma to react with and remove the surface of a material on the wafer. Details of each technique can be found in Chapter 2.

Doping of specific impurities is used to create piezoresistive elements into the silicon crystalline structure. Depending on the dopant and its concentration, electrical properties of cantilever can be effectively tuned for many different uses. Incorporation of boron, aluminum, or gallium (Group 13) into silicon is referred to as p-type (positive) doping. These elements have one less electron than silicon and thus create "holes" in the structure due to their ability to accept electrons. Incorporation of phosphorus or arsenic (Group 15) into silicon is referred to as n-type (negative) doping. These elements have one more valence electron than silicon and thus leave a free electron inside the structure [11]. These "holes" and "free electrons" serve as charge carriers that move about inside the silicon lattice (for doping see Chapter 4, Section 4.3.1).

Silicon is the most common material used in microcantilever fabrication due to the unique electrical properties and ease of fabrication (Fig. 6.12). Additionally, the abundance of silicon compounds makes it one of the cheapest materials to fabricate devices from. Oxygen and silicon are the two most common elements in the Earth's crust, and together they form silica and silicates that can be used to extract pure silicon for fabrication. The abundance of the element leads to a low cost of production, an important factor for commercialization.

Silicon is suitable for cantilever beams also for both its mechanical and thermal properties. According to Table 6.1 [13], silicon's Young's modulus is comparatively low, allowing for the elasticity needed for static and dynamic deflection modes.

Polymer Based Cantilevers

Since the late 1990s, an increasing amount of work has been reported on the fabrication of polymer-based cantilevers. One advantage of using polymers is that Young's modulus is

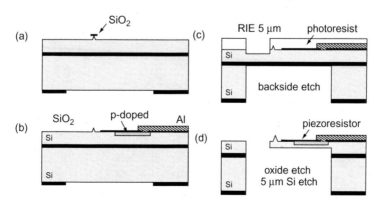

Figure 6.12: Fabrication of Silicon Piezoresistive Cantilever [12]. The p-doped Area in (b) Becomes the Piezoresistor in (d).

Table 6.1: Mechanical Properties of Common Materials

	Yield Strength (10^{10} dyne/ cm^2)	Knoop Hardness (kg/mm^2)	Young's Modulus (10^{12}dyne/cm^2)	Density (gr/ cm^3)	Thermal Conductivity (W/cm°C)	Thermal Expansion (10^{-6}/°C)
*Diamond	53	7000	20.35	3.5	'20	1.0
*SiC	21	2480	7.0	3.2	3.5	3.3
*TiC	20	2470	4.97	4.9	3.3	6.4
*A1,0,	15.4	2100	5.3	4.0	0.5	5.4
*Si$_3$N$_4$	14	3486	3.85	3.1	0.19	0.8
*Iron	12.6	400	1.96	7.8	0.803	12
SiO$_2$ (fibers)	8.4	820	0.73	2.5	0.014	0.55
*Si	7.0	850	1.9	2.3	1.57	2.33
Steel (max. strength)	4.2	1500	2.1	7.9	0.97	12
W	4.0	485	4.1	19.3	1.78	4.5
Stainless Steel	2.1	660	2.0	7.9	0.329	17.3
Mo	2.1	275	3.43	103	1.38	5.0
Al	0.17	130	0.70	2.7	2.36	25

Source: [13].

typically two orders of magnitude lower than for conventional silicon-based materials [9]. Because of the lower stiffness, more variation in deflection occurs, leading to a more sensitive device. The cantilever's stiffness can be further tailored simply by changing materials. Another advantage of polymer-based cantilever is the relatively cheap raw material and fabrication [9].

In 1994 Pechman et al fabricated an early polymer microcantilever with a standard Novolak-based photoresist as the material [14]; Genolet et al utilized the negative epoxy photoresist SU-8 to build AFM-cantilevers in 1999 [15]. Other polymers such as fluoropolymer [16], polyimide [17], polystyrene [18], polypropylene [11], polyethylene therephtalate [19], and parylene [20] have been used and evaluated.

Fabrication Techniques

Polymeric microcantilevers are fabricated in a variety of ways depending on the type of polymer to be used. Unlike the bulk micromachining of silicon cantilevers, the polymeric cantilever structure is fabricated by depositing layers on the surface of a base substrate. The procedure is shown in Fig. 6.13. (a) First, a sacrificial layer is formed, followed by deposition of a cantilever material. Polymer deposition is usually achieved by spin-coating an organic solution. Parylene is a type of polymer formed by

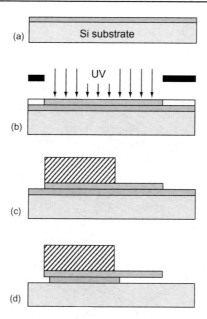

Figure 6.13: Summary of Microfabrication Steps for Polymer-Based Microcantilevers.
From [9].

CVD of di-para(xylylene) [21]. (b) UV-photolithography is then used to pattern the polymer. The figure shows an example of using photosensitive, or photocurable polymer such as SU-8 or a type of polymide. (c) After the cantilever is formed, a thick polymer layer may be added on top to facilitate handling of the fragile cantilevers after device release. (d) Finally, the cantilevers are released from the frontside of the wafer. The sacrificial layer is removed by a selective etching resulting in cantilevers suspended some micrometers above the silicon substrate. Either wet etching or dry etching may be used for this step.

An interesting alternative to classical surface micromachining is the soft lithography technique (see Chapter 2 Section 2.5.2). Injection molding of cantilevers has been used for polystyrene, polypropylene and nanoclay polymer composite [11]. As seen in Figure 6.14, in this process, a molten polymer is forced under pressure into a steel cavity. The mold are heated to the temperature of the molten polymer to ensure mold filling. Any thermoplastic polymer can be formed into microcantilevers. Injection molding is cost effective and suitable for mass production (Fig. 6.14). Other techniques including laser ablation, microstereolithography and multi-photon-absorption polymerization [11] present a high degree of freedom for design prototypes, although they are not suitable for a large scale production.

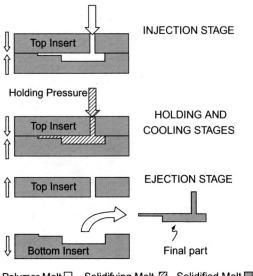

Polymer Melt ☐   Solidifying Melt ☒   Solidified Melt ■

Figure 6.14: General Injection Molding Process [11].

Advantages and Disadvantages

Advantages in polymer cantilevers are evident. The lower Young's modulus allows fabrication of cantilevers with high degrees of sensitivity. Also, the mechanical and chemical properties are readily tailorable [22]. Raw materials and fabrication techniques are cheaper than traditional silicon cantilevers [23].

One of the main drawbacks is the stability of the devices and the materials. Due to the small spacing between the substrate and the cantilever, the risk of adhesion of the cantilever to the substrate increases and the phenomenon called stiction might occur [11]. Polymer cantilevers present moisture absorption in liquids or degassing in vacuum that can result in unstable output signal [11]. Other drawbacks include fatigue, ageing or bleaching, which can affect the long-term stability of cantilevers. Metal coating is also needed for optical read-out since reflectivity of polymers is insufficient.

6.2.4 Principles of Sensing

There are two main principles of sensing with cantilevers: static and dynamic. Static sensing involves measurement of a fixed and stable change in the position of the cantilever. For biomolecular detection, surface stress-induced deflection is commonly used for static sensing. Dynamic sensing involves determination of the effect of analyte binding on cantilever motion. Measurement of resonance frequency is the preeminent method of dynamic sensing.

> **Sensing Principles**
>
> 1. **Static Measurement**
>   Measures deflection induced by surface stress
> 2. **Dynamic Measurement**
>   Measures change in resonant frequency

Static Measurement:

For sensor systems where analyte binding induces surface stress, flexible cantilevers functionalized on one side are used to transduce that stress into a measurable deflection. The operating principles and physics behind deflection-based microelectromechanical systems (MEMS) and nanoelectromechanical systems (NEMS) cantilever sensors will be explored.

Principles

Stress change in a thin film on the surface of a cantilever is illustrated in Fig. 6.15. On the left, the cantilever bends downward and expands until the stress created in the cantilever balances the expanding film on top. The stress on the top side of the supporting cantilever is tensile, because it is pulled by the film. In the case of a contracting film the cantilever bends upwards creating a compressive stress on the top side.

Stoney's equation is used to quantify the surface stresses generated in molecular thin films from different molecular recognition events. Stoney's equation for a cantilever is written as:

$$\sigma_{surface} = \frac{Et^2}{6R} \tag{6.61}$$

where R is the radius of curvature, E is the Young's modulus, t is the thickness of the cantilever, σ is the surface stress generated. When a substance binds to the surface of the cantilever, it will induce a deflection in the cantilever that will signify detection. Deflection

Tensile stress from the film   Compressive stress from the film

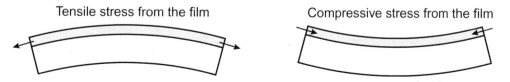

Figure 6.15: Expansion and Contraction.

based cantilevers have a sensitivity that is dependent on the radius of curvature. Therefore materials with a lower Young's modulus will achieve a greater sensitivity. Whether the cantilever experience expansion or contraction depends on the sign of the surface stress induced by substance. Figure 6.16 shows both cases of negative (left) and positive (right) surface stresses [23]. In the left panel, the cantilever surface is deposited with a gold thin film. When 10 μL of 5.85 mg/mL IgG is injected, a negative surface stress is observed. Figure 6.16(b) and (c) show two models explaining the phenomenon. The corresponding concentration is calculated to be 0.195 mg/mL. In Fig. 6.16(b) proteins try to "spread" or expand the surface. In Fig. 6.16(c), relatively immobile proteins deform or "flatten" to expand the surface. In the right panel, 10 μL of 6 mg/mL BSA is injected and positive stress is observed. Figure 6.16(e) describes the model where relatively mobile proteins pack together. As they try to contract, a tensile stress arises on top part of the substrate.

The origin of the surface stresses that give rise to nanomechanical bending in cantilevers has been much debated. Experiments involving DNA hybridization appear to indicate that the deflection is produced by a combination of various intermolecular forces including electrostatic charge attraction/repulsion, hydrophobic/hydrophilic interactions, hydrogen bonding, Van der Waals forces, and other events that attempt to achieve a thermodynamically favorable state [9]. In these instances, coating the surface with gold

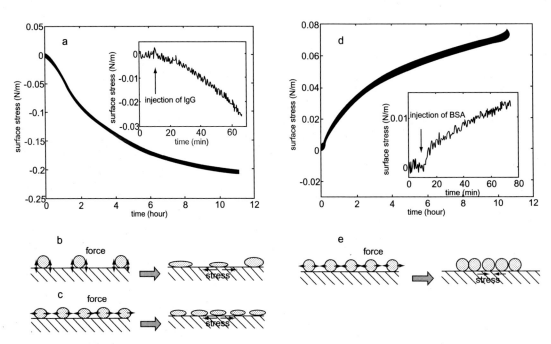

Figure 6.16: Negative and Positive Surface Stresses and Molecular Adsorption. (a-c) Injection of IgG Induces Negative Surface Stress. (d-e) Injection of BSA Induces Positive Stress.
From [23].

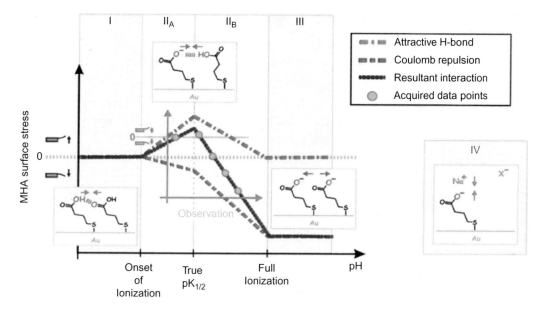

Figure 6.17: Surface Stress and Surface Functionalization [24].

appeared to dramatically improve the magnitude of deflection for a given concentration of analyte [9]. Specifically, the movement of charge from the alkanethiol to the gold layer apparently induces the vast majority of additional surface stress during binding events [3,9]. Figure 6.17 shows another example of competing forces at a cantilever's surface and how they affect the displacement of the cantilever. Changes in surface stress were measured as surface-attached carboxyl groups transit from a fully protonated state at a pH of 4.5 (Phase I) to a fully ionized state at a pH of 9.0 (Phase III). Attractive hydrogen bonding induces tensile stress whereas electrostatic repulsion produces compressive stress.

Practical Considerations and Limitations

In recent years, deflection-based cantilevers have shown potential for widespread use as biological sensors. These devices are versatile in that they can be operated in air or transparent liquid, allowing real-time analysis, provided that there are no turbulences or significant fluctuations in temperature of the dielectric constant of the liquid [3].

Some limits in deflection-based cantilevers are the need for surface functionalization, which almost always involves a gold cantilever bilayer structure to take advantage of the gold-biomolecule linking chemistry [3]. The gold layer can increase background noise, since it is a good conductor and the cantilever structure is very sensitive to temperature fluctuations. For the static deflection methods reference cantilevers may be used to subtract background and improve their detection limits.

Another limit is that to achieve greater cantilever sensitivity, in order to have larger deflection, an increase in the length or decrease in the thickness of the cantilever is needed (see Stoney's formula, Eqs 6.36 and 6.40). To create such cantilevers, bulk micromachining is needed to overcome the risk of stiction during fabrication and operation. Another limit to this technique is that it is limited to the near monolayer regime. Single molecule adsorption cannot generate a measurable deflection in a realistic device. For instance, the lower limit detection of DNA hybridization was found to be about 2×10^{10} hybridized molecules per square millimeter [3].

Dynamic Measurement

Principles

Most dynamic sensing methods are based on measurement of resonant frequencies. When a cantilever is initially actuated, it will oscillate at a particular frequency, known as its natural frequency. When analyte molecules are adsorbed to the cantilever, the additional mass induce a change in the resonant frequency. Theoretical details were described earlier. Here, we remind the readers the **responsivity** of a mechanically resonating system:

$$Rsp = -\frac{f}{2m_{\text{eff}}} = \frac{\Delta f}{\Delta m} \qquad (6.62)$$

where f and m_{eff} are the resonant frequency and the effective mass of the system, respectively, and Δm and Δf are changes in the mass and the frequency, respectively. **Sensitivity** is defined as the minimum input level required to produce an output that overcomes the threshold, which is typically chosen just above the noise level.

Practical Considerations and Limitations

As Eq 6.62 indicates, the responsivity of the system can be improved by either increasing the resonant frequency or reducing the mass. Reducing mass can be accomplished in two ways: reducing the dimensions of the cantilever beam and reducing the density of the material used to fabricate the cantilever. Reducing mass not only reduces the effective mass of the cantilever but also increases the resonant frequency. Nanoscale resonators that we will describe in Section 6.2.8 are good examples of systems that take advantage of an extremely small mass to obtain higher resolution.

A critical issue for a mechanical resonating system is dissipation of kinetic energy as a result of damping. Dissipation is defined as the ratio of energy lost per cycle to the stored energy of the system [9]. It is found as the inverse of quality factor. This dissipation broadens the resonant peak and introduces noise that increases the minimum detectable frequency change. Damping can be either intrinsic or extrinsic. Intrinsic damping is the damping caused by the cantilever material or anchor loss [25–28], while extrinsic damping is the damping caused by viscous and other forces from the surrounding media. In general, extrinsic damping dominates in silicon cantilevers and intrinsic damping dominates in polymer cantilevers [9]. Because of this, the damping of silicon cantilevers can be greatly reduced by placing the cantilevers in a vacuum or a low-viscosity medium, although such environments limit the applications of the sensors. Reducing damping in polymer cantilevers is more difficult since the source of the damping is from the material itself.

Independent of the materials used, In order to increase surface stress sensitivity, the material should have low internal damping and the cantilever geometry should allow for a high quality (Q) factor. From reports, the Q factor of a silicon microcantilever normally ranges from 500 to 10 000, with some cases as low as 50. This is still larger compared to the Q factors of polymer-based cantilever sensors, whose ranges are between 10 and 25 [9].

6.2.5 Principles of Transduction

The precision of the electrical transduction or sensing readout presents another limiting factor to the overall sensitivity of a cantilever based sensor system. Specialized instruments are used to accurately measure changes in deflection or resonant frequency at the micro- and nanoscale [1]. We describe the following four important types of readout techniques.

- Laser reflection
- Interferometry
- Piezoresistance
- Capacitance.

The first two methods are laser-based optical measurements. The other two are categorized as electrical transduction. Types of cantilevers that use **piezoelectricity** such as quartz tuning forks will be described in Section 6.3. Other transduction techniques specific to nanomaterials such as carbon nanotubes or graphene are described in Section 6.2.8.

Laser Reflection

Working Principles

Laser reflection is used for most commercially available AFMs [29]. In this method, a laser beam is focused at a point on the surface of the cantilever. The beam is then reflected onto a position-sensitive photodetector. Typically the photodetector is a two-segment photodiode capable of converting the incoming laser light into a voltage or current. Figure 6.18 shows

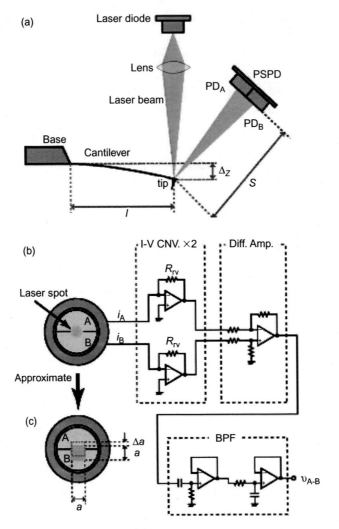

Figure 6.18: Laser Reflection-based Sensing Cantilever [30].

an example of laser reflection measurement [30]. As the cantilever changes position, the laser is reflected differently. The incoming laser light then changes its position on the photodetector, A differential amplifier is used to amplify the current or voltage difference between the two photosensitive areas to generate an electrical signal that corresponds to the position of the light spot (see Figure 6.18(b) and (c)). In resonant frequency sensing modes, a temporal component should also be measured to determine the resonant frequency.

Obtaining a reflective surface suitable for laser reflection-based readouts is relatively straightforward for silicon cantilevers. Typically, the default silicon surface is reflective enough for a laser. Issues arise with the use of polymer cantilevers, which tend to be less reflective than their silicon counterparts. In these cases, a thin film of gold, chromium, or some other reflective material is applied to the surface of the cantilever to ensure that the laser is reflected properly. This causes other issues; adding a gold layer to polymer cantilevers leads to unwanted bending caused by thermal-expansion or molecular adsorption that must be accounted for when performing sensor calibration [9]. The substance needs to be chemically inert in order to avoid any interaction with the media; at the same time, it may need to be functionalized by the attachment of receptor molecules to the surface. For these reasons, gold is the most practical choice for metallization of polymer based cantilevers.

Advantages and Disadvantages

Due to the relatively simple fabrication techniques and ease of use, the reflected laser beam is one of the most popular methods for transduction. No additional circuitry needs to be embedded in the cantilever greatly simplifying the instrumentation required. Furthermore, photodiodes can be manufactured to be extremely sensitive. Deflection resolutions in sub-nanometer and even sub-ångström regimes have been regularly achieved [5].

The largest drawback of this method is necessity of the precise optical alignment. Each time a new cantilever is employed in the system, the laser spot has to be carefully realigned to obtain correct signals. This issue also makes it difficult to construct a massively parallelized system because it requires a laser-photodiode system for each cantilever, massively increasing the size and cost of the device [5]. Several solutions have been proposed. IBM Zurich developed a system incorporating eight laser emitters, with each emitter paired with a single cantilever [31−33]. The lasers were angled in such a way that all reflected beams fell upon a single photodiode. The photodiode was able to track individual beam movements, allowing for an enhanced degree of parallelization. Another technique developed by Majumdar et al at the University of California, Berkeley, enabled the use of a two-dimensional array of cantilevers [34−36]. An expanded and collimated laser beam illuminates every cantilever of the array simultaneously. The reflections are imaged by a high-resolution charged-coupled device camera (CCD). It enables positional resolutions of approximately one nanometer, allowing for massive parallelization; 720 cantilevers have been simultaneously used in this type of configuration.

Other shortcomings include thermal issues associated with the use of a laser in certain media. In a liquid environment for example, thermal management can be an issue as the laser beam heats the liquid. Additionally, the liquid may deflect or scatter the beam, which introduces errors in the measurement.

Interferometry

Working Principles

When two coherent beams of light are superimposed, they create an interference pattern that can be measured and characterized. Figure 6.19 illustrates the schematic of an interferometer-based sensor [37]. A light beam is focused on the backside of a cantilever. The beam reflected from the lever and a reference beam reflected from the optical flat creates an interference pattern which is projected on a photodiode. The interference pattern is highly dependent on the path length of the reflected light. When a cantilever changes position, the intensity of the interference pattern also changes. This change is determined by the photodetector. Chapter 5, Problem 5.7 describes a simplified model of interferometry.

In [38], a diode laser and fiber optics are used instead of the microscope objective to create a compact interferometry system. The end of a fiber-optic waveguide is brought into close proximity to the surface of the cantilever. A laser is emitted from the waveguide, and the beam is reflected back into the waveguide. The laser is also partially reflected at the end of the waveguide. Those two beams create an interference pattern to be measured [38]. Since

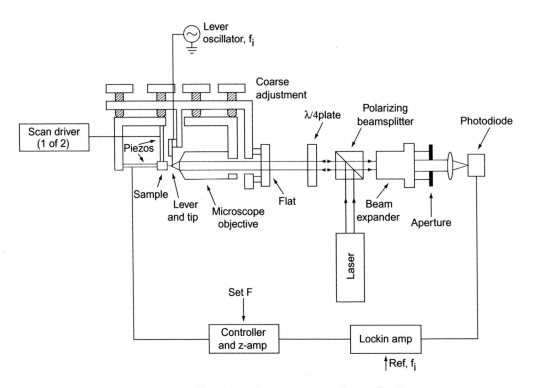

Figure 6.19: Schematic of Interferometer-Based Cantilever Sensor.
From [37].

interferometry relies on laser reflection from the surface of the microcantilever, basically the same fabrication techniques can be employed to create cantilevers for interferometry sensing and laser reflection sensing.

Advantages and Disadvantages

Interferometry is a highly sensitive sensing technique; it is capable of detecting displacements of 0.1 Å [39]. Many of the benefits conferred by this sensing mode mirror those of laser reflection-based position sensing. Similarly, the drawbacks are highly similar: thermal management and scattering in liquid environments are practical concerns that limit the utility of interferometric sensing [39]. This becomes particularly troublesome when attempting to perform biosensing, which is frequently performed in a liquid environment. Furthermore, the waveguide must be carefully positioned in order to achieve accurate measurement.

Piezoresistance

Another practically very important sensing method used in cantilevers is through piezoresistivity. A piezoresistive material is a type of material that changes its electrical resistivity as a result of mechanical stress. Doped silicon exhibits a strong piezoresistive effect in which the resistance of the doped region changes reliably when it experiences a stress due to deflection. Therefore, piezoresistive cantilever sensors are usually constructed of doped silicon. Piezoresistive elements are embedded at the top surface where the largest strain is induced. As mechanical deflection of the cantilever occurs, the resulting stress variation will produce strain that can be measured as a change in the material's resistivity. Sensing the cantilever deflection by integrating piezoresistors on the cantilever was initially developed by two groups for AFM imaging: Tortonese et al at Stanford University and Rangelow et al at Kassel University in 1993 and 1995, respectively [40,41].

Working Principles

The principle behind piezoresistive cantilevers is that molecular adsorption generates a surface stress due to interaction between the molecules and the cantilever surface. The difference in surface stresses will be balanced via cantilever bending. In Section 6.2.2, we described relationship between cantilever bending and the stress and strain distribution. The relative change in resistance due to an applied strain is linearly proportional to the longitudinal and transversal strain. This relationship is conceptualized by the gauge factor (GF). The GF of a material is used to characterize its strain sensitivity and is defined by:

$$\text{GF} = \frac{\Delta R}{R} \cdot \frac{L}{\Delta L} = \frac{\Delta R}{R} \cdot \frac{1}{\varepsilon} \tag{6.63}$$

or

$$\frac{\Delta R}{R} = \text{GF} \cdot \varepsilon \qquad (6.64)$$

where ΔR is change in resistance, R is the initial resistance, ΔL is change in length, and L is the initial length. In Eq 6.64, the change in the resistance is linearly correlated with the strain.

The stress distribution is a function of the cantilever's thickness and length and is distributed in such a way that it is greatest at both surfaces near the supporting end of the cantilever (see Fig. 6.6). It is near these surfaces where a piezoresistive material is most sensitive. Therefore, piezoresistive materials are usually localized very near the supporting surface of the beam. The beam can then be used as the arm of a Wheatstone bridge to calculate, through a common voltage divider, the change in resistivity when the beam is bent. The configuration is very similar to that of a piezoresistive pressure sensor we described in Chapter 2, Section 2.4.1. Figure 6.20 illustrates a basic schematic of a piezoresistive microcantilever [40].

Fabrication Techniques

Figure 6.21 shows a typical fabrication procedure for a silicon piezoresistive cantilever [42].

(a) A bare silicon substrate is coated with a silicon oxide or nitride layer deposited by thermal oxidation or low-pressure chemical vapor deposition (LPCVD). A substrate deposited with an additional thin silicon layer, which will construct the cantilever body, may be used instead of a bare wafer. A backside silicon nitride or oxide window are patterned with RIE (reactive ion etching), followed by wet-etching or DRIE (deep RIE) patterning.

(b) The top oxide or nitride layer is patterned using RIE or HF wet etching. Regions of p-type doping are created by thermal diffusion or ion implantation.

(c) Another layer of silicon oxide or nitride is deposited to insulate the doped region. A tip may be integrated depending on the application.

(d) A metal (Au/Cr or Al) layer is deposited and patterned for electrical connection.

(e) Additional protective layer may be deposited to cover the metal wiring.

In this example, a Wheatstone bridge consisting of symmetric four p-type resistors is implanted on the surface near the cantilever base where maximum surface stress σ arises (Figure 6.21(f)).

After the cantilever is fabricated, Additional Au/Cr layers may be deposited on the surface for an immobilization layer for proteins or nucleic acid interactions. The piezoresistive cantilevers can be further integrated with electronics, such as CMOS circuitry, to accomplish electrical amplification and processing such as those in commercial accelerometers and pressure sensors.

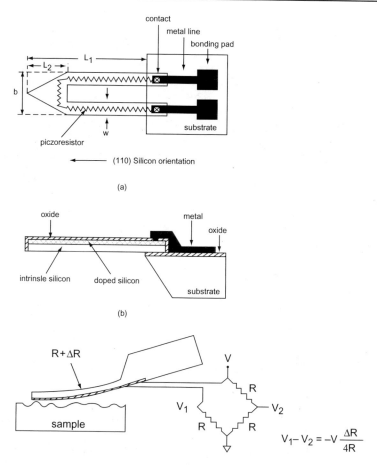

Figure 6.20: Schematic of Piezoresistive Cantilever [40].

Advantages and Disadvantages

The main advantage of a piezoresistive cantilever is that the measurement system can be integrated on the same chip without the need for optical alignment. This allows for the fabrication of large arrays of microcantilevers that can be read out simultaneously [3]. It is also possible to integrate a compact system suitable for an in situ analysis. They work well under liquid environments [43].

The main problem associated with a piezoresistive system is thermally induced conductance fluctuation, or thermal drift [8]. Piezoresistive cantilevers are fundamentally sensitive to temperature changes, since the conductivity of a semiconductor relies on transportation of thermally excited electrons (see Chapter 4). This thermal sensitivity will introduce fluctuation in the measurements and output signal, and thus thermal detection is desired to

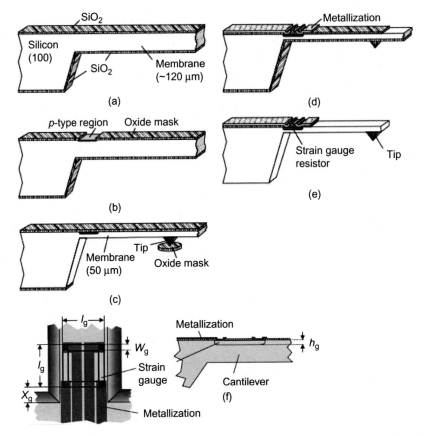

Figure 6.21: (a)−(e) Fabrication Process of a Piezoresistive Microcantilever. (f) Wheatstone Bridge Implemented on the Cantilever Surface [42].

better calibrate the measurements. In the case of piezoresistive cantilevers, the thermal drift can be reduced by using the Wheatstone bridge configuration. This is because the thermal variation would cause changes to all resistances in the Wheatstone bridge in the same way. The temperature variation would cause an effect both in the numerator and the denominator of the voltage dividing equation on the Wheatstone bridge. Another type of thermal effect is known as the Johnson−Nyquist noise, which arises from the thermal agitation of the electrons in the piezoresistor. The Johnson−Nyquist noise cannot be eliminated by the Wheatstone bridge. It defines the theoretical sensitivity limitation of semiconductor based sensors (see Chapter 1, Section 1.4.2 for discussion).

Applications

There are various applications that use piezoresistive microcantilevers such as commercial accelerometers and pressure sensors. Specific commercial products that utilize this type of sensing technique is the CantiChip biosensor produced by Cantion A/S, Denmark [3]. Cantion A/S was one of the first companies which offer commercially available piezoresistive cantilevers. Their technology involves cantilevers placed in arrays of either 4 or 16. Such arrays have been used, for example, to monitor consolidated bioprocessing by detecting ethanol and glucose and saccharide [9]. Research-grade cantilevers have been fabricated in numerous experiments and used for a wide variety of applications. These include micromachined piezoresistive thin-film self-sensing micro-cantilever array sensors, polydimethylsiloxane (PDMS) fluidic cells for bioassay studies, polycarbonate liquid chambers [1], NEMS-based cantilevers for DNA-binding assays [2], microscale air flow sensors [44] and tactile sensors for robotic applications [45].

Capacitance

Working Principles

Capacitive cantilevers may be operated in both resonant and static modes. In the resonant mode, the cantilever is subjected to a voltage that vibrates the beam at its resonant frequency. Oscillation amplitude, phase, and frequency of the cantilever are modified depending on interactions with the sample, and these changes can be measured to quantify analytes specifications [22].

Capacitive detection of cantilever deflection is based on the basic principles of the parallel-plate capacitor model we discussed in Chapter 2. The cantilever itself acts as one of the two plates, and adsorption of an analyte creates deflection that alters the distance between them [46]. Since capacitance is inversely proportional to distance between the plates, a decrease in distance results in a measurable increase in capacitance. Knowledge of the surface area of the cantilever plate and the permittivity of the dielectric medium allows for calculation of absolute displacement, making this technique potentially highly sensitive [22]. The equations describing the capacitance, the voltage and the charge in a parallel-plate model are explained in Chapter 2, Section 2.4.2. The capacitance is proportional to the surface area S and is inversely proportional to the gap x. (see also Eq. 2.4):

$$C = \varepsilon_0 \frac{S}{x} \qquad (6.65)$$

Since capacitance is defined as the ratio of charge Q on each conductor to the voltage V between them:

$$Q = CV \qquad (6.66)$$

While the simplicity of the associated electronics makes capacitive detection advantageous, the technique is not commonly used due to a number of limitations. In order to accurately record cantilever deflection, the dielectric material between the conductive plates must be

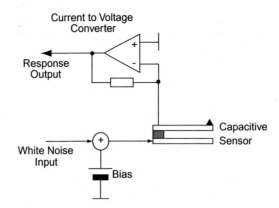

Figure 6.22: Capacitive Sensing Based Cantilever [47].

known or remain constant throughout the experiment. This is problematic if analytes or medium goes inside the gap, effectively changing dielectric constant [46]. Thus, capacitive cantilevers are suited for gas sensing and function poorly with aqueous solutions. Capacitive detection is also not suitable for measuring large displacements. If the cantilever deflects too closely to the plate beyond the pull in limit, the opposing charges will cause the two plates to stick together (see discussions in Chapter 2, Section 2.4.2.). Furthermore, scaling down the size of the capacitive cantilever will lower its overall sensitivity because capacitance of the system is directly proportional to its surface area [46]. Figure 6.22 shows a schematic diagram of the setup used in detecting change in capacitance [47]. The cantilever beam is virtually grounded by the current to voltage converter, minimizing the effects of stray capacitances.

Fabrication Techniques

The main components of a capacitive cantilever system consist of two adjacent, highly doped single-crystal silicon beams forming the two plates of a capacitor. These two beams are referred to as the cantilever and the counter electrode, and are anchored to a larger silicon substrate. Typical cantilever dimensions are 60 μm wide, 5−6 μm thick, and 275−350 μm long. In [47], the experimental capacitive cantilevers are fabricated by silicon bulk micromachining (Fig. 6.23). The fabrication begins with a 15 μm-thick n-type (1 0 0) oriented silicon film bonded on a 400 μm n-type (1 0 0) oriented silicon substrate, with a 1 μm silicon dioxide layer in between. Next, two 1.7 μm silicon dioxide layers are thermally grown on the outside surfaces of the silicon wafers. Photolithography and wet etching are then used to define patterns for the cantilever system in the top oxide layer. The counter electrode is etched by RIE and the exposed middle oxide layer is subsequently removed by wet etching. A 1 μm-thick layer of aluminum is next deposited to form the ohmic contacts of both cantilever and its substrate. These ohmic contacts serve as a junction point between the metal leads of wires and

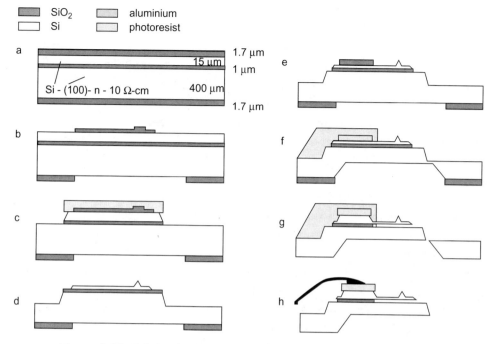

Figure 6.23: Fabrication Process of Capacitive Microcantilever [47].

the semiconductive material. An appropriately patterned photoresist cover layer protects the aluminum bonding pads from being attacked in the subsequent fabrication steps. Finally, the cantilever and counter electrode are released from one another using both RIE and wet etching to ensure proper removal of the oxide layer [47].

Applications

Capacitive sensing is one of very basic techniques for microstructures. It is used for displacement sensing of many MEMS based resonators [48,49] and gyroscopes [50–52]. However, as previously mentioned, it is not commonly used in detection methods for microcantilevers. Many capacitive cantilevers produced are mainly for research purposes.

Li et al. at National Tsing Hua University reported electric actuation of the cantilever which produced a resonant frequency of 396.46 kHz and a quality factor of 2600 at 10 mTorr. Deposition of a 0.1 μm silicon layer caused a 140 Hz shift in resonant frequency. They reported that the system detected on average 353 fg/Hz, and a resolution of 5×10^{-18} g [46]. Verd discussed similar results using an integrated circuit and capacitive cantilever, reporting both high sensitivity (10^{-18} g) and high spatial resolution (300 nm) [53]. For gas sensing applications, Amirola et al used capacitive detection of volatile organic compounds and found the limit of detection below 50 ppm for toluene and 10 ppm for octane [54].

6.2.6 Molecular Sensing with Cantilever-Based Sensors

An early study of microcantilevers used for molecular sensing were reported by Thundat et al at Oakridge National Laboratory in 1994 [55]. They measured the deflection of scanning force microscope cantilevers influenced by both thermal heating and variation in relative humidity. The amount of vapor adsorbates in the cantilever can be measured by monitoring the change in the resonant frequency [55]. Barnes et al of IBM, Zurich used the properties of cantilevers to perform sensitive photothermal spectroscopy. They reported on a cantilever sensor capable of detecting temperature changes in the order of picojoules [56]. In 1996, Butt et al measured the increase of the surface stress of the cantilever in aqueous medium with increasing pH and unspecific binding of proteins to hydrophobic surface [57]. Lang et al of IBM Zurich in 1998 developed a system with eight cantilevers that could distinguish different alcohols due to their different adsorption rates in the polymer-coated cantilevers [31]. In the field of biomolecule detection, Fritz et al. in 2000 developed the cantilever-based sensor that discerns single-base variation in DNA strands without using fluorescent labels [58]. Transduction of DNA hybridization is measured via surface stress changes of a cantilever. They were able to detect even a single base mismatch between two oligonucleotides. Figure 6.24 shows differential signal of a hybridization experiment. Two cantilevers were functionalized with sequences that are different only in one base: 5'-CTATGTCAGCAC-3' and 5'-CTATGTAAGCAC-3'. Injection of the first complementary oligonucleotide as shown in Fig. 6.24(B) increased the differential signal (interval II in Fig. 6.24A). Injection of the second complementary oligonucleotide as shown in Fig. 6.24(C) decreased the differential signal (interval III in Fig. 6.23A). They further studied detection of immunoglobins and other proteins, suggesting a number of possible new applications based on biomolecule detection [58].

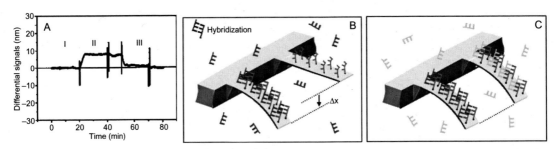

Figure 6.24: Measurement of DNA Hybridization [58]. (a) Differential Signal of Two Cantilevers Functionalized with Sequences Different in One Base. (b) Injection of the First Complementary Oligonucleotide Increased the Difference. (c) Injection of the Second Complementary Oligonucleotide Decreased the Difference.

Other interesting applications include the use of cantilevers as cancer-detecting microchips and glucose biosensing [39]. In the case of cancer detection, Majumdar et al [59] functionalized the cantilever by coating the surface with antibodies specific to prostate-specific antigen (PSA), a prostate cancer marker found in the blood of a patient with prostate cancer. When the coated microcantilever interacted with the sample solution added with PSA molecules, antigen−antibody complexes were formed and the cantilever bent as a consequence of the adsorption. Figure 6.25 shows measured surface stresses as a geometry-independent parameter for assaying PSA. Surface stresses measured with cantilevers with different geometries fit onto a single curve as a function of free PSA concentration.

The bending of the cantilever was detected using low-power laser reflection with subnanometer precision [39]. In the case of glucose biosensing, Pei et al [60] reported a technique for micromechanical detection of biologically relevant glucose concentrations by immobilization of glucose oxidase onto the microcantilever surface. The microcantilever undergoes bending due to a change in surface stress induced by the reaction of glucose oxidase with glucose present in solution.

6.2.7 Discussion on Silicon Cantilever Sensors

One of main advantages of microcantilever sensors is the compatibility with the silicon integrated circuits.

Overall, microcantilever sensors are highly sensitive devices capable of detecting minute changes within a medium. Surface stresses induced by binding events can alter the shape and conformation of a cantilever, and the mass of bound analytes can affect the resonant frequency of the cantilever. From a material standpoint, silicon is the dominant bulk

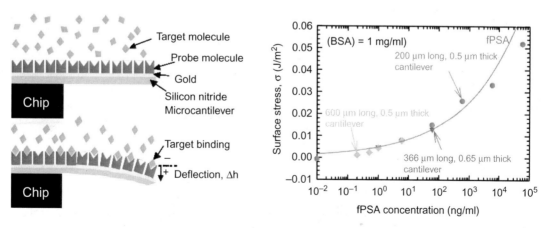

Figure 6.25: PSA Detection Using Microcantilevers.
From [59].

Table 6.2: Comparison of Cantilever Materials and Fabrication Methods

Material	Si[a]	SiN[b]	SiO[c]	SU-8	Topas	Polystyrene
Cantilever Fabrication						
Fabrication method	Etching	Etching	Etching	UV	NIL	Injection
Fabrication costs	High	High	High	Medium	Medium	Low
Influence of metal coating deposition	None	None	None	Bending	Bending	Bending
Cantilever properties	Theoretical values based on typical dimensions (L = 500 μm. w = 100 μm)					
Thickness h	500 nm	500 nm	1 μm	2 μm	4.5 μm	5 μm
Youngs' modulus E	180 Gpa	290 Gpa	85 GPa	4 GPa	2.6 G Pa	3 Gpa
Poisson's ratio v	0.28	0.27	0.25	0.22	0.26	0.34
Density ρ	2.3 g cm^{-3}	3.0 g cm^{-3}	2.7 g cm^{-3}	1.2 g cm^{-3}	1.0 g cm^{-3}	1.0 g cm^{-3}
Spring constant k	4.5 mN m	7.3 mN m	17.0 mN m	6.4 mN m	47.4 mN m	75.0 mN m
Resonance frequency f_o ($f_o = \omega_o/2\pi$)	2.8 kHz	3.2 kHz	3.7 kHz	2.4 kHz	4.6 kHz	5.5 kHz
Surface stress sensitivity $\Delta z/\Delta\sigma$	12.0 m^2 N^{-1}	7.6 m^2 N^{-1}	6.6 m^2 N^{-1}	36.6 m^2 N^{-1}	10.5 m^2 N^{-1}	6.6 m^2 N^{-1}
Mass responsivity $\Delta f/\Delta m$	24.4 Hz ng^{-1}	21.2 Hz ng^{-1}	13.8 Hz ng^{-1}	10.0 Hz ng^{-1}	10.1 Hz ng^{-1}	10.6 Hz ng^{-1}
Measurements	Qualitative comparison of sensor performance					
Damping (quality factor Q)	Viscous	Viscous	Viscous	Material	Material	Material
Thermal stability	High	High	High	Medium	Low	Low
Moisture absorption in bulk	Low	Low	Low	High	Medium	Medium
Time-stability	Years	Years	Years	Months	Months	—
Reflection of optical beam without metal	High	High	High	Low	Low	Low

[a]Crystalline Si.
[b]LPCVD nitride.
[c]PECVD oxide.
Source: [9]

material for cantilever fabrication due to its abundance and good thermal, electrical, and mechanical properties. Table 6.2 shows the different material properties for each possible bulk material as well as its fabrication methods and costs. Note the variability in each material's Young's modulus, their minimum dimensions, density, spring constant, and other values critical to developing effective cantilever sensors.

Additionally, each transduction method confers unique benefits and detriments that affect their usefulness in a particular sensing application. Laser reflection and interferometry are frequently used due to high familiarity with techniques due to the development of AFM, but suffer from difficulty in parallelizing processes. Piezoresistive cantilevers allow for massive parallelization but are more difficult to fabricate. Capacitive sensing is difficult to implement in liquid environments owing to the variable dielectric constant in a changing fluid.

6.2.8 Nanomaterial-Based Cantilever Sensors

Here we describe molecular sensors based on nanometer scale mechanical cantilevers. Most nanoscale cantilevers are resonant sensors that take advantage of the small mass (see the discussion earlier). We introduce mechanical resonators based on carbon nanotubes, silicon nanowires, graphene, and other materials.

Carbon Nanotubes as Mechanical Resonators

Carbon nanotubes (CNTs) can be used as a mechanical resonator, which allow for the detection of molecules or other changes on the nanometer scale [61–63]. The principle is basically the same as we discussed in Section 6.2.2 for silicon oscillating cantilevers. A nanotube resonates at a particular frequency due to the size and shape of the tube. If mass is loaded onto this tube, the resonant frequency will change. Due to the orders of magnitude in size difference between CNTs and other MEMS sensors, CNT based sensors show much higher resonant frequencies and sensitivities to smaller masses.

Figure 6.26 shows a typical experimental setup for carbon nanotube resonator based molecular sensors, which is prepared in an ultra high vacuum (UHV) chamber $(1 \times 10^{-10}$ torr) [61]. A radio-frequency (RF) electric field is applied to the CNT to forces it into resonance. The field emission current from the nanotube to the counter electrode is measured to monitor the nanotube resonance. In this case, gold atoms evaporated from a resistive heater attach to the nanotube. A quartz crystal membrane sensor is used to stabilize the evaporation rate. The setup can be considered as a special type of evaporation chamber (see Chapter 2 for evaporation). Double walled CNTs with the outside diameter of $\sim$2 nm, length of $\sim$250 nm, and mass of $\sim$2 $\times$ 10^{-21} kg were used and operated at the resonant frequency of $\sim$300 MHz. Responsivity of 0.104 MHz/zg and the sensitivity 1.3×10^{-25} kg/Hz$^{1/2}$, which is equivalent to 0.40 gold atoms per Hz$^{1/2}$, were demonstrated [61]. In a similar study, vibration is measured by monitoring the change in the current through the field-effect transistor composed of the nanotube oscillating at $\sim$170 MHz [62]. They measured deposition of Cr atoms and demonstrated the responsivity of 11 Hz/yg $(=0.011$ MHz/zg$^1)$. The dimensions of the nanotube used were diameter $d = 1.2$ nm and the length $L = 900$ nm.

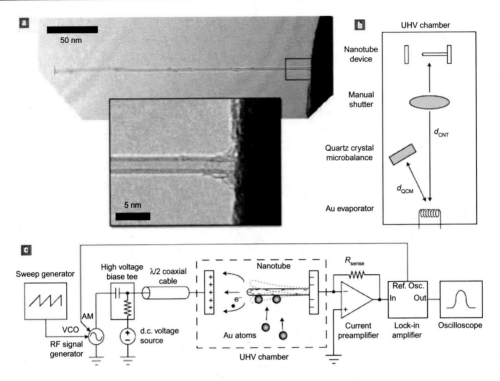

Figure 6.26: Carbon Nanotube-Based Resonator.
From [61].

A larger effective stiffness of the resonator leads to increase in the sensitivity. In [64], a nanotube with a diameter of 1.7 nm is supported at both ends with a suspended section of ~150 nm, which is much shorter than previous studies. The resonator vibrates at almost 2 GHz with the estimated effective mass of $m_{eff} = $ ~3×10^{-19} g, which corresponds to a responsivity of ~3.3 MHz/zg. The demonstrated sensing resolution of 1.7×10^{-24} g allowed for detection of naphthalene molecule adsorption and measurement of the binding energy of a xenon atom.

One critical issue with nanometer scale resonators is that as the mass of a resonator becomes smaller, the dumping effect from the surrounding medium becomes dominant. The fact that most CNT based resonators must be operated in a vacuum limits the applications of such nanometer-scale resonators.

Graphene Nanoelectromechanical Resonators

Studies on graphene have seen explosive growth recently, fueled by the 2010 Nobel Prize awarded to the researchers who first characterized the material. It has been

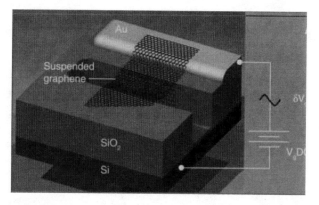

Figure 6.27: Schematic of a Graphene NEMS Resonator [65].

considered an ideal material for use in NEMS. The material's properties such as low density and stiffness make it attractive to be used in applications such as force and mass sensing. Current studies have shown that graphene flakes that are on the micrometer scale can be used as NEMS resonators.

As shown in Fig. 6.27, a graphene NEMS resonator often consists of a silicon wafer with a silicon dioxide layer on top of it. The graphene layers are then suspended over the silicon dioxide substrate via van der Waals forces. These graphene sheets are exfoliated mechanically over trenches that were etched into the silicon dioxide substrate through dry plasma etching. The metallic electrodes, often made of gold, are made through photolithographic processes. Vibrations with fundamental frequencies are then applied across the graphene layer and detected [65].

One application of NEMS resonators includes mass loading. Acting in principle similar to other acoustic wave-based sensors, NEMS resonators are able to detect an addition in mass by there being a shift in the frequency at which the graphene sheet would vibrate. This shift can then be detected allowing for mass detection. As shown in Fig. 6.28, there is an increase in frequency shift as mass is added [66]. Graphene based sensors are able to detect on almost an atomic level changes in mass.

In one experimental testing the mass loading sensitivity was observed to be about 2 zg. This result is slightly larger than the type of sensitivity demonstrated by nanotube but the sensitivity may be improved by reducing the noise of the detecting systems [67]. Graphene resonators also have the advantages of having a larger surface area that can capture more incoming mass as well as sensitive electrical properties. Also graphene has the advantage of being able to withstand high strains, which means there is a less of a chance of fracture when it is mass loaded [67].

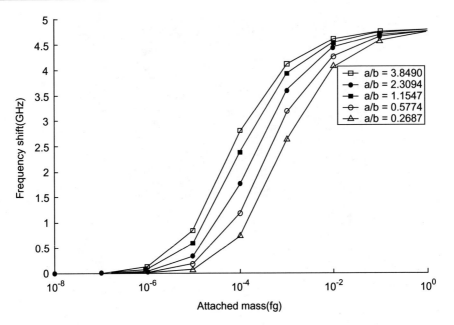

Figure 6.28: Plot Showing Frequency Dependency on Mass.
From [66].

Other Nanoresonators

Nanoscale mechanical resonators have been designed using other materials such as nanobelts, which can be composed of a number of metal oxides including ZnO, SnO_2, In_2O_3, Ga_2O_3, CdO, and PbO_2 [68]. Silicon nanowires have also been found to be effective resonators [69].

Conclusions

In order to demonstrate advantages over present commercial sensing technologies such as quartz crystal microbalance (QCM) and surface plasmon resonance (SPR) sensors, microcantilever sensors need to provide greater compatibility with existing silicon based digital technology. In addition, the microcantilever sensing system should be easy to operate and field deployable. Based on these criteria and the state of current technology, future research should focus on more-stable sensing materials, techniques to increase sensitivity, and techniques to lower response time. Cantilever sensing is still a relatively new field of research, but has provided several interesting opportunities for integrated sensors. Novel research concepts and implementations continue to proliferate, broadening the already varied list of potential and realized applications. Work to further characterize cantilever behavior has the potential to make available additional methods of exploiting this particular micromechanical device, providing exciting new developments in the realm of miniaturized sensors.

6.3 Acoustic Molecular Sensors: TSM, SAW, FPW, APM, and Tuning Forks

6.3.1 Introduction

In the 1970s, a class of sensor that utilizes acoustic waves was invented which steadily expanded into many applications [70,71]. Currently, acoustic wave sensors are capable of measuring properties in solid, liquid, and gaseous phases such as density, viscosity, elastic moduli, conductivity, concentration, and displacement with competitive sensitivities compared to other sensor designs [70,71]. In this section we will introduce the theory, design, fabrication, and multifaceted applications of devices utilizing acoustic waves including traditional bulk and surface wave propagation devices.

Types of Acoustic Sensors

A seminal review paper written in 1993 by Jay Grate, Stephen Martin, and Richard White broadly outlines four basic types of acoustic wave sensors [70,71], which are summarized in the list below. In addition to the four types of sensors, we will also introduce mechanical resonators based on a quartz tuning fork.

1. Thickness-shear-mode (TSM) sensors

    A TSM sensor is composed of a piezoelectric substrate that is placed in between two electrodes. It utilizes shear waves that propagate throughout the bulk in a direction perpendicular to plates. Mass loading alters the wave frequency. It is also called bulk acoustic wave (BAW) sensors. Operating frequencies ranges $5-10$ MHz. Wavelengths are typically ~ 500 μm.

2. Surface acoustic wave (SAW) sensors

    SAW sensors utilize the modulation of surface acoustic waves to sense a physical phenomenon. They use interdigitated transducers (IDTs) to generate Rayleigh waves that travel along the surface. Operating frequencies ranges $30-300$ MHz. Wavelengths are typically ~ 20 μm.

3. Flexural plate wave (FPW) sensors

    FPW sensors are composed of plates or membranes much thinner than an acoustic wavelength. IDT electrodes are used. Operating frequency range is $2-7$ MHz. Wavelengths are typically ~ 100 μm.

4. Acoustic plate mode (APM) sensors

    APM sensors are similar in form factor to SAW devices but feature thin plates that are only a few acoustic wavelengths thick. It shows properties of both the TSM and SAW devices. Operating frequencies are 25–200 MHz. Wavelengths are ~50 μm.

5. Quartz tuning fork resonators

A quartz tuning fork is a type of resonator that utilizes the piezoelectric effect. It shares many characteristics and implementation methods with other types of acoustic sensors. Operating frequencies range from 20–200 kHz.

6.3.2 Principles of Piezoelectric materials

Basics of Piezoelectric Materials

Fundamentally, many existing acoustic wave devices rely on the piezoelectric effect and it is prudent to provide a brief review here. In short, piezoelectric effect refers to the production of electric polarity by applying a mechanical stress to a special class of crystals, such as quartz and lithium tantalate. The inverse piezoelectric effect refers to the phenomena that stress produced in a crystal as a result of an applied potential difference, as shown in Fig. 6.29.

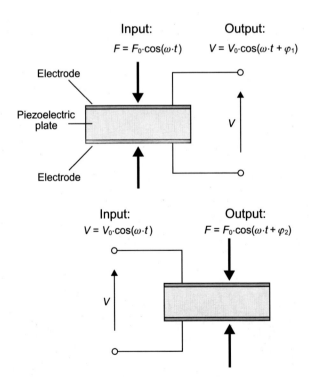

Figure 6.29: Piezoelectric (Top) and Inverse Piezoelectric Effect (Bottom).

Quantitatively, the relation between the force **F** and the induced charge **Q** is described by Eqs 6.67 or 6.68.

$$\mathbf{Q} = \mathbf{dF} \tag{6.67}$$

$$\begin{bmatrix} Q_x \\ Q_y \\ Q_z \end{bmatrix} = \begin{bmatrix} d_{11} & d_{12} & d_{13} \\ d_{21} & d_{22} & d_{23} \\ d_{31} & d_{32} & d_{33} \end{bmatrix} \begin{bmatrix} F_x \\ F_y \\ F_z \end{bmatrix} \tag{6.68}$$

where **d** is the coupling matrix describing the intensities of charges that are accumulated in response to a force applied with a particular direction. For most materials, forces only induce charges on directly opposing faces. That is to say,

$$\begin{bmatrix} Q_x \\ Q_y \\ Q_z \end{bmatrix} = \begin{bmatrix} d_{11} & 0 & 0 \\ 0 & d_{22} & 0 \\ 0 & 0 & d_{33} \end{bmatrix} \begin{bmatrix} F_x \\ F_y \\ F_z \end{bmatrix} \tag{6.69}$$

As we discussed, this mechanism also works in reverse as the inverse piezoelectric effect (i.e., application of electrical charge can induce mechanical motion). Accordingly, a piezoelectric material can be induced to vibrate by application of a sinusoidally varying electric signal thereby producing an acoustic wave. This effect can be harnessed to transduce between electrical and mechanical signals. Acoustic wave devices exploit this effect to measure a wide variety of measurands by careful selection of materials and structures whose mechanical properties vary with the measurand of interest. A simple but practically very useful case is that mechanical stress in certain crystalline materials induces charge formation on opposing surfaces of the material in one direction of Eq. 6.69 as:

$$Q_z = d_{33} \cdot F_z \tag{6.70}$$

This forms the basis of a large class of acoustic sensors. Acoustic wave sensors essentially function by piezoelectrically generating, passing, and monitoring a mechanical sinusoidal wave across a medium of interest. The difference between the input and output waves in terms of phase, amplitude, and frequency can be used to measure properties of the medium. By pretreating the medium with different materials, the medium can be sensitized to a variety of measurands including biomolecules, mass, radiation, magnetic fields, and temperature, among others [72]. Acoustic wave sensors rely on piezoelectric materials to transduce between electrical and mechanical waves and can be effectively modeled by the construction of equivalent circuits [73].

Equivalent Circuit of a Piezoelectric Mechanical System

Piezoelectric and mechanical systems can often be analyzed by means of an equivalent circuit which models the physical system being discussed. Acoustic sensors based on

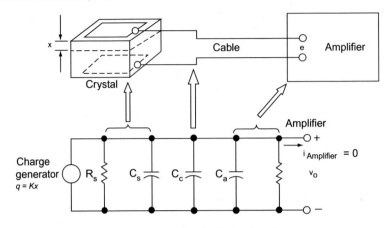

Figure 6.30: Equivalent Circuit.

piezoelectric materials has very high DC output impedance, therefore it can be modeled as a voltage source v_0. As we learned from the previous section, the output voltage v_0 is proportional to the applied force (or strain). Just as we perform signal transductions using transfer functions in electrical engineering, the sensor output signal is related to the input mechanical force as if it had passed through an equivalent circuit, as shown in Fig. 6.30.

In the equivalent circuit, x is the strain induced in the crystal by the external mechanical force, and K is the piezoelectric coefficient. The capacitance C_s consists of two parts: one represents the static capacitance of the sensor, the other part is inversely proportional to the mechanical elasticity of the sensor. The resistance R_s represents the leakage resistance of the sensor. In practice, the acoustic sensor is always connected to external systems, such as cables (represented as capacitance C_c), amplifiers (represented as capacitance C_a), and the subsequent systems which can be represented in general using load resistance.

Piezoelectric materials are not generally very good dielectrics. As represented as a parallel resistance R_s as shown in the equivalent circuit, a charge placed on a pair of electrodes gradually leaks away. The equivalent circuit provides the time constant for the retention of a voltage on the piezoelectric sensor after the application of a force. Figure 6.31 shows displacement response (a) with an ideal dielectric material with no leakage and (b) with actual piezoelectric material with different time constant. The time constant of such RC circuits dependent on: the combined capacitance of the sensor and the leakage resistance. When the time constant is smaller, the material leaks faster. For many popular piezoelectric materials, the time constants are of the order of 1 second. Due to such leakage, piezoelectric sensors are not very useful for the detection of static quantities, such as the weight of an object.

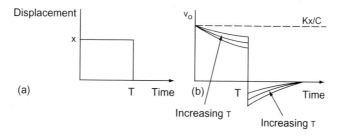

Figure 6.31: Piezo Leakage. (a) Ideal Displacement Response in an on-off Operation without Any Leakage (b) Actual Displacement Responses with Different Time Constants.

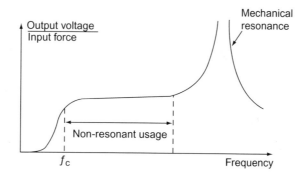

Figure 6.32: Piezomaterial Frequency Response.

For the general dynamic operations, a typical frequency response of a piezoelectric material is shown in Fig. 6.32, where the output voltage vs. applied force is shown as a function of frequency. The flat region of the frequency response, between the high-pass cutoff and the resonant peak, is typically used for non-resonant applications. Note that if the material is connected to a load resistance, the high-pass cutoff frequency will be increased. In other words, the load and leakage resistance need to be large enough that low frequencies of interest are not lost.

6.3.3 Thickness-Shear-Mode (TSM) Sensors

Working Principles of TSM Sensors

The TSM resonators are the oldest, simplest, and most prevalent type of acoustic wave devices. TSM resonators rely on the use of bulk acoustic waves, in which the propagated waves travel throughout the bulk of the substrate as opposed to just the surface. TSM resonators are composed of a piezoelectric substrate, typically quartz that is sandwiched

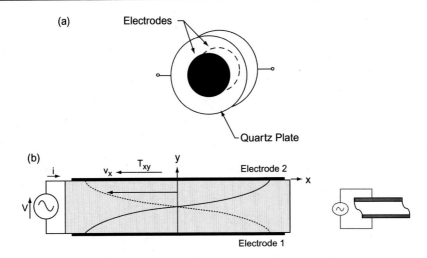

Figure 6.33: (a) Diagram of TSM Structure. (b) Vibrational Mode of a TSM Sensor.

between two thin metal electrodes. The use of quartz has often led TSM to be referred to quartz crystal microbalance (QCM).

Depending on the orientation of the crystal material, different modes of vibration can be induced. TSM resonators rely on shear waves that propagate throughout the substrate. The most commonly used for TSM sensors is an AT-cut thin disk, where a quartz mineral is cut at an angle of about 35° to the optic axis of the crystal [74]. A typical design of a TSM resonator is shown in Fig. 6.33(a). The diameter is typically about 1.5 cm (0.5 inch) and the thickness is ~100−200 μm. When an electric field is applied between the electrodes, it induces tangential deformation where the top and bottom surfaces move in parallel but opposite directions, resulting in acoustic waves that propagate through the crystal material. The vibrational mode is shown in Fig. 6.33(b).

The frequency of the oscillation is dependent on the thickness of the plate, as plate thickness decreases the frequency will increase [70]. The oscillation is sensitive to mass loading, and there is a linear relationship between changes in resonance frequency and mass accumulation. When there is a change in mass on the surface, a change in resonance frequency will also occur, which in turn will allow for detection.

The Sauerbrey equation describes analytical relation between the amount of adsorbed mass and the resonant frequency shift of the oscillating quartz, and is given as:

$$\Delta f = -\frac{2f_0{}^2}{A\sqrt{\rho_q\mu_q}}\Delta m \qquad (6.71)$$

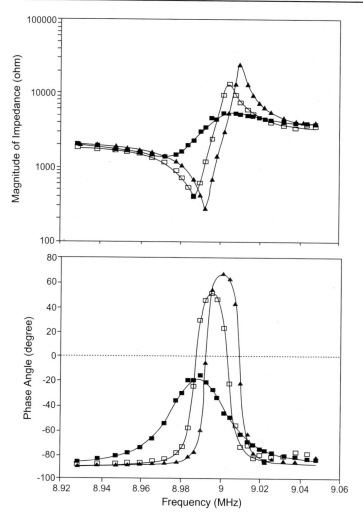

Figure 6.34: TSM Sensor Frequency Response.
Points are typical responses in experiments where one side of the crystal in contact with methanol
(▲), water (□), and cyclohexanol (■). Lines are equivalent circuit simulations. *From [75].*

where, f_0 is resonant frequency (Hz), Δf is frequency change (Hz), Δm is mass change (g), A is the area between electrodes (cm^2), ρ_q is density of quartz ($\rho_q = 2.648$ g/cm^3), μ_q is the shear modulus of quartz for AT-cut crystal ($\mu_q = 2.947 \times 10^{11}$ g/cm/s^2).

The equation forms the basis for measuring the resonant frequency and its changes by using the TSM as the mass deposition on the sensor surface. The response of TSM has been studied to measure liquid properties, such as the density and viscosity of liquids. Figure 6.34 shows the amplitude and phase characteristics of impedance for a 9-MHz TSM device. Points are typical measurements where one side of the crystal in contact with methanol (▲), water (□), and cyclohexanol (■). Lines are equivalent circuit simulations. The resonant peak shifts to smaller values as the product of the viscosity and density of the liquid increases [75]. The capability of measuring liquid properties extends the applications of TSM sensors to in situ monitoring of molecular changes occurring at the fluid—solid interface.

Applications of TSM Sensors

Deposition Rate Monitor

TSM resonators, or QCM sensors, have the ability to act as a sensor for a variety of parameters. One common application of QCM sensors is to monitor the deposition in thin film deposition systems under vacuum. The resonant frequency of a QCM is continuously measured to observe the rate of material deposition in e-beam/resistive evaporation tools. Density of the deposited material needs to be known to find the deposition rate in terms of thickness per time. For a thick film deposition, the elasticity of the film has to be also taken into consideration. See Chapter 2 for the actual use of a QCM sensor in metal deposition apparatuses.

Epinephrine Biosensor

Figure 6.35 shows an example of a molecular sensor. There are various components that can affect the oscillation of TSM resonators in this system. Such components include hydrophilicity/hydrophobicity, slip, mass, thickness of the crystal, coupling, and roughness. These factors allow for TSM to become a possible sensor for applications in corrosion detection, chemical sensing, viscosity sensor, and others [76].

Epinephrine is a neurotransmitter that is able to affect nerve impulses. There are several life phenomena that are correlated to the concentration of epinephrine found in blood, and medically it has also been used as a frequent emergency healthcare medicine. This is why much attention has been gathered towards the determination of epinephrine concentration. Current techniques include cation exchange; however, this method has rather low sensitivity. Another method, immunoassays, has the desired high sensitivity, but is time consuming and the materials to perform it are expensive. Also they have poor reproducibility and stability [77]. This is why a more convenient method for the direct determination of epinephrine is sought. TSM devices are one of the more favorable transducers because they are sensitive to variations in properties of their surrounding medium, and they are able to operate in a liquid. The surface of the TSM resonator was functionalized to promote specificity. In this application, the polymer coating that contains specific binding sites was added through spin coating. When the functionalized sensor is immersed in a solution with epinephrine, the neurotransmitter binds to the sensor, resulting in a shift in the resonant frequency. A direct relationship between the concentration of epinephrine and the frequency shift was observed, as shown in Fig. 6.36.

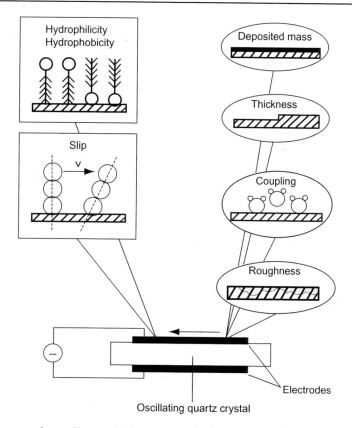

Figure 6.35: Factors that Affect TSM Oscillation Include Hydrophilicity/hydrophobicity, Slip, Mass, Thickness of the Crystal, Coupling, and Roughness [76].

Discussion on TSM Resonators

TSM Resonators carry the advantage that they have a low production cost due to the fact that quartz is abundant and inexpensive. The surface of the sensor can be easily functionalized (examples of functionalization methods for QCM sensors are discussed in Chapter 2, Section 2.5.3). It also has the ability to endure under liquid, or harsh environments, good sensitivity to mass deposition, and temperature stability. Since TSM resonators are able to detect and measure liquids they are extremely good candidates for biosensors. However, out of the four types of acoustic wave sensors, TSM are thought to be the least sensitive [78].

6.3.4 Surface Acoustic Wave Sensors

SAW technology allows for the production of versatile sensors that can be tailored to the application. They provide real-time data and have many applications [70]. The technology

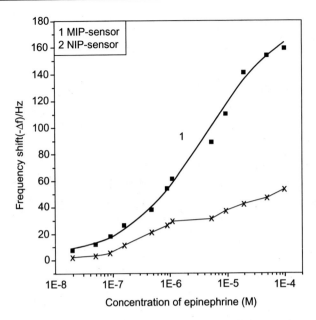

Figure 6.36: Relationship Between Concentration of Epinephrine and Frequency Shift [77].

operates on shear Rayleigh waves that traverse the surface of a piezoelectric substrate between one or more IDTs. Using changes in the waveform that include characteristics such as velocity, amplitude, frequency, and delay time, the signal can be transduced into electric signals to be analyzed [79]. There are several materials that can utilized for substrate, but all must have piezoelectric properties. Biomedical applications include vapor sensors, liquid biosensors, and others [71,72,80,81]. Through the 1980s the technology drew attention and has been studied widely. The first generation of commercial products is emerging on the market.

Working Principles

SAW devices use IDTs to convert electric signals to acoustic waves, which can be altered by sensing event and sent to a computational device for analysis. The IDTs generate a type of shear wave, called a Rayleigh wave, which propagates across the surface of the piezoelectric substrate [70]. Since these waves have transverse motion even though they are located at the surface, they must penetrate a depth of around one wavelength. The finger spacing and length of the IDTs primarily determine the shape of the waves. The wavelengths are typically ~20 μm.

Love wave sensors are a category of SAW devices that utilize surface wave with shear horizontal particle displacement to function. The sensors are typically composed of a piezoelectric substrate material with a guiding layer stretched across the surface. The metal

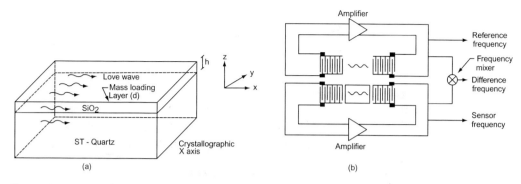

Figure 6.37: Generic Love Wave Sensor. (a) Geometry of the Love Waveguide. (b) Schematic Diagram of the Love-wave Dual-channel Delay Line Oscillator. The Sensing Channel Is Coated with A Chemically Selective Layer.
From [82].

layers of the interdigitated electrodes are sandwiched between the two layers of the substrate and the guiding layer, as shown in Fig. 6.37. These elements have variations and advantages that give rise to a wide range of devices with different sensitivities and applications. Generally the substrate is a piezoelectric material that consists of lithium, oxygen, and a group five metal such as tantalum or niobium, the thin guiding layer is typically made of a polymer (usually PMMA (poly(methyl methacrylate))) or silicon dioxide. The way these devices use surface acoustic waves is fairly simple in concept. The characteristics of the wave change due to changes in stiffness, electric and dielectric properties, as well as the mass density [82]. The sensing event or molecular adsorption initiates a physical or chemical change near the guiding layer. This change will cause a variation in the velocity, frequency, or amplitude of the acoustic waves that are traversing the guiding layer. A positive feedback loop created by the delay line will send the vibrations to the IDTs. As they are transducing these changes into electric signals by oscillation, they are mirrored into the new signals. A circuit can then process the data and present it in a readable fashion. Acoustic waves travel across surface, the addition of a guiding layer provides protection from the sensing environment. An example shown in Fig. 6.37(b) has a reference signal and a sensing signal to compare the sensing event.

Implementation Method

SAW sensors are built using basic photolithographic techniques, making them both inexpensive to produce and maintain a high level of quality, which will ensure they have high signal-to-noise ratios and consistent functionality. We discuss several applications of sensors that include their specific manufacturing methods including techniques and materials used. Quartz, GaAs, $LiNbO_3$, and $LiTaO_3$ are usually used as the piezoelectrical material.

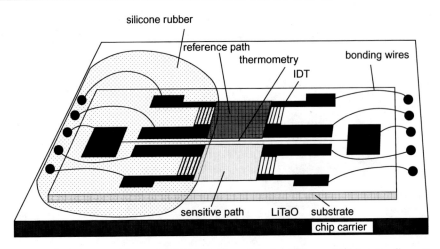

silicone rubber

reference path
thermometry
IDT

bonding wires

sensitive path   LiTaO   substrate

chip carrier

Figure 6.38: Typical Composition of a SAW Device with the Silicone Rubber Sealing. For Clarity the Sealing is Shown for One Side of the Device Only. The Sensitive Area Between the Transducers Is 1.4 × 3 mm². The Thin Metal Strip Between the Delay Paths Is Used for Thermometry.
From [83].

Lithium tantalate, $LiTaO_3$, is regarded as one of the ideal materials for biological sensing applications given its high electromechanical coupling coefficient, low temperature variability, and high dielectric constant [83]. For the IDTs, gold is generally preferred with its high conductivity and corrosion resistance which is particularly desirable when contacting volatile biological mediums. Fabrication generally utilizes photolithography. After patterning of the spin-coated photoresist layer, the unprotected part of the gold is removed by wet-etching. Finally, the remaining photoresist is removed leaving the gold interdigitated transducers on the substrate. The IDTs can be connected to the control circuitry in a variety of ways including wire bonding, mechanical pressurization, or capacitive coupling. A diagram of a representative assembly of an antibody based SAW biosensor is illustrated in Fig. 6.38.

Most designs feature two opposing IDTs separated by an open plane of lithium tantalate. The substrate is coated with antibody bonding proteins which provides excellent adhesion to the underlying lithium tantalate substrate. The protein layer can then be exposed to an aqueous antibody solution to induce the binding of antibodies on the substrate. After drying a homogenous monomolecular layer, uniformly aligned antibodies is obtained [84].

For integration onto miniaturized microfluidic devices such as lab-on-a-chip, several modifications for fabrication have been proposed and experimentally evaluated. SAW biosensors are integrated with a sampling chamber by using a UV-curing process, where a three dimensional structure is constructed with layer-by-layer exposure [83]. The result of this process is a fully encapsulated, disposable SAW biosensor that is suited for high throughput biological analyses.

Applications

SAW-based devices have a variety of applications that include filters, resonators, pressure and temperature sensors, oscillators, mass detectors and many others [70,81].

Chemical Sensors

The ability to detect and analyze chemical agents is of vast of importance in many areas. Some of these areas include: testing the air quality in living areas, monitoring toxicity levels in industrial plants, as well as checking for chemical agents in war zones. The development of a universal system that has quick detection abilities and high sensitivity is needed and is also relatively low in weight and cost.

SAW molecular sensors have been designed for the detection of minute biomolecule concentrations. Generally, these devices share some common characteristics. First, they are based on traditional SAW devices which include IDTs at both the input and output. The output signal of the device is interpreted to determine how it changed from the input signal in terms of frequency, phase, amplitude, or other characteristics. Second, in order to achieve molecular selectivity, SAW biosensors include a specifically sensitive thin-film that can change one or more of its properties in response to the presence of the targeted biomolecule. When binding targeted molecules, the thin-film increases its apparent mass which consequently affects the output signal of the device.

Some SAW chemical vapor sensors are currently being sold commercially. Microsensor Systems Inc. at the Naval Research Laboratory developed the first SAW chemical vapor sensor in 1999 called VaporLab. The portable units provide wireless transmission of data and are capable of identifying a wide spectrum of types of chemical gases. These devices also offer long-term stability and good sensitivity. Also the device has a response time of about 30 seconds, and its data storage allows up 200 vapor patterns as well as up 100 discrete measurements. The product's ability to be able to detect numerous vapor patterns is essential when in use; it ensures that multiple potential harmful chemical gases can be detected. It can be is handheld and easy to carry around to various locations. Hazmatcad™ developed by Microsensor Systems Inc. is a commercially available SAW sensor designed to detect chemical warfare agents (CWAs) and toxic industrial chemicals (Fig. 6.39). Hazmatcad and other types of commercially available devices are compared in the literature [85].

A typical vapor sensor consists of a polymer layer, which acts as an active layer whose physical properties are altered when they come into contact with the molecules or chemicals [86]. The surface acoustic wave sensor then transduces these physical changes to an electrical response; An acoustic wave sensor primarily includes a wave generating input transducer, a sensing film that lies on top of the piezoelectric material, and an output wave detector.

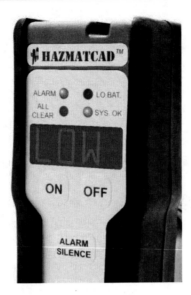

Figure 6.39: Hazmatcad™ Hazardous Material Chemical Agent Detector.
© *MSA 2000−2011 (http://us.msasafety.com/).*

When a voltage is applied, the input transducer generates a surface acoustic wave which moves across the piezoelectric substrate that has the chemical sensing film coating. The sensing film is a polymer surface to which desired chemical analytes bind. The output transducer will then detect changes in resonances frequency. The changes are then measured by an internal microcomputer that can then determine the presence of chemical agents as well as their concentration.

According to [87], the minimum detectable levels (MDL) of concentration were tested for tabun (GA), sarin (GB), and mustard (HD) and measured to be 1.37 mg/m^3 (0.056 ppm) in 44−57 s, 0.22 mg/m^3 (0.032 ppm) in 30−34 s, and 0.85 mg/m^3 (0.14 ppm) in 43−46 s, respectively. The immediate danger to life or health (IDLH) values for GB and HD when exposed up to 30 min are both 0.2 mg/m^3 (0.03 ppm).

Biomolecular Sensors

A promising future application of SAW biosensors is in high throughput biomolecular analyses intended for patient diagnosis. Many diseases such as cancer manifest with distinct although subtle changes in blood composition. Generally, the concentrations of tens to hundreds of unique biomolecules in the blood can be affected [88]. In order to accurately and inexpensively determine disease progression and regression in patients, a device which can automatically test for the presence of multiple biomolecules is desired. The small size, high sensitivities, and the modularity of SAW biosensors make them an excellent candidate for future cancer testing.

Outside of clinical and research usage, SAW biosensors hold a promise for general pathogen detection. For instance, researchers have developed a sensor capable of detecting *Salmonella* in the liquid phase by using antibodies specific to antigens found on the membranes of the bacteria [89]. This application could find widespread deployment in the food industry as a safeguard against contamination.

Länge et al overview sensors that are capable of detecting *Escherichia coli*, *Legionella*, the anthracis simulant B8 *Bacillus thuringiensis* (B8), and M13 bacteriophage (M13) [72]. All of these sensors use the basic principle of mass loading; however they incorporate antibodies that attach to the guiding layer which serve to entrap the molecules of interest as opposed to allowing gravity to hold them down. Another example for Love wave sensors is an oscillator since that is the core of how the device transduces information to begin with. The oscillator is composed of a Love wave sensor in combination with an amplifier acting as a positive feedback loop. The current direction for Love wave sensors has to do with improving their performance and versatility. In liquid applications, the viscosity of the medium can reduce the accuracy and sensitivity of the sensor. This is a measure that should be compensated for. The positioning of elements such as the IDTs may help elucidate new avenues of application [72].

The "sam X" biosensor system produced by SAW Instruments is an example of a commercially available sensor utilizing SAWs to detect pathogens and is shown in Fig. 6.40. Specifically, it provides for automated, simultaneous testing of a single analyte for multiple target biomolecules including those commonly found on the exterior of pathogens. First released in 2011, the sam X consists of a bench top apparatus that accepts sensor chips, which are manufactured with a variety of coatings. Researchers and laboratory personnel using the device can attach multiple targeting molecules of their choosing to the coating. Following attachment, the sensor chips and samples are loaded into the apparatus and data can be gathered in real time as the sample passes through the sensor's microfluidic channels and slowly deposits molecules of interest onto the sensor's delay line. In cases where the samples contain whole cells, the entire cell can be adhered and characterized. Depending on the system configuration, up to 8 discrete sensors can tested at once with minimal operator interference. Samples may be arbitrarily distributed among the sensors.

Chemical sensors that are based off acoustic wave technology often offer good sensitivity as well as good stability. The devices are able to perform sensing functions by the application of a sensing layer that will react to any desired substance. In comparison to competing technologies for chemical vapor sensors, such as optical and electrochemical, acoustic waves are able to detect a smaller change in mass loading giving them better sensitivity. However, with surface acoustic wave sensors, there exists the issue of overlapping sensitivities. The surface waves are also often influenced by changes in pressure, temperature, and viscoelasticity. These can be sources for potential error [84]. Therefore it is necessary to take these potential issues into consideration during the development of a SAW chemical sensor, such that they do not cause a change in frequency that could affect the output of the sensor.

Love wave-based biosensors exhibit improved accuracy over alternative devices such as surface plasmon resonance based biosensors [90]. Specifically, frequency sensitivities as high as 1500 Hz/(ng/mm^2) have been reported. Given the ability of current electronic analysis technology to distinguish between nearby frequencies, this corresponds to an overall mass sensitivity of 20 pg/mm^2). Accordingly, just 20 pg of target molecules need to bind to the each square millimeter of the device in order to produce a measurable signal. Theoretically, the highest frequency sensitivity achievable with traditional SAW biosensor designs has been estimated as 110 Hz/(ng/mm^2) [84].

SAW Sensors for Position Sensing

Here we introduce interesting application other than molecular sensing. Love waves can be harnessed to improve upon touchscreen technology due to their high sensitivity [73]. Consider a rectangular surface in which SAW waves are constantly propagating across. If a pulse is launched across the surface, it will have a characteristic amplitude and speed. If a

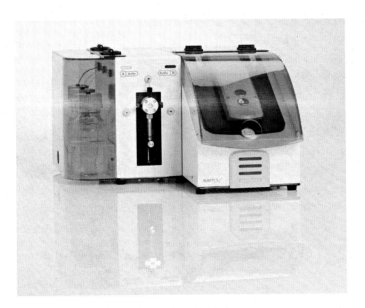

Figure 6.40: Sam X Biosensor.
© 2013 SAW Instruments GmbH (http://saw-instruments.com/products/samx.php).

finger applies pressure to the surface, the wave will be altered by attenuation and velocity change. Using this output data, the location and intensity being applied by the finger can be calculated [91].

6.3.5 Flexural Plate Wave Sensors

Working Principles

Flexural plate wave (FPW) sensors rely on a very thin delay line which is only a fraction of an acoustic wavelength thick, typically $2-3$ μm [80]. IDTs are positioned on the backside of the device and generate Rayleigh waves that propagate on the surface of the device. Interference between Rayleigh waves across the delay line plate produces Lamb waves. These waves can be conceptualized as a hybrid between surface and bulk waves. Figure 6.41 illustrates Lamb wave propagation. A primary advantage of this setup is that the IDTs can be isolated from the fluid medium which reduces noise and short circuiting effects. Additionally, when operating at low frequencies, comparatively high sensitivities are maintained [92]. Wavelengths are typically ~ 100 μm.

Implementation Method

FPW sensors can be fabricated using traditional photolithography methods similar to that of SAW sensors, given that both devices make use of IDTs to transduce electrical signals. Selection of an appropriate thin-film for a given application may be more complex. Factors to take into consideration include temperature dependence, thin-film expansion in response to adsorbed target molecules, and specificity against nontarget molecules [92].

Applications

FPW have found several applications. Molecular sensing is possible by immobilizing antibodies within a polymer bed. Specifically, the platinum substrate surface was functionalized with 3-aminopropylsilane and poly(acrylic acid). Antibodies were prepared by first coupling them to 1-ethyl-3-(3-dimethylaminiopropyl) carbodiimide and then linking them to the poly(acrylic acid) bed with glycine, as illustrated in Fig. 6.42 [93].

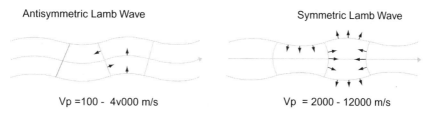

Antisymmetric Lamb Wave   Symmetric Lamb Wave

Vp =100 - 4v000 m/s   Vp = 2000 - 12000 m/s

Figure 6.41: Lamb Wave Propagation [80].

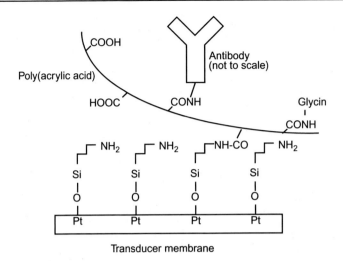

Figure 6.42: Flexural Plate Wave Sensor for Molecular Sensing.

From [93].

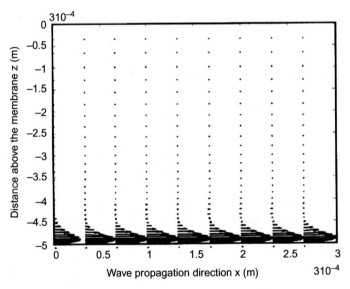

Figure 6.43: Velocity Field of Induced by a Flexural Plate Wave Pump [94].

FPW sensors have been configured into microfluidic streaming actuators. Bypassing an acoustic wave through the thin-film of the delay line, directional mechanical energy leaks out from the thin-film and drives fluid motion. Interestingly, near the surface of the thin-film, the average velocity is zero. Instead, the device can drive fluid motion at a distance as shown in Fig. 6.43. This technology holds promise for both microfluidic transport and mixing processes which are critical in lab on chip applications. One potential concern with this method of pumping is that it has currently only been demonstrated in gaseous mediums such as air [94].

Discussion

In summary, FPW devices have been demonstrated as different types of sensors and actuators. The advantages of FPW sensors include the ability to isolate the IDTs from the analyte medium. This has the effect of reducing both signal noise and artifacts and prevents device short circuiting which would confound measurement. Further, they can operate at relatively high sensitivities despite low frequency. A drawback is that FPW sensors require a somewhat involved double-sided fabrication method that hampers mass fabrication to a degree. Further, they are restricted to gentle applications given the fragility of the sensitive thin-film.

6.3.6 Acoustic Plate Mode Sensors

Working Principles

Acoustic plate mode (APM) sensors consist of two metal IDTs, often aluminum, and a piezoelectric material such as quartz or $LiTaO_3$ [80]. They are similar in their form to SAW devices; however, they have thin plates that are only a few acoustic wavelengths thick. APM sensors have waves that travel through both the surface and throughout the bulk of the material; Rayleigh waves propagate through the surface, while shear horizontal waves will move through the bulk. The frequencies of APM devices range from 25−200 MHz and wavelengths are ~50 μm. [70], Similar to TSM, decreasing plate thickness increases plate mode frequencies.

These devices hone in on a particular plate mode by adjusting the number of fingers and spacing of the IDTs. To remove the SAW frequencies (i.e., Rayleigh waves), APM sensors employ a band-pass filter. The modes are typically separated by several MHz which is adequate in keeping them well separated from one another.

Applications

Advantages of APM include that it has a scalable fabrication method, is usable in both gas and liquid phases, and is potentially cheaper to produce in comparison to the other types. It can also be configured to isolate IDTs from analyte fluid. However, APM does also have a few disadvantages such as that it has comparatively low mass sensitivity and larger dimensions. APM sensors can be utilized for long-term biosensing such as in vivo blood sugar measurement.

The cell monitor device reported in [95] uses APM to allow for underside IDTs; cells are grown on surface of sensor and differences in mass between cell cycles allows the constant monitoring of cell growth over time [95]. This sensor is produced using ST-cut quartz which dimensions of 20 mm by 1.5 mm. The IDTs are produced with E-beam deposition which include 20 finger pairs. Using this method, an assessment of cell adhesion can be performed in vitro.

APM sensors can be used as a biosensor like many other acoustic wave sensors. Gizeli et al. have combined Love waves with APM devices and a polymer guiding layer to create a new mass sensor [90]. The device was produced from single crystal Y-cut quartz, with IDTs patterned on the surface with basic photolithographic techniques. Then a PMMA layer was applied by spin coating to the surface of the sensor.

Discussion

APM sensors allow for the placement of the electrodes to be on the opposite surface from the sensing element. This opens up possibilities for applications; however, the waves cannot be confined to the surface due to the inherent width required to propagate the wave through the bulk. Attempts have been made to mitigate this problem by combining different technologies to find protein adsorption.

6.3.7 Quartz Tuning Fork Resonators

Working Principles

Quartz tuning forks, or crystal oscillators as they are more generally known, are fundamentally transducers that convert between electrical and mechanical energy. Most commonly, they are found in applications such as watches, clocks, and digital electronics where they convert electrical energy into precisely timed mechanical and electrical frequencies which can be used to keep time. For sensing applications, mechanical or electrical perturbation of the device can be transduced into a measurable electrical signal. Events such as atomic contact and binding generate subtle differences in the output electrical signal in terms of frequency, phase, or amplitude [96].

The mechanism behind crystal oscillators centers upon the piezoelectric crystal substrate that makes up the bulk of the device, which provides the basic functionality of converting mechanical deformation into electrical current. By shaping the material into a tuning fork shape, a set of resonant frequencies is established. Particularly, the first resonance mode frequency is approximately described by

$$\omega_1 = \frac{1.76t}{L^2} \sqrt{\frac{E}{\rho}}$$

(6.72)

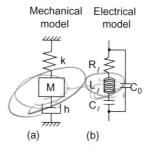

Mechanical   Electrical
model   model

(a)   (b)

Figure 6.44: Mechanical and Electrical Models of a Quartz Tuning Fork [96].

where ω_1 is the angular frequency of the first resonance mode, L is the length of the tuning fork prongs, t is the thickness of the tuning fork prongs, E is the Young's modulus of the material, and ρ is the density of the material [96]. Quartz is widely regarded as the ideal crystal for most crystal oscillator applications given its high stiffness, small device dimensions, and low cost.

Mathematically, a crystal oscillator can be modeled through both mechanical and electrical means as a damped oscillator as illustrated in Fig. 6.44.

Applications

Atomic Force Microcopy

A type of quartz tuning fork with a resonant frequency of 32.768 kHz is commonly used for watches. This type of cheap, commercially available device is used in several interesting applications for nanoscale force measurements.

One example is AFM, in which a 3D topographic image of surface is achieved by monitoring atomic-scale forces such as electrostatic repulsion and van der Waals attraction with the aid of a very sensitive probe which can produce a signal in response to "touching" a surface. Traditionally, silicon cantilevers have been employed for this purpose, as we discussed in Section 6.2. However, quartz tuning forks have been successfully utilized in atomic force microscopes and have shown the potential to surpass silicon cantilevers in several applications, namely in noncontact situations where the material of interest can be damaged by physical contact and should be avoided [97,98]. It is important to note that although quartz tuning forks are the transducers in such applications, the actual contacting probe is typically an additional structure such as glass fiber tip (see Fig 6.45) [98] or silicon tip [99].

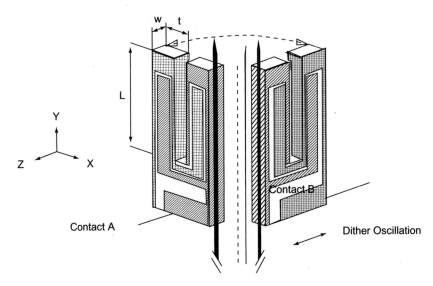

Figure 6.45: Quartz Tuning Fork for Atomic Force Microscopy [98].

Biosensing

In addition to microscopy, quartz tuning forks have also been adapted to biosensing applications. The basic principle utilizes an antibody specific to a target biomolecule that is immobilized onto the prongs of the tuning fork [100,101]. Mass loading upon binding of the target biomolecule manifests as measurable changes in the resonant frequencies, primarily by altering the effective density and modulus of the prongs.

Discussion

Quartz tuning forks are remarkably ubiquitous in everyday life as basic timekeeping devices. This contributes to their low cost and very well-characterized properties. Studies have extended their use into biomolecular assays for both research and potentially clinical use.

6.3.8 Conclusion

Acoustic wave devices come in a wide variety of shapes, layouts, and materials. This lends itself to an expansive number of studies on potential applications. We examine four types of acoustic wave sensors as well as a tuning fork sensor. Though the applications of these devices are similar, subtle nuances in the structure of the sensing element allow room for the optimization of sensitivity and signal-to-noise ratios of the devices.

6.4 Thermal Sensors

Since temperature is related to mechanical activities of molecules, thermal sensors are used in a variety of applications of molecular sensing. This section focus on the principles of thermal sensing, biomedical applications, commercial products, and current research of thermal sensors related to molecular sensing.

Many types of thermal sensors transduce a thermal signal directly into an electrical signal. Such type of sensors include thermocouples, thermistors, and transistors. Other types of thermal sensors convert temperature change into mechanical deflection that can be transduced into electrical signal.

6.4.1 Introduction

The sensors that this section will cover include thermocouples, thermistors, transistors, and mechanical cantilevers. Other types of temperature sensing such as infrared (IR) imaging is briefly discussed. The following summarizes the sensors that will be discussed in this section.

1. Thermocouple

    A thermocouple consists of two metals in contact. Due to the thermoelectric effect, or Seebeck effect, it produces a voltage proportional to the change in temperature. An important application of thermocouple sensors is a calorimetric sensing, where changes in temperature on absorption or desorption of analyte molecules are measured as a sensing effect.

2. Thermistor

    A thermistor is a type of resistor whose resistance changes significantly with temperature and can be utilized as a thermal sensor.

3. CMOS device

    CMOS (complementary metal−oxide−semiconductor) device is a device fabricated by a standard silicon microfabricaiton process. The carrier concentration of silicon is highly dependent on the operation temperature, which can be found by measuring voltage-current (*V-I*) characteristics.

4. Bilayer cantilever

    A bilayer cantilever consists of two materials with different thermal expansion coefficients. A change in temperature results in a mechanical deflection of a cantilever.

5. IR imaging

    Infrared energy emitted from the surface of an object directly correlates with the temperature of the object itself. Temperature distribution of a surface can be measured by recording images of IR emission.

6.4.2 Thermocouples

Working Principle

Thermocouples are based on a phenomenon called the Seebeck effect, also known as the thermoelectric effect, which is a physical phenomenon that takes the differences in temperature and converts them into differences in electrical voltages. Figure 6.46 shows a schematic of typical temperature measurement using a thermocouple. When we apply temperature difference ΔT between measuring point P and reference points Q and R, electrons flow toward cold side inducing voltage differences along P-Q for metal A and P-R for metal B. Since metal A and B have different temperature-voltage characteristics, we observe a potential difference between the two reference points Q and R.

In order to precisely measure the temperature of the measuring point, the temperatures of the reference points should be precisely known to serve as the reference temperatures. The two junctions are usually kept at 0°C by ice water and are called cold junctions. Another common method is cold junction compensation, where the temperature of the reference point is measured by other thermal sensors that can measure the absolute temperature at reference point. The use of two types of thermal sensors may sound redundant. This configuration is needed when the temperature of a small area needs to be measured. A thermocouple can be implemented into a very small junction that can be used as a measuring probe.

General Applications of Thermocouple Sensors

Thermocouples are widely used both in science fields and industry. The metallic thermocouple probes are made from several alloys depending on the desired sensitivity and range. For example, a Type E thermocouple is constructed with chromel and constantan

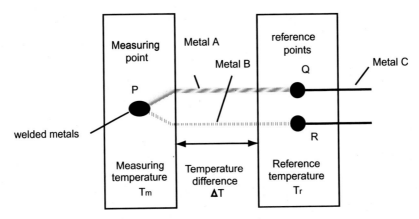

Figure 6.46: Temperature Measurement with a Thermocouple.

alloyed probes. They are simply produced by enclosing the two alloy probes into insulation. The advantages of thermocouple sensors include the facts that they can be used multiple times, they are well characterized, simple to construct and inexpensive. Some disadvantages are limited accuracy, the requirement of reference temperature point and their narrow temperature range.

Thermocouples are versatile devices that can be used for a variety of applications aside from their commonly known utilization as thermal sensors. They can also be used as power generators and even coolers [102]. There are several types of thermocouples available based on the desired temperature range for the application [103]. Thermocouples can be used in very high temperature applications such as furnaces, turbines, and nuclear reactors. Thermocouples are used in biomedical applications as well. For instance, a type of thermocouple is used as a probe for open heart surgery applications. The device is a Cu/CuNi thermocouple that has an accuracy of $\pm 0.1°C$ in physiological temperature ranges (15–45°C). The device is also autoclavable for sterilization; it also has a response time of 0.3 seconds that allows for consistent accurate reading of the patient during thoracic surgery.

Calorimetric Molecular Sensors

Calorimetry is a method where enthalpy changes induced by absorption, i.e. heat of condensation, or desorption, i.e. heat of vaporization, of analyte molecules in the sensing site are measured in terms of variations in temperature. Thermocouple sensors are commonly used for integrated calorimetric sensors because of the ease of fabrication.

The calorimetric sensing process typically includes the following four principal steps [10]:

1. Absorption of the analyte and subsequent partitioning,
2. Liberation or abstraction of heat resulting from (1)
3. Temperature changes resulting from (2)
4. Thermovoltage induced by (3).

The calorimetric sensor detects only transients, meaning only changes in the analyte concentration.

Figure 6.47 shows the structure of a calorimetric sensor reported in [104,105]. Thermopiles consisting of a p-type silicon/aluminum thermocouple are implanted on a silicon wafer. An enzyme layer is attached below the thermopile. As illustrated in Fig. 6.47, calorimetric sensors are usually suspended membranes in order to maximize the sensing effect and isolating heat flux between the sensor and the main structure. Figure 6.47(b) shows an experimental setup where glucose solution is introduced through an integrated fluidic system. The thermopile sensitivity to a temperature change is assumed to be about 55 to 60 mV/K, which gives the measurement with a rate of 8 V/W in practice. Based on a

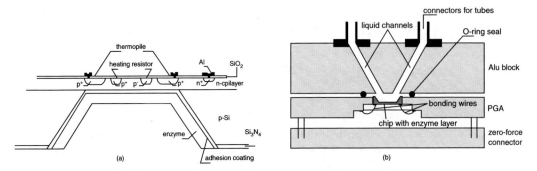

Figure 6.47: Structure of a Calorimetric Sensor. (a) Cross Section of a Microcalorimeter and (b) Experimental Set up of Calorimetric Sensing in a Fluidic System (from [105]).
(a) From [104]; (b) from [105].

sensitivity of 45 kV/mmol and a noise level of below 1 pV, the minimum measurable concentration was estimated to be in the order of 20 pmol.

A similar calorimetric sensor integrated with a microfluidic channel has been reported in [106]. Their device consists of a freestanding p-type polysilicon/gold microthermopile integrated with a glass microfluidic reaction chamber. The sensor measured a head reaction of glucose oxidase, catalase, and urease on glucose, hydrogen peroxide, and urea with sensitivities of 53.5 μV/M for glucose, 26.5 μV/M for hydrogen peroxide and 17 μV/M for urea. Since the chamber volume is 15 nl, the detection limit of the amount of molecules is calculated to be $\sim$2 mM (30 pmol).

An interesting approach for colorimetric sensing reported in [107] is the use of gold nanoparticles instead of thermocouple sensors. The method uses the change in the absorbance spectrum of a self-assembled monolayer of gold particles on glass, as a measure of molecular binding to the surface of the immobilized colloids. The spectrophotometric sensor demonstrated a detection limit of 16 nM for streptavidin (for general discussions on absorption spectroscopy and optical characteristics of gold nanoparticles, see Chapter 5).

Discussions on Thermocouples

Thermocouples as mentioned previously are highly applicable in a variety of thermal sensing applications from measuring temperatures in high temperature environments such as nuclear reactors to being able to accurately and quickly gauge the temperature inside of human heart during surgery. Thermocouples are well developed and understood in modern times. Therefore the sensitivities of the device are high, except for at extreme temperatures where the accuracy of temperature isn't as important. Many digital thermometers used in the medical field are all based on thermopiles, which are composed of several thermocouples connected usually in series. Thermocouples are excellent for this application because they are able to accurately gauge temperature change while they are also cheap and easy to manufacture. Thermocouples are also able to be reused several times without being replaced making them economical, which is also important for devices in the medical field.

6.4.3 Thermistors

Working Principles

The resistance of any conductive material is dependent on temperature to some extent. The resistance of a thermistor changes significantly with temperature and can be utilized as a thermal sensor. Thermistors can be used in several different applications. Some thin-film resistors are used in industry and research for cryogenic temperature measurements, monitoring of industrial components, and even solid state physics experiments such as the superconductor linear collider [108]. Furthermore, thermistors can be used in biomedical applications such as continuous body and tissue temperature measurements and temperature control for fluid flow. The advantages of thermistors are high sensitivity and precision, a wide range of resistance values for specific uses, the ability to shape the sensor into different forms, and fast thermal response times. Some disadvantages of this technology is the fact that the relationship between resistance and temperature is nonlinear and that the semiconducting material decalibrates at high temperatures [109].

Thermistors work on a principle which is easy to utilize in a sensing system. The selected sensor material changes its resistance with change in the temperature. The sensor system measures this temperature-dependent change in resistance and measures the temperature of the object or environment of interest based on a previously calibrated standard curve of resistance vs. temperature [109].

There are two general types of thermistors, negative temperature coefficient (NTC) thermistors and positive temperature coefficient (PTC) thermistors. The resistance of a NTC thermistor decreases with an increase in temperature while that of a PTC thermistor increases [109]. Most PTC thermistors function as a kind of "switch," whose resistance rises suddenly at a certain critical temperature. NTC thermistors generally have a relatively gradual change in resistance compared to PTC thermistors.

The resistance of NTC thermistors can be modeled as a nonlinear relationship to temperature as expressed in Eq. 6.73, where R_T is the final resistance at temperature T, R_0 is the initial resistance, T_0 is the initial temperature, and B is a constant.

$$R_T = R_0 \exp\left[B\left(\frac{1}{T} - \frac{1}{T_0}\right)\right] \qquad (6.73)$$

This is based on the fact that the conductivity σ of a semiconductor is proportional to the carrier concentration n, which is proportional to $\exp(1/2kT)$, namely:

$$\sigma \propto n \propto \exp(1/2kT), \tag{6.74}$$

where k is Boltzmann's constant. More general discussion on resistivity, conductivity, carrier concentration, and temperature is made in Chapter 4, Section 4.3.1.

In practice, linear approximation is often used to easily find the temperature value. The linear coefficient α (% per °C) of resistance change can be defined in the following way:

$$\alpha = \frac{1}{R_0} \times \left. \frac{dR_T}{dT} \right|_{T=T_0} \times 100 \tag{6.75}$$

An example is given in problem 6.11.

Materials and Fabrication

Thermistors are typically made of semiconductor materials that have been selected due to their temperature-dependent resistance change. These materials usually consist of a mixture of different oxides of metals such as cobalt, copper, magnesium, manganese, titanium, and nickel or doped ceramics. These metal oxides and doped ceramics are manufactured in a controlled atmosphere through the process of sintering. Sintering is the process of taking a mixture of powders in a mold and heating it up past the melting temperature to allow the different materials to mix. This method allows thermistors to be shaped into different forms such as beads, disks, rods, washers, or flakes [109].

In order to tailor the response, thermistors can be placed in series or parallel as well as changing the mixture of oxides. These packages of resistors can be encapsulated in different packaging materials, such as glass, or epoxy resin, to increase its robustness [110]. The electrical leads can then be connected using sputtering and photolithography. Furthermore, the sensitivities and total range of measured temperatures can be changed by varying the ratio of metal oxides or doped ceramics. Figure 6.48 shows an example of a thin film thermistor. In addition, in the case of these particular thin-film thermistors, the comb and finger part of the thermistor are varied to change the properties of the thermistor. These thin-film resistors are made from a mixture of zirconium oxy-nitrides and the ratio of conducting zirconium nitride to zirconium oxide can be changed to tailor the resistivity [110]. Thermal sensor based on a thermistor is categorized as a type of conductometric sensor (see Chapter 4, Section 4.2).

Applications

General Applications

There are many applications for thermistors as they are simple, robust devices that are also versatile. NTC thermistors can have temperature coefficients as large as several percent per °C, which allows for an increased sensitivity over other temperature sensors.

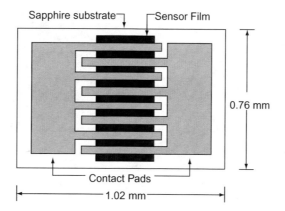

Figure 6.48: Schematic of a Thin-film Thermistor.
From [110].

Thermistors are not only used in industrial applications, but are also utilized in biomedical applications as well. Thermistors are widely used for medical procedures. These thermistors are matched pairs of small, glass encapsulated thermistor beads that are connected in series or parallel circuits and are housed in different ways depending on the use. The thermistors are used to provide continuous body and tissue temperature measurements in patients and can be used to monitor and subsequently control the temperature of fluid flow.

Thermistors are also used in fever thermometers, skin sensors, blood analyzers, incubators, heart catheters, hypodermic needle sensors, esophageal stethoscopes, and respiration monitors due to their small size, accuracy, and reliability. A specific example is the thermodilution catheters that are used during cardiac procedures for both diagnostic and corrective purposes. The catheters have small thermistor beads inserted into the tips and the whole catheter is then inserted into a hypodermic needle. These needles are extremely important in open-heart surgery as they are put into the myocardium muscle to monitor the temperature of the heart throughout the procedure. Another example is the use of thermistors in the care of premature babies. The thermistor sensor is exposed to the closed environment of the baby and continually monitors both the environment's temperature and air circulation to ensure both are optimized for the baby.

Temperature Control in Microfluidic Systems

One excellent application area of miniaturized fluidic system is polymerase chain reaction (PCR), because PCR does not require a large amount of DNA to run a reaction, in fact, theoretically it only needs a single fragment of double stranded DNA to initiate PCR. PCR systems require careful monitoring of temperature so as to ensure optimum operating conditions for denaturation, elongation and termination to occur (see Chapter 3,

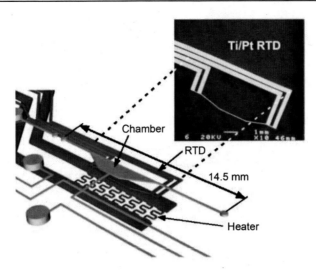

Figure 6.49: Microfluidic PCR System.
From [111].

Section 3.3.3 for details). In the microfluidic device (Fig. 6.49) reported in [111], PCR and electrophoresis are combined, such that both the reaction and analysis are on a single chip, and have reduced the reaction time to 10 minutes with 20 cycles [111].

To monitor the temperature of the heat being delivered, resistance temperature detectors (RTD) are used; these sensors are placed in the PCR reaction chamber so as to provide accurate temperature measurements throughout the reaction of the high sensitivity PCR. RTDs were used in this device for their ability to respond quickly to a temperature change and their accurate temperature measurement.

Microfabrication was used to form the RTD sensor on a glass wafer. The wafer was first coated with a layer of titanium and platinum using vapor deposition. The metal layer was then etched by hot aqua regia (mixture of one part nitric acid and three parts hydrochloric acid) to create the elements of the resistance temperature detector [111].

Discussion

Although thermistor technology is well-established, there are some issues related to it. One of the major disadvantages of thermistors is the fact that the semiconductor materials used for the sensors are susceptible to decalibration at high enough temperatures. The standard operating temperature range for thermistors are usually under a few hundred degrees, which may not be as large an issue for biomedical applications. Another issue for thermistors is their tendency to overheat themselves due to their small size and low thermal response time.

6.4.4 CMOS Temperature Sensors

Working Principle

A CMOS device is a device fabricated by a standard silicon microfabrication process. CMOS is a style of circuit design that is integrated into circuit chips (see Chapter 4, Section 4.3.1 for CMOS). CMOS temperature sensor can be categorized as a special type of thermistor that uses silicon as the sensing material. Since the carrier concentration of silicon is dependent on the temperature, any type of electrical element integrated onto a silicon including resistor, diode, and transistor can be potentially utilized as a temperature sensor. The temperature can be found by measuring the temperature-dependent voltage-current (*V-I*) characteristics.

Most CMOS temperature sensors are bipolar transistors because the base-emitter voltage and saturation current have good temperature characteristics [112]. Temperature is determined on the signals generated from the base-emitter voltage and the saturation current [113]. All the integration is done on the surface of a microchip not only making it extremely portable but also mass producible.

Applications

CMOS temperature sensing systems are versatile because of their well-understood, mass-producible characteristics. They are used in a variety of appliances because of their compatibility with other digital circuitry. They are used in cell phones, portable AV equipment, palm top computers, etc. A typical CMOS sensor product has a temperature range ranging from −30°C to 80°C. They are used in several devices not only because of their ability to be mass produced on a small scale but also because of their power conserving capabilities.

6.4.5 Bilayer Cantilever Sensors

Working Principle

A bilayer cantilever consists of two layers of different materials. The temperature is found by measuring the deflection of the cantilever, which results from the different thermal expansion coefficients of the top film and the probe body (see Fig. 6.50). If a piezoelectric material is a part of the bimaterial cantilever, the deflection will result in a measurable voltage produced. These cantilevers are also useful in the application of actuators.

Let us consider a bilayer cantilever where a thin film is deposited on a relatively thick cantilever body to form bilayers. If the temperature and the thermal expansion coefficients of the two materials are known, the thermal deflection can be found in the following way [114].

At the initial temperature of $T = T_0$, the lengths of both the film and the body equal to L. When the temperature increases to $T = T_0 + \Delta T$, the two materials expand with

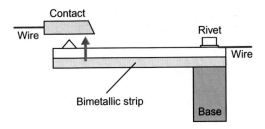

Figure 6.50: Schematic Diagram of a Bimaterial Cantilever Sensor.

different expansion rates. We can find the free lengths (i.e. the lengths when they are separate and no external force is applied) of the film $L + \Delta L_f$ and the cantilever body $L + \Delta L_b$ based on their thermal expansion coefficients. Let us assume a case where $\Delta L_f > \Delta L_b$ for an example. When they are bonded back to back, the lengths will be forced to be the same $L + \Delta L$. With our assumption, the forced length becomes somewhere between the free lengths of the film and the cantilever body, namely $L + \Delta L_f > L + \Delta L > L + \Delta L_b$.

The internal forces should satisfy the force equilibrium equation:

$$F_f = F \tag{6.76}$$

where F_f is the force induced in the film, and F is the force induced in the body. Since F_f and F can be correlated with the mismatch lengths $\Delta L - \Delta L_b$ and $\Delta L_f - \Delta L$ and Young's modulo of the two materials, we can find $\Delta L_f - \Delta L$ and thus the induced force F_f. Now we can find the radius of curvature R by recalling Eq. 6.33.

$$F_f = \frac{Et^2 w}{6R} \tag{6.77}$$

where E and t are the Young's Modulus and the thickness of the cantilever body. See also problem 6.13.

The advantages of cantilevers are that they do not require a power source to effect temperature sensing, they have a high sensitivity to changes in temperature, a large operating range, are easy to produce in a large scale, and can be tailored to a specific application by changing the shape, size, or biomaterials used. On the other hand some disadvantages are that inaccuracy of temperature readings can result due to the fact that the deformation may not be within the plane that is desired; this could be a problem when the goal is to close a circuit as a result of a temperature change. Finally, the current cantilevers' sensitivity is lowered at extremely high or low temperatures [115].

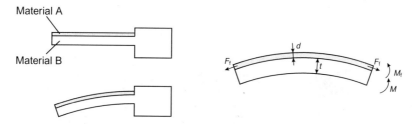

Figure 6.51: Example of Bi-metallic Thermostat.

Applications

There are a number of different applications in which a thermal cantilever can be used; they are used when one wants to effect the closing, or opening, of a circuit as a result of a temperature change. This type of actuation is most commonly used in thermostats for climate control. A thermostat is a device that senses ambient temperature and helps in its regulation. In the case of the bi-metallic thermostat there are two thermally different metals stuck together. So when it is cold or cool the contacts are closed and current is allowed to pass through the circuit. When the metal starts to heat up it will begin to expand causing the other bonded metal strip to bend up or down opening the contact and stopping the current from flowing (Fig. 6.51). This type of system creates an on/off action that allows for the monitoring of the temperature with a way to signal that the temperature is not what is desired. There have been interesting studies to create microscale thermostats based on MEMS techniques [116,117].

6.4.6 IR Imaging

Background

IR light was first discovered by Fredrick William Herschel in the 1800s when he was measuring the temperature of different colors. He found that the temperature became warmer as the light got closer to the color red and even warmer past red; this is what he described as infrared [118]. IR is a form of electromagnetic radiation, as are of radio waves, UV light rays, visible light, X-rays, and gamma rays. All of these forms of electromagnetic waves travel at the speed of light. The wavelength range of visible light ranges from 390–750 nm, while infrared ranges from 0.78–1000 μm, much longer than that of the light in the visible spectrum. IR radiation is classified into near infrared, mid infrared and far infrared, as seen in Fig. 6.52 [118].

Infrared imaging, or **infrared thermograph**, is used in various applications. For instance, night-vision devices use IR radiation to allow humans to see in the dark. IR imaging can also be used in biological applications such as observing the changing in blood flow as

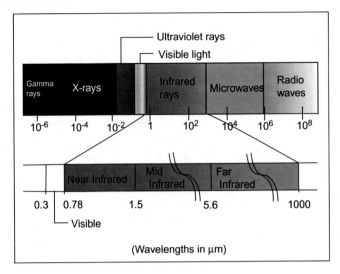

Figure 6.52: Spectrum of Electromagnetic Radiation [118].

thermal monitoring and regulation. An IR imaging device uses a lens to focus the IR signal onto a detector that converts the energy into an electrical signal. This allows for the temperature of an object to be taken from a distance, giving it an advantage over other thermal sensors that require direct contact with the object of interest.

Working Principles

IR energy is emitted by the motion of atoms and molecules on the surface of an object. The release of the energy from the surface of the object directly correlates with the temperature of the object itself. **Planck's law** describes the electromagnetic radiation emitted by a black body in thermal equilibrium:

$$u(T) = \frac{8\pi hc}{\lambda^5} \cdot \frac{1}{\exp\left[\frac{hc}{\lambda kT}\right] - 1},$$ (6.78)

where u is the spectral energy density (energy per unit wavelength per unit volume) of the surface of the black body, T is its absolute temperature, λ is its wavelength, k is the Boltzmann constant, h is the Planck constant, and c is the speed of light. Planck's equation allows us to correlate the relationship between spectral emissivity, temperature and radiant energy, which is the fundamental of the IR imaging.

By finding the solution to

$$\frac{\partial u}{\partial \lambda} = 0,$$ (6.79)

the peak wavelength λ_{max} for the radiation is found to be

$$\lambda_{max} = \frac{2,897 \text{ nm} \cdot \text{K}}{T} \tag{6.80}$$

This equation relates blackbody radiation to the temperature and is called **Wien's displacement law**, named after Wilhelm Wien, who derived the relationship before Planck's law was developed in 1900.

Instrumentation

Special cameras can be used to detect IR radiation emitted from a body. The higher is an objects temperature, the more IR radiation is emitted. This allows for the ability to see objects in total darkness or through smoke. IR cameras usually have a single color channel that distinguishes between intensities of certain wavelengths in the IR spectrum. There are also more advanced cameras that allows the acquisition of different wavelengths in the IR spectrum that can display a multicolor image [119].

Currently there are two types of thermography cameras, cool and uncooled IR image detectors. Cooled IR are typically contained in a vacuum seal and cryogenically cooled. More precise measurements are made possible by operating the detectors under a known reference temperature, which ranges from 4 K to just below room temperature. Uncooled IR detectors operate at ambient temperatures that are able to detect IR radiation by the change of resistance, voltage, and current in the infrared detector. The changes are then compared to that of the operating temperature of the detector. These detectors tend to have lower resolution and image quality than that of cooled detectors. These detectors are based off pyroelectric and ferroelectric materials or microbolometer technology [119].

A basic IR thermography consist of several components; a lens that collects the energy emitted by the target, a transducer that converts the energy into an electrical signal, an emissivity adjustment, and compensation that ensures that the ambient temperature is not transferred to the final output. Advances in selective filtering of the incoming IR signal have been made by the availability of more sensitive detectors and more stable signal amplifiers.

Applications in Biomedical Science

IR thermography can be used in many different applications such as night vision, civil engineering, and material testing. Here we will focus on the use of IR imaging in the field of medicine and in biomedical applications [120].

Measurements of skin temperature to evaluate patient health is useful as it can be an indicator of body dysfunction and can be associated with inflammation, infection, or even the growth of tumors [120,121]. This is due to the fact that skin temperature depends on

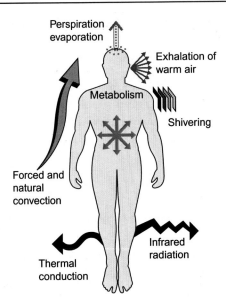

Figure 6.53: Diagram of the Dependence of Skin Temperature on a Plurality of Factors [121].

multiple factors that could be affected by disease. Figure 6.53 shows a diagram of the dependence of skin temperature on a plurality of factors such as metabolism, heat transfer through conducting and convection, perspiration, and breathing.

IR imaging allows the creation of a "body surface temperature map" to produce meaningful information through a non-invasive, non-radiating, and painless way [122]. In the past, IR imaging was not heavily used in widespread medical applications due to poor performance combined with prohibitively high costs, but in the last two decades IR imaging has seen a resurgence in biomedical applications as improved technologies, e.g. image processing, graphics cards, have allowed more accurate analysis of IR images and lowered costs [121].

There are two general methods of IR imaging, static and dynamic. The static method consists of creating a single thermographic image in which the spatial distribution of temperature across a surface is mainly analyzed. The dynamic method involves the recording and evaluation of both spatial and temporal behavior of skin temperature. This method can be broken down into two specific methods, active and passive. Active methods measure the body's reaction to an applied external stress. An example of such a stress is immersing a patient's limb in hot or cold water. Passive methods do not use external stress, but can include an internal stress such as the body's temperature after exercise [121].

One of the earliest applications of IR technology in medicine was the detection of breast cancer in patients. The basis of this application is that the skin temperature that is overlying the tumor gives off an increased amount of infrared radiation. This is due to the increased metabolic activity, blood flow, and the vascular proliferation due to angiogenesis around the tumor as the cancer grows [122,123]. Although the method alone does not provide sufficient information for diagnosis, IR thermography can be utilized in breast cancer screening as one part of a multimodality-screening environment. Despite the inability of IR breast thermography in locating the exact size and placement of a breast lesion, it is still useful in evaluating the presence of breast cancer in a patient with an average sensitivity and specificity of 90% [122]. Not only can IR imaging be used for detection of the presence of breast cancer, but it can also be used in the detection of melanoma in any part of the body [122].

Another application of IR imaging is in open-heart surgery. One specific application is a type of thermoangiography where the heart temperatures are monitored during coronary bypass surgery. The IR imaging is utilized for real time evaluation of the cooling effects of the cardioplegia solutions given to patients during the surgery, the flow of blood through the heart and its arteries, and coronary perfusion after the surgery. In such a procedure an IR camera is mounted on an adjustable mechanical arm above the patient bed and the thermograph actively transmitted to a display that the surgeon can observe during the procedure. This allows the surgeon to better monitor patient health as the IR image can be used for early detection of a narrow anastomosis or other reasons for blood flow restriction in the heart that could lead to complications [124].

Discussion on IR Imaging

The main limiting factor on this technology is the sophistication of other technologies that couple with IR sensors. Although technological improvements have allowed for better temperature, temporal and spatial resolution, there is still much room left for improvement in how the large data acquired by the system should be interpreted as useful information [120]. Advanced methods of IR image processing for computer aided detection (CAD) are needed to simplify and help reduce diagnostic ambiguities. Another reason why IR thermal sensing was not adopted was due to prohibitively high costs. The development of these technological tools, increased innovations and inventions, and the advancement of more efficient methods would decrease the cost of IR imaging, making it a viable strategy in thermal imaging in biomedical science as well as other applications.

6.4.7 Discussions on Thermal Sensors

We introduced several types of thermal sensors related to molecular sensing and biomedical applications. Each of these different sensors has their own approaches and methods that allow them to be used in specific applications.

Thermocouples and thermistors use electrical changes to measure the temperature. CMOS thermal sensors are useful for digital devices because of the compatibility with other silicon microdevices. Bimetallic cantilevers are able to detect the temperature without a power source due to their sensing basis in mechanical properties. IR thermal sensors can create heat maps of the temperature of interest.

Although different techniques can be interchangeably used in many applications, unique characteristics of each method such as size, operating temperature range, response time, cost, lifetime, and energy consumption should be discussed in selecting types of sensors to be employed.

6.5 Conclusions

Current micromachining techniques can mass-produce mechanical structures on silicon wafers, and use these to produce multi-target molecular sensor arrays on a single chip. The improvement in technology opens new avenues of incorporating micro mechanical sensors into microsystems consisting of electronic, mechanical, and fluidic devices. The low cost, low power consumption, and label-free detection characteristics of mechanical sensors make them very appealing for potential applications.

Micromachined devices have made it possible to analyze complex biosamples utilizing biochemical binding events, DNA hybridization assays, antibody−antigen assays, among many others. Development of new mechanical sensors is underway utilizing silicon as a solid technology platform and microlithography techniques borrowed from chip fabrication facilities. State-of-the-art silicon fabrication techniques enable an increase in the sensitivity and decrease of background signals. The great advantage of these sensors is the ability to mass manufacture these devices on silicon wafers in a very cost-effective and modular fashion while still attaining tremendous sensitivity and specificity. This quality allows for the integration of micromechanical sensors to be easily incorporated into many existing technologies.

Problems

P6.1 Piezoresistivity

The resistor implanted in the cantilever has the length $L = 20\,\mu m$ and the cross sectional area $A = 10\,\mu m \times 1\,\mu m$ as shown in Fig P6.1. The resistivity of the resistor is $\rho = 0.5\,\Omega$ cm. Answer the following questions.

1. What is the resistance R of this resistor?
2. Because of the bending, the total length L becomes longer by $\Delta L = 0.1\,\mu m$. Calculate the strain ε.
3. Gauge factor of this material is GF $= 30$. Find the change in the resistance.

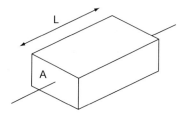

Figure P6.1

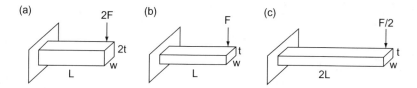

Figure P6.2

P6.2 Cantilever Bending

Consider the three cantilevers as shown in Fig. P6.2:

1. With which of the three cantilevers will we observe the largest deflection?
2. Where and with which cantilever will we will observe largest strain?

P6.3 Surface Stress, Normal Stress, and Deflection

The three cantilevers shown in Fig. P6.3 are experiencing the same intensity of surface stress induced by molecular adsorption.

1. With which of the three cantilevers will we observe the largest deflection?
2. Where and with which cantilever will we will observe largest strain?

P6.4 Silicon Cantilever Basics

Consider a silicon cantilever beam with the length $L = 40\ \mu m$, width $W = 30\ \mu m$, thickness $T = 5\ \mu m$.

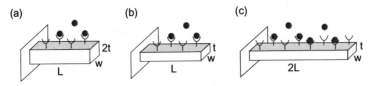

Figure P6.3

1. What is the effective spring constant? Use Young's modulus of silicon
   $E = 130 \times 10^9 \text{ m/kg} \cdot \text{s}^2$.
2. A force is applied at the end of the cantilever and it is deflected by 10 μm. What is the
   value of maximum strain? Where does it occur?
3. Find the resonant frequency of the beam. Use the density of silicon
   $\rho = 2.33 \times 10^3 \text{kg/m}^3$.
4. What is the effective mass of the cantilever?
5. What is the actual mass of the cantilever?

P6.5 Cantilever Actuation

A silicon cantilever shown in Fig. P6.5 is electrostatically actuated.

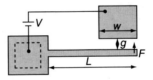

Figure P6.5

1. Find the effective stiffness of this cantilever in the direction of actuation. The thickness
   of the cantilever is t, and the Young's modulus is E.
2. Calculate the electrostatic force F based on assumption of a parallel plate actuator.
3. Find the pull in voltage V, assuming the force F is applied at the end of the cantilever.
   Use vacuum permittivity ε_0.

P6.6 Cantilever Resonance

We have a silicon cantilever that measures length $L = 350$ μm, width $w = 30$ μm, and
thickness $t = 340$ nm.

1. What is the effective mass of the cantilever? Calculate from the dimensions. Use
   density of silicon $\rho = 2.33 \times 10^3 \text{kg/m}^3$.

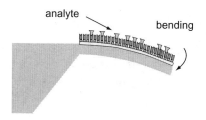

Figure P6.7

2. When a small bead with a mass of 14 ng is attached to the tip of the cantilever, the resonant frequency changed from 2370 kHz to 987 kHz. Find the effective spring constant of the cantilever.

3. When we attach another object with unknown mass at the tip, the resonant frequency changed from 987 kHz to 634 kHz. Find the mass of the object.

P6.7 Cantilever-Based Biosensors

We have a piezoresistive cantilever based molecular sensor (Fig. P6.7). The cantilever parameters are given as follows:

Thickness $T = 5\ \mu m$
Length $L = 2000\ \mu m$
Width $w = 10\ \mu m$
Young's modulus $E = 2 \times 10^{11}$ N/m
Gauge factor GF $= 100$.

If the resistance of the piezoresistive cantilever changes 0.01% due to the bending caused by binding events on the cantilever surface, find (a) the strain within the cantilever; (b) the surface stress caused by the analyte molecule binding on the surface.

P6.8 Microfabrication of Cantilever Biosensors

Cantilever is a popular form of molecular sensor. Answer the following questions:

1. Describe how a cantilever could be used as a molecular sensor in terms of the three basic sensor elements.

2. Draw a diagram showing the fabrication procedure of a cantilever made from a single crystal silicon wafer. Name two key techniques in such fabrication.

3. Name four methods to measure the deflection of silicon cantilever.

P6.9 QCM Acoustic Biosensors

For a quartz crystal microbalance (QCM) based biosensor for molecular recognition, the following design parameters are given: fundamental resonant frequency of 5 MHz, sensor diameter of 1 cm, shear modulus of 2 g/(cm s^2) and density of 8 g/cm^3. Estimate the change in molecules mass loading corresponding to the shift of 1 MHz in resonant frequency.

P6.10 MEMS Acoustic Biosensors

Draw and label the features of a surface acoustic wave (SAW) sensor. Explain the difference between a SAW sensor and a quartz crystal microbalance (QCM) sensor.

P6.11 Thermistor

A thermistor has a resistance of 1.50 kΩ at 25°C. The linear coefficient of resistance change for this thermistor change is given as:

$$\alpha = \frac{1}{R_0} \times \frac{dR_T}{dT}\bigg|_{T=T_0} \times 100 = -4.5\%/°C$$

If the resistance of this thermistor is measured to be 1.53 kΩ, what is the temperature?

P6.12 Temperature Measurement

1. You want to measure the inside temperature of a furnace. The temperature ranges 25−500 °C. Discuss what type of sensor can be used.
2. You want to develop a system that continuously monitors your body temperature and sends the data to your cell phone. Discuss what type of sensor should be used. Compare with other types of temperature sensors.

P6.13 Thermal Deflection of a Bilayer Cantilever

The linear coefficient of thermal expansion α of a cantilever is defined in the following equation.

$$\Delta L = \alpha \cdot L \cdot \Delta T,$$

where L, ΔL, and ΔT are original length, change in length, and change in temperature.

Consider a bimetal cantilever with the dimensions shown in Fig. P6.13. The linear coefficients of thermal expansion of the film and the body are α_1 and α_2, respectively.

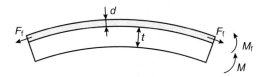

Figure P6.13

1. If the film and the body are separate, find the changes in length ΔL_f, and ΔL_b for the film and the body, respectively, at a temperature change of ΔT.
2. When they are bonded as a bimetal cantilever, What will be the actual change ΔL in length? Young's modulo of the film and the body are E_f, and E_b, respectively. Assume $\alpha_1 > \alpha_2$.
3. Find the induced force F and the radius of curvature R at a temperature change of ΔT.

References

[1] Alvarez M, Lechuga LM. Microcantilever-based platforms as biosensing tools. Analyst 2010;135:827−36.

[2] McKendry R, Zhang J, Arntz Y, Strunz T, Hegner M, Lang HP, et al. Multiple label-free biodetection and quantitative DNA-binding assays on a nanomechanical cantilever array. Proceedings of the National Academy of Sciences 2002;99:9783−8.

[3] Raiteri R, Grattarola M, Butt H-J, Skládal P. Micromechanical cantilever-based biosensors. Sensors and Actuators B: Chemical 2001;79:115−26.

[4] Ziegler C. Cantilever-based biosensors. Analytical and Bioanalytical Chemistry 2004;379:946−59.

[5] Boisen A, Thundat T. Design & fabrication of cantilever array biosensors. Materials Today 2009;12:32−8.

[6] Binnig G, Quate CF, Gerber C. Atomic force microscope. Physical Review Letters 1986;56:930−3.

[7] Albrecht T, Akamine S, Carver T, Quate C. Microfabrication of cantilever styli for the atomic force microscope. Journal of Vacuum Science & Technology A: Vacuum, Surfaces, and Films 1990;8:3386−96.

[8] Wolter O, Bayer T, Greschner J. Micromachined silicon sensors for scanning force microscopy. Journal of Vacuum Science & Technology B: Microelectronics and Nanometer Structures 1991;9:1353−7.

[9] Boisen A, Dohn S, Keller SS, Schmid S, Tenje M. Cantilever-like micromechanical sensors. Reports on Progress in Physics 2011;74:036101.

[10] Hierlemann A, Brand O, Hagleitner C, Baltes H. Microfabrication techniques for chemical/biosensors. Proceedings of the IEEE 2003;91:839−63.

[11] McFarland AW, Colton JS. Chemical sensing with micromolded plastic microcantilevers. Journal of Microelectromechanical Systems 2005;14:1375−85.

[12] Brugger J, Despont M, Rossel C, Rothuizen H, Vettiger P, Willemin M. Microfabricated ultrasensitive piezoresistive cantilevers for torque magnetometry. Sensors and Actuators A: Physical 1999;73:235−42.

[13] Petersen KE. Silicon as a mechanical material. Proceedings of the IEEE 1982;70:420−57.

[14] Pechmann R, Kohler J, Fritzsche W, Schaper A, Jovin T. The Novolever: a new cantilever for scanning force microscopy microfabricated from polymeric materials. Review of Scientific Instruments 1994;65:3702−6.

[15] Genolet G, Brugger J, Despont M, Drechsler U, Vettiger P, Rooij NDe, et al. Soft, entirely photoplastic probes for scanning force microscopy. Review of Scientific Instruments 1999;70:2398−401.

[16] Lee L, Berger S, Liepmann D, Pruitt L. High aspect ratio polymer microstructures and cantilevers for bioMEMS using low energy ion beam and photolithography. Sensors and Actuators A: Physical 1998;71:144−9.

[17] Wang X, Ryu KS, Bullen DA, Zou J, Zhang H, Mirkin CA, et al. Scanning probe contact printing. Langmuir 2003;19:8951−5.

[18] McFarland AW, Poggi MA, Bottomley LA, Colton JS. Production and characterization of polymer microcantilevers. Review of Scientific Instruments 2004;75:2756−8.

[19] Zhang XR, Xu X. Development of a biosensor based on laser-fabricated polymer microcantilevers. Applied Physics Letters 2004;85:2423−5.

[20] Yao T-J, Yang X, Tai Y-C. BrF_3 dry release technology for large freestanding parylene microstructures and electrostatic actuators. Sensors and Actuators A: Physical 2002;97:771−5.

[21] Kramer P, Sharma A, Hennecke E, Yasuda H. Polymerization of para-xylylene derivatives (parylene polymerization). I. Deposition kinetics for parylene N and parylene C. Journal of Polymer Science: Polymer Chemistry Edition 1984;22:475−91.

[22] Goeders KM, Colton JS, Bottomley LA. Microcantilevers: sensing chemical interactions via mechanical motion. Chemical Reviews 2008;108:522.

[23] Moulin A, O'Shea S, Badley R, Doyle P, Welland M. Measuring surface-induced conformational changes in proteins. Langmuir 1999;15:8776−9.

[24] Watari M, Galbraith J, Lang H-P, Sousa M, Hegner M, Gerber C, et al. Investigating the molecular mechanisms of in-plane mechanochemistry on cantilever arrays. Journal of the American Chemical Society 2007;129:601−9.

[25] Lifshitz R, Roukes ML. Thermoelastic damping in micro-and nanomechanical systems. Physical Review B 2000;61:5600.

[26] Mohanty P, Harrington D, Ekinci K, Yang Y, Murphy M, Roukes M. Intrinsic dissipation in high-frequency micromechanical resonators. Physical Review B 2002;66:085416.

[27] Stemme G. Resonant silicon sensors. Journal of Micromechanics and Microengineering 1991;1:113.

[28] Photiadis DM, Judge JA. Attachment losses of highQ oscillators. Applied Physics Letters 2004;85:482−4.

[29] Meyer G, Amer NM. Novel optical approach to atomic force microscopy. Applied Physics Letters 1988;53:1045−7.

[30] Fukuma T, Kimura M, Kobayashi K, Matsushige K, Yamada H. Development of low noise cantilever deflection sensor for multienvironment frequency-modulation atomic force microscopy. Review of Scientific Instruments 2005;76: pp. 053704-053704-8

[31] Lang H, Berger R, Battiston F, Ramseyer J-P, Meyer E, Andreoli C, et al. A chemical sensor based on a micromechanical cantilever array for the identification of gases and vapors. Applied Physics A: Materials Science & Processing 1998;66:S61−4.

[32] Lang H, Baller M, Berger R, Gerber C, Gimzewski J, Battiston F, et al. An artificial nose based on a micromechanical cantilever array. Analytica Chimica Acta 1999;393:59−65.

[33] Lang H, Berger R, Andreoli C, Brugger J, Despont M, Vettiger P, et al. Sequential position readout from arrays of micromechanical cantilever sensors. Applied Physics Letters 1998;72:383−5.

[34] Lim S-H, Raorane D, Satyanarayana S, Majumdar A. Nano-chemo-mechanical sensor array platform for high-throughput chemical analysis. Sensors and Actuators B: Chemical 2006;119:466−74.

[35] Yue M, Stachowiak JC, Lin H, Datar R, Cote R, Majumdar A. Label-free protein recognition two-dimensional array using nanomechanical sensors. Nano Letters 2008;8:520−4.

[36] Yue M, Lin H, Dedrick DE, Satyanarayana S, Majumdar A, Bedekar AS, et al. A 2-D microcantilever array for multiplexed biomolecular analysis. Journal of Microelectromechanical Systems 2004;13:290−9.

[37] Erlandsson R, McClelland G, Mate C, Chiang S. Atomic force microscopy using optical interferometry. Journal of Vacuum Science & Technology A: Vacuum, Surfaces, and Films 1988;6:266−70.

[38] Rugar D, Mamin H, Guethner P. Improved fiber-optic interferometer for atomic force microscopy. Applied Physics Letters 1989;55:2588−90.

[39] Vashist SK. A review of microcantilevers for sensing applications. Journal of Nanotechnology 2007;3:1–18.

[40] Tortonese M, Barrett R, Quate C. Atomic resolution with an atomic force microscope using piezoresistive detection. Applied Physics Letters 1993;62:834.

[41] Linnemann R, Gotszalk T, Hadjiiski L, Rangelow I. Characterization of a cantilever with an integrated deflection sensor. Thin Solid Films 1995;264:159–64.

[42] Behrens I, Doering L, Peiner E. Piezoresistive cantilever as portable micro force calibration standard. Journal of Micromechanics and Microengineering 2003;13:S171.

[43] Onoe H, Gel M, Hoshino K, Matsumoto K, Shimoyama I. Direct measurement of the binding force between microfabricated particles and a planar surface in aqueous solution by force-sensing piezoresistive cantilevers. Langmuir 2005;21:11251–61.

[44] Takahashi H, Aoyama Y, Ohsawa K, Tanaka H, Iwase E, Matsumoto K, et al. Differential pressure measurement using a free-flying insect-like ornithopter with an MEMS sensor. Bioinspiration & Biomimetics 2010;5:036005.

[45] Noda K, Hoshino K, Matsumoto K, Shimoyama I. A shear stress sensor for tactile sensing with the piezoresistive cantilever standing in elastic material. Sensors and Actuators A: Physical 2006;127:295–301.

[46] Li Y-C, Ho M-H, Hung S-J, Chen M-H, Lu MS. CMOS micromachined capacitive cantilevers for mass sensing. Journal of Micromechanics and Microengineering 2006;16:2659.

[47] Blanc N, Brugger J, De Rooij N, Durig U. Scanning force microscopy in the dynamic mode using microfabricated capacitive sensors. Journal of Vacuum Science & Technology B: Microelectronics and Nanometer Structures 1996;14:901–5.

[48] Pourkamali S, Hashimura A, Abdolvand R, Ho GK, Erbil A, Ayazi F. High-Q single crystal silicon HARPSS capacitive beam resonators with self-aligned sub-100-nm transduction gaps. Journal of Microelectromechanical Systems 2003;12:487–96.

[49] Lee JE, Zhu Y, Seshia A. A bulk acoustic mode single-crystal silicon microresonator with a high-quality factor. Journal of Micromechanics and Microengineering 2008;18:064001.

[50] Alper SE, Akin T. A single-crystal silicon symmetrical and decoupled MEMS gyroscope on an insulating substrate. Microelectromechanical Systems, Journal of 2005;14:707–17.

[51] Xie H, Fedder GK. Vertical comb-finger capacitive actuation and sensing for CMOS-MEMS. Sensors and Actuators A: Physical 2002;95:212–21.

[52] Wu J, Fedder GK, Carley LR. A low-noise low-offset capacitive sensing amplifier for a $50\text{-}\mu g/\sqrt{Hz}$ monolithic CMOS MEMS accelerometer. Solid-State Circuits, Journal of IEEE 2004;39:722–30.

[53] Verd J, Abadal G, Teva J, Gaudó MV, Uranga A, Borrisé X, et al. Design, fabrication, and characterization of a submicroelectromechanical resonator with monolithically integrated CMOS readout circuit. Journal of Microelectromechanical Systems 2005;14:508–19.

[54] Amírola J, Rodríguez A, Castañer L, Santos J, Gutiérrez J, Horrillo M. Micromachined silicon microcantilevers for gas sensing applications with capacitive read-out. Sensors and Actuators B: Chemical 2005;111:247–53.

[55] Thundat T, Warmack R, Chen G, Allison D. Thermal and ambient-induced deflections of scanning force microscope cantilevers. Applied Physics Letters 1994;64:2894–6.

[56] Barnes J, Stephenson R, Welland M, Gerber C, Gimzewski J. Photothermal spectroscopy with femtojoule sensitivity using a micromechanical device. Nature 1994;372:79–81.

[57] Butt H-J. A sensitive method to measure changes in the surface stress of solids. Journal of Colloid and Interface Science 1996;180:251–60.

[58] Fritz J, Baller M, Lang H, Rothuizen H, Vettiger P, Meyer E, et al. Translating biomolecular recognition into nanomechanics. Science 2000;288:316–8.

[59] Wu G, Datar RH, Hansen KM, Thundat T, Cote RJ, Majumdar A. Bioassay of prostate-specific antigen (PSA) using microcantilevers. Nature Biotechnology 2001;19:856–60.

[60] Pei J, Tian F, Thundat T. Glucose biosensor based on the microcantilever. Analytical Chemistry 2004;76:292−7.

[61] Jensen K, Kim K, Zettl A. An atomic-resolution nanomechanical mass sensor. Nature Nanotechnology 2008;3:533−7.

[62] Lassagne B, Garcia-Sanchez D, Aguasca A, Bachtold A. Ultrasensitive mass sensing with a nanotube electromechanical resonator. Nano Letters 2008;8:3735−8.

[63] Chowdhury R, Adhikari S, Mitchell J. Vibrating carbon nanotube based bio-sensors. Physica E: Low-dimensional Systems and Nanostructures 2009;42:104−9.

[64] Chaste J, Eichler A, Moser J, Ceballos G, Rurali R, Bachtold A. A nanomechanical mass sensor with yoctogram resolution. Nature Nanotechnology 2012.

[65] Bunch JS, Van der Zande AM, Verbridge SS, Frank IW, Tanenbaum DM, Parpia JM, et al. Electromechanical resonators from graphene sheets. Science 2007;315:490−3.

[66] Sakhaee-Pour A, Ahmadian M, Vafai A. Applications of single-layered graphene sheets as mass sensors and atomistic dust detectors. Solid State Communications 2008;145:168−72.

[67] Chen C, Rosenblatt S, Bolotin KI, Kalb W, Kim P, Kymissis I, et al. Performance of monolayer graphene nanomechanical resonators with electrical readout. Nature Nanotechnology 2009;4:861−7.

[68] Wang ZL. Functional oxide nanobelts: materials, properties and potential applications in nanosystems and biotechnology. Annual Review of Physical Chemistry 2004;55:159−96.

[69] Feng X, He R, Yang P, Roukes M. Very high frequency silicon nanowire electromechanical resonators. Nano Letters 2007;7:1953−9.

[70] Grate JW, Martin SJ, White RM. Acoustic wave microsensors. Analytical Chemistry 1993;65:940A−8A.

[71] Grate JW, Martin SJ, White RM. Acoustic wave microsensors. Part II. Analytical Chemistry 1993;65:987A−96A.

[72] Länge K, Rapp BE, Rapp M. Surface acoustic wave biosensors: a review. Analytical and Bioanalytical Chemistry 2008;391:1509−19.

[73] Ballato A. Modeling piezoelectric and piezomagnetic devices and structures via equivalent networks. Ultrasonics, Ferroelectrics and Frequency Control, IEEE Transactions on 2001;48:1189−240.

[74] Ferreira GN, da-Silva A-C, Tomé B. Acoustic wave biosensors: physical models and biological applications of quartz crystal microbalance. Trends in Biotechnology 2009;27:689−97.

[75] Yang M, Thompson M, Duncan-Hewitt WC. Interfacial properties and the response of the thickness-shear-mode acoustic wave sensor in liquids. Langmuir 1993;9:802−11.

[76] Thompson M, Kipling AL, Duncan-Hewitt WC, Rajaković LV, Čavić-Vlasak BA. Thickness-shear-mode acoustic wave sensors in the liquid phase. A review. Analyst 1991;116:881−90.

[77] Liang C, Peng H, Zhou A, Nie L, Yao S. Molecular imprinting polymer coated BAW bio-mimic sensor for direct determination of epinephrine. Analytica Chimica Acta 2000;415:135−41.

[78] Drafts B. Acoustic wave technology sensors. IEEE Transactions on Microwave Theory and Techniques 2001;49:795−802.

[79] Jakoby B, Vellekoop MJ. Properties of Love waves: applications in sensors. Smart Materials and Structures 1997;6:668.

[80] Rocha-Gaso M-I, March-Iborra C, Montoya-Baides Á, Arnau-Vives A. Surface generated acoustic wave biosensors for the detection of pathogens: a review. Sensors 2009;9:5740−69.

[81] Campbell CK. Applications of surface acoustic and shallow bulk acoustic wave devices. Proceedings of the IEEE 1989;77:1453−84.

[82] Du J, Harding G, Ogilvy J, Dencher P, Lake M. A study of Love-wave acoustic sensors. Sensors and Actuators A: Physical 1996;56:211−9.

[83] Länge K, Blaess G, Voigt A, Götzen R, Rapp M. Integration of a surface acoustic wave biosensor in a microfluidic polymer chip. Biosensors and Bioelectronics 2006;22:227−32.

[84] Welsch W, Klein C, Von Schickfus M, Hunklinger S. Development of a surface acoustic wave immunosensor. Analytical Chemistry 1996;68:2000−4.

[85] Smith JP, Hinson-Smith V. The new era of SAW devices. Analytical Chemistry 2006;78:3505−7.

[86] Frye GC, Martin SJ, Velocity and attenuation effects in acoustic wave chemical sensors, in Proceedings IEEE 1993 Ultrasonics Symposium, 1993, pp. 379−383.

[87] Cajigas JC, Longworth TL, Davis N, Ong KY, Testing of HAZMATCAD Detectors Against Chemical Warfare Agents: Summary Report of Evaluation Performed at Soldier Biological and Chemical Command (SBCCOM), DTIC Document, 2003.

[88] Soper SA, Brown K, Ellington A, Frazier B, Garcia-Manero G, Gau V, et al. Point-of-care biosensor systems for cancer diagnostics/prognostics. Biosensors and Bioelectronics 2006;21:1932−42.

[89] Pathirana S, Barbaree J, Chin B, Hartell M, Neely W, Vodyanoy V. Rapid and sensitive biosensor for *Salmonella*. Biosensors and Bioelectronics 2000;15:135−41.

[90] Gizeli E, Bender F, Rasmusson A, Saha K, Josse F, Cernosek R. Sensitivity of the acoustic waveguide biosensor to protein binding as a function of the waveguide properties. Biosensors and Bioelectronics 2003;18:1399−406.

[91] Adler R, Desmares PJ. An economical touch panel using SAW absorption. IEEE Transactions on Ultrasonics, Ferroelectrics, and Frequency Control 1987;34:195−201.

[92] Vellekoop M, Lubking G, Sarro P, Venema A. Integrated-circuit-compatible design and technology of acoustic-wave-based microsensors. Sensors and Actuators A: Physical 1994;44:249−63.

[93] Pyun J, Beutel H, Meyer J-U, Ruf H. Development of a biosensor for *E. coli* based on a flexural plate wave (FPW) transducer. Biosensors and Bioelectronics 1998;13:839−45.

[94] Nguyen N-T, Meng AH, Black J, White RM. Integrated flow sensor for in situ measurement and control of acoustic streaming in flexural plate wave micropumps. Sensors and Actuators A: Physical 2000;79:115−21.

[95] Pan L, Liang Y, Tseng F, Leou K, Chen L, Lai Y, A novel application of acoustic plate mode sensor in tissue regeneration, in Proceedings of the IEEE-EMBS Special Topic Conference on Molecular, Cellular and Tissue Engineering, 2002. 2002, pp. 143−144.

[96] Friedt J-M, Carry E. Introduction to the quartz tuning fork. American Journal of Physics 2007;75:415.

[97] Edwards H, Taylor L, Duncan W, Melmed AJ. Fast, high-resolution atomic force microscopy using a quartz tuning fork as actuator and sensor. Journal of Applied Physics 1997;82:980−4.

[98] Karrai K, Grober RD. Piezoelectric tip-sample distance control for near field optical microscopes. Applied Physics Letters 1995;66:1842−4.

[99] Hoshino K, Rozanski LJ, Vanden Bout DA, Zhang X. Near-field scanning optical microscopy with monolithic silicon light emitting diode on probe tip. Applied Physics Letters 200892:131106-131106-3

[100] Su X, Dai C, Zhang J, O'Shea SJ. Quartz tuning fork biosensor. Biosensors and Bioelectronics 2002;17:111−7.

[101] Zhang J, O'Shea S. Tuning forks as micromechanical mass sensitive sensors for bio- or liquid detection. Sensors and Actuators B: Chemical 2003;94:65−72.

[102] Van Herwaarden A, Sarro P. Thermal sensors based on the Seebeck effect. Sensors and Actuators 1986;10:321−46.

[103] Riffat SB, Ma X. Thermoelectrics: a review of present and potential applications. Applied Thermal Engineering 2003;23:913−35.

[104] Van Herwaarden A. Overview of calorimeter chips for various applications. Thermochimica Acta 2005;432:192−201.

[105] Van Herwaarden A, Sarro P, Gardner J, Bataillard P. Liquid and gas micro-calorimeters for (bio) chemical measurements. Sensors and Actuators A: Physical 1994;43:24−30.

[106] Zhang Y, Tadigadapa S. Calorimetric biosensors with integrated microfluidic channels. Biosensors and Bioelectronics 2004;19:1733−43.

[107] Nath N, Chilkoti A. A colorimetric gold nanoparticle sensor to interrogate biomolecular interactions in real time on a surface. Analytical Chemistry 2002;74:504−9.

[108] Heine G, Lang W. Magnetoresistance of the new ceramic Cernox thermometer from 4.2 K to 300 K in magnetic fields up to 13 T. Cryogenics 1998;38:377−9.

[109] Childs P, Greenwood J, Long C. Review of temperature measurement. Review of Scientific Instruments 2000;71:2959−78.

[110] Courts S, Swinehart P. Stability of Cernox® Resistance temperature sensors. Advances in cryogenic engineering 2000;45:1841−8.

[111] Lagally ET, Emrich CA, Mathies RA. Fully integrated PCR-capillary electrophoresis microsystem for DNA analysis. Lab on a Chip 2001;1:102−7.

[112] Meijer GC, Wang G, Fruett F. Temperature sensors and voltage references implemented in CMOS technology. IEEE Sensors Journal 2001;1:225−34.

[113] Blauschild RA, Tucci PA, Muller RS, Meyer RG. A new NMOS temperature-stable voltage reference. IEEE Journal of Solid-State Circuits 1978;13:767−74.

[114] Timoshenko S. Analysis of bi-metal thermostats. Journal of the Optical Scoiety of America 1925;11:233−55.

[115] Lai J, Perazzo T, Shi Z, Majumdar A. Optimization and performance of high-resolution micro-optomechanical thermal sensors. Sensors and Actuators A: Physical 1997;58:113−9.

[116] Shin DD, Lee D-G, Mohanchandra KP, Carman GP. Thin film NiTi microthermostat array. Sensors and Actuators A: Physical 2006;130:37−41.

[117] Wijngaards D, De Graaf G, Wolffenbuttel R. Single-chip micro-thermostat applying both active heating and active cooling. Sensors and Actuators A: Physical 2004;110:187−95.

[118] Davis A, Lettington A. Principles of thermal imaging. Applications of Thermal Imaging 1988;:1−34.

[119] Rogalski A. Infrared Detectors. Boca Raton: CRC Press; 2010.

[120] Jones BF. A reappraisal of the use of infrared thermal image analysis in medicine. IEEE Transactions on Medical Imaging 1998;17:1019−27.

[121] Jiang L, Ng E, Yeo A, Wu S, Pan F, Yau W, et al. A perspective on medical infrared imaging. Journal of Medical Engineering & Technology 2005;29:257−67.

[122] Meola C, Carlomagno GM. Recent advances in the use of infrared thermography. Measurement Science and Technology 2004;15:R27.

[123] Wang J, Chang K-J, Chen C-Y, Chien K-L, Tsai Y-S, Wu Y-M, et al. Evaluation of the diagnostic performance of infrared imaging of the breast: a preliminary study. Biomedical Engineering Online 2010;9:1−10.

[124] Kaczmarek M, Nowakowski A, Siebert J, Rogowski J. Infrared thermography: applications in heart surgery. Optoelectronic and Electronic Sensors III 1999;:184−8.

[125] Hinterdorfer P, Baumgartner W, Gruber HJ, Schilcher K, Schindler H. Detection and localization of individual antibody-antigen recognition events by atomic force microscopy. Proc Natl Acad Sci 1996;93 (8):3477−81.

[126] Lee GU, Chrisey LA, Colton RJ. Direct measurement of the forces between complementary strands of DNA. Science 1994;266(5186):771−3.

[127] Florin E-L, Moy VT, Gaub HE. Adhesion forces between individual ligand-receptor pairs. Science 1994;264(5157):415−7.

[128] Lee GU, Kidwell DA, Colton RJ. Sensing discrete streptavidin-biotin interactions with atomic force microscopy. Langmuir 1994;10(2):354−7.

Further Reading

Mechanics of Cantilevers

Boresi AP, Chong K, Lee JD, Boresi AP, Chong KP, Elasticity in Engineering Mechanics. Wiley, New York, 3rd edition, December 21, 2010.

Hibbler RC, Hibbler RC, Engineering Mechanics: Statics and Dynamics. Prentice Hall, Upper Saddle River, 13th edition, July 15, 2012.

CHAPTER 7

Implantable Sensors

Chapter Outline

7.1 Introduction

One of the most vital application areas of the molecular sensors is measurement of activities in human body. Among such tools for biomedical sensing applications, implantable sensors raise unique problems regarding the design and realization of device structures that fulfill the requirements for the use in the human body. The use of implantable sensors has several advantages over other monitoring paradigms. Implantable sensors enable self-monitoring systems in which the patient acts as a part of the sensing system to detect changes, consciously or unconsciously, of the health conditions. Such devices may remove, to a certain degree, the subjectivity, variable patient ability, and offers sufficient sensitivity by design for detecting subtle changes.

Implantable molecular sensors and devices would provide real-time feedback that allow for treatments that are safer, more efficient, less painful, and more convenient. Not so obviously, treatment would also be cheaper, as there would be a reduction in the amount of disposable materials such as needles, along with an increase in more useful and accurate data measurements [1]. The continual breakthroughs in microfabrication methods have enabled engineers to implant smaller and smaller devices into the human body. The best example of how microfabrication has advanced the development of implantable devices may be the pacemaker. The first practical implantable pacemaker was made possible by the invention of the transistor in 1947. The size of pacemakers has shrunk as fabrication methods have become more refined. Today, a modern pacemaker can be as small as a quarter, or about an inch (2.5 cm), and weighs less than 15 g [2]. Modern pacemakers have been a true success story in the application of nano- and microfabrication technology. They have shown that the refinement and downsizing of microfabrication have profound effects in the field of implantable sensing.

In this chapter, we describe implantable molecular sensors and related devices. We define an implantable molecular sensor or device to be a sensor or device that functions in the body under physiological conditions for some significant period of time. We first introduce the pacemaker as a model implantable device, which integrates several essential sensing elements required for implantable applications. Then we discuss elements used in implantable systems, including electrodes for electrical and chemical sensing and mechanical pressure sensors. We also describe recent advancements in drug delivery, energy harvesting, and other issues related to implantable sensing.

7.2 Sensing in Cardiac Pacemakers

A cardiac pacemaker is mainly used to treat arrhythmias, or irregular heartbeats. It stimulates the heart with an electrical pulse to restore proper rhythm. They have played an important role in the lives of millions of patients around the world. Between 1993 and

2009, 2.9 million people received permanent cardiac pacemakers in just the United States alone. Overall use of pacemakers has risen over 50% in that time period [3]. As one of the most influential medical sensing devices, it is important to understand more about cardiac pacing and the technology behind it.

Medtronic, Inc. has played a critical role in the development of cardiac pacemakers. The first battery-operated, external pacemaker was invented in 1957 by E. Bakken, the co-founder of Medtronic, Inc. It was not long after that when the first commercially produced implantable pacemaker was invented by Medtronic, Inc. partners W. Chardack and engineer W. Greatbatch in 1960. The next major breakthrough in pacemaker technology came in 1976 with the invention of the lithium batteries. These longer-lasting batteries were critical to the future success of the pacemaker and are still used in many pacemakers today. In 1986, Medtronic, Inc. introduced a rate-responsive pacemaker that adjusted the pacing rate to match a patient's level of physical activity. Improvements continued over time, focusing on size, functionality, and use with magnetic resonance imaging (MRI). Now, pacemakers can take on multiple roles and can be as small as a few centimeters. We will first focus on the method of operation of a popular type of pacemaker, before describing other types as well. We will end this section by discussing future research in the field of cardiac pacemakers.

7.2.1 Components of Pacemakers

A pacemaker consists of two main parts, the electrodes attached to pacemaker leads and the pulse generator (Fig. 7.1). Additional sensors are attached to rate-responsive pacemakers. These additional sensors can be split into two main categories: activity sensors and metabolic sensors.

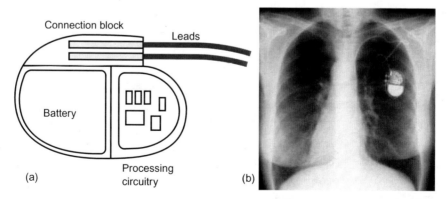

Figure 7.1: (a) Various Parts of the Pacemaker. (b) Implanted Pacemaker in Human Body.
(b)   http://www.medtronic.com/for-healthcare-professionals/products-therapies/cardiac-rhythm/pacemakers/pacing-leads/index.htm#tab2.

Pacemaker Electrodes

The heart has a natural pacemaker that controls the heart rate and rhythm. A normal heartbeat begins in the right atrium. Located here is the sinoatrial (SA) node. The SA node generates regular electrical impulses that first contract the atria and then after a short delay caused by the impulse passing through the atrioventricular (AV) node, the ventricles as well. The typical frequency of these pulses is around 60–80 times per minute for a resting individual. The voltage generated from the depolarization and repolarization of the SA node is measured by an ECG (electrocardiogram) to be around 2.5 mV. However, the internal depolarization and repolarization of the cells and their voltage generation is beyond the scope of this book.

The electrodes of the pacemaker monitor this signal's frequency in addition to other typical waveforms and pass this information to the computer in the pulse generator to determine the patient's heart rate and rhythm. In other words, part of the electrical signal the heart generates is passed through the conductive electrode to the pulse generator. Signal sampling rates can vary and are typically in the order of 0.01–1.0 ms. From here, the computer in the pulse generator determines if it is necessary to "reset" the heart rhythm or adjust the heart rate according to programmed base levels. The electrodes also deliver the stimuli to the heart muscle. Each pulse causes the heart to contract. The leads that carry the current from the pulse generator to the electrodes are insulated wires with low resistance values typically around 40 Ω.

The electrodes typically tend to be conducting materials, such as platinum, titanium, or various alloys. They typically have a hook or anchor of some sort that helps hold the electrode into place in the heart muscle to ensure proper electrical stimulation. The outer layer may be a porous material (with pores in the micron size range) that contains drugs to prevent tissue fibrosis. The surface area of these electrodes varies, but typical sizes range from 1 to 6 mm^2 (see Table 7.1). The number of electrodes that the pacemaker contains varies with the type of pacemaker. These different types are discussed in a later section.

Table 7.1: Size of Pacing Leads and Electrodes

	CapSure® SP Novus 5092, 5592	CapSure Sense 4074, 4574	CapSure Z Novus 5054, 5554
Tip electrode	Platinized platinum	Titanium nitride (TiN) coated	Platinized platinum
Lead body size	6 Fr (=2 mm)	5.3 Fr (=1.8 mm)	6 Fr (=2 mm)
Electrode surface area	Tip: 5.8 mm^2	Tip: 2.5 mm^2	Tip: 1.2 mm^2

1 Fr (French catheter scale) = 1/3 mm

Adapted from Medtronic website (http://www.medtronic.com/for-healthcare-professionals/products-therapies/cardiac-rhythm/pacemakers/pacing-leads/index.htm#tab2).

Pacemaker Pulse Generators

The pulse generator contains the battery and a small computer as well. Usually, a rate-responsive sensor can also be found with the pulse generator system. When the heart has an abnormal heartbeat, the generator sends an electrical pulse to the heart to "reset" its heartbeat rhythm. Now having explained the general operation of a pacemaker, we will further discuss the two main sensors associated with detecting the heart's electrical activity.

Activity Sensors

Rate responsive sensors can measure a variety of attributes including muscle vibrations, breathing rate, blood temperature, and body acceleration to provide "rate-responsive therapy". There are two main types of rate-responsive sensors: activity sensors and metabolic sensors. The activity sensor draws a relation between the patient's physical activity level and heart rate. This type of sensor is older and more common. Activity level can be measured by various things including a piezoelectric crystal (see Chapter 6, Section 6.3 for basics of piezoelectric effects) that recognizes muscular pressure waves and an accelerometer that identifies changes in posture that correspond to physical activity (see Fig. 7.2). These sensors can be located in either the pulse generator chamber or the electrode tip. Activity sensors provide a rapid response to the body's need to pump additional blood because they directly measure physical activity. In [4], an accelerometer integrated pacemaker was evaluated in 11 patients. The accelerometer demonstrated improved rate stability and higher rate response to activities of jogging and standing, although significant improvement was not observed for responses to other daily activities and treadmill exercise.

There are other factors that play a role in the patient's heart rate that are not related to physical activity. This is where metabolic sensors can help [5].

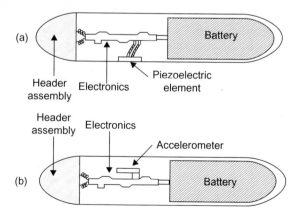

Figure 7.2: (a) Piezoelectric-based Vibration Activity Sensor and (b) Accelerometer-based Activity Sensor.
From [6].

Metabolic Sensors

Metabolic sensors can measure QT interval, minute ventilation, and peak endocardial acceleration. The metabolic sensor helps adjust to physical and mental stress requirements. Most of the metabolic sensors are those that measure minute ventilation that can be derived from measuring the variations in the transthoracic impedance signal. Minute ventilation is the product of respiratory rate and tidal volume and is shown to be correlated with metabolic demand and heart rate. The pacing rate is adjusted based on the patient's minute ventilation rate. Ventilation rates may change $\sim 10-50$ L/min depending on physical activities [7]. A metabolic sensor measures ventilation rates based on a change in resistance [8]. The impedance across the thorax region is measured with low-level electrical signals to avoid unwanted stimulation of cardiac cells. For example, a 1 mA current with a narrow pulse width of $7-30$ µs at 20 Hz has been used [9]. A trend is emerging that combines a piezoelectric crystal-based activity sensor and a minute ventilation sensor with various algorithms to closer replicate the SA node's behavior [10].

7.2.2 Types of Pacemakers

There are three main types of pacemakers in clinical use today: single chamber, dual chamber, and biventricular. The dual chamber has become the most prevalent, replacing the single chamber pacemakers. The biventricular pacemaker is used in patients with heart failure or a damaged electrical system in the heart. Each type of pacemaker varies in the number of leads used and is named after how many chambers of the heart have pacemaker leads.

Single Chamber Pacemaker

The single chamber pacemaker has only one lead that can be placed in one chamber of the heart. They typically carry electrical impulses from the pulse generator to the right ventricle. However, in some occasions, it can also have its one lead in the right atrium. This type of pacemaker was popular when pacemakers were originally being developed. However, since only one chamber can be stimulated, it could lead to an improper timing between contractions of the atrium and ventricle. In recent years, the single chamber pacemaker has been losing popularity and represents about 10% of all pacemaker implantations.

Dual Chamber Pacemaker

The dual chamber pacemaker has two leads. It has leads in the right atrium and the right ventricle to help control the timing between the contractions of different chambers. Dual chamber pacemakers are the most popular, with over 80% of implants being dual chamber.

Biventricular Pacemaker

The biventricular pacemaker is used to make the heart pump more efficiently by contracting both the right and left ventricles at the same time. Because this system resets the heart's original pumping mechanism, it is also called cardiac resynchronization therapy (CRT) [11]. Figure 7.3 illustrates the locations of the leads we have discussed.

7.2.3 Recent Developments and Future Research

While the original pacemaker was a mid-20th century invention, developments to improve the experience for patients continue. Beyond developing longer lasting batteries, pacemakers have seen drastic improvements in reduced interference from electromagnetic devices, remote monitoring, and pacemakers requiring only minimally invasive surgery. We also see a trend for multi-tasking pacemakers that provide other monitoring benefits. Further, we have also seen developments in harvesting energy from pacemaker leads to power the pacemaker battery.

Future Sensing Components in Pacemakers

We are seeing many new implantable sensors being incorporated onto pacemakers to record a patient's vital information. Already, we have seen pacemakers being combined with miniature, implantable cardioverter defibrillators (ICDs) to improve the functional status of the pacemaker and to help treat many life-threatening arrhythmias [12]. Another major research area has focused on incorporating drug pumps and hemodynamic monitoring as alternative methods to detect and treat irregularity in the heart's rhythms [12].

In the coming years, we are likely to see new sensors for blood pressure, glucose levels, oxygen concentration, and heart volume output being incorporated into pacemakers for

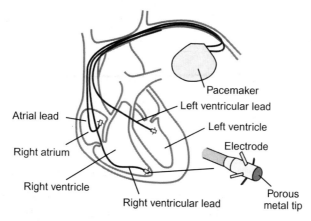

Figure 7.3: Location of a Pacemaker and its Leads in the Body.

permanent implants. Doctors will be able to access this information to provide a better overall patient experience. Basic principles for many of these sensors are discussed in other chapters of this book.

Recent Pacemaker Advancements

Until recently, pacemaker patients experienced major effects from electromagnetic interference. In 2011, Medtronic, Inc. invented the first cardiac pacemaker that was useable with many MRI environments. MRI is a critical diagnostic tool in today's clinical environment and until now, pacemaker patients were unable to undergo MRI diagnosis of medical conditions [13]. However, some problems still exist for pacemaker patients from even basic items such as cellphones and many other household devices.

Another major breakthrough in recent pacemaker technology was the invention of wireless transmission of data collected by the pacemaker directly to the clinician without patient interaction. This pacemaker, invented by St. Jude Medical, is a single chamber pacemaker that allows clinicians to continually monitor device performance without frequent patient check-ups.

The next development could potentially be considered one of the biggest advances in pacemaker technology. While this device is still in the research stage, Medtronic, Inc. is in the process of developing a pacemaker the size of a vitamin capsule. With a pacemaker so small, they hope to avoid major surgery for implantation completely by having the device inserted directly to the heart via a catheter. This product is thought to be coming to market in the next few years from Medtronic Inc.

Energy Harvesting

About 5−10 years after an initial pacemaker implantation surgery, the battery requires replacement and the entire pulse generator must be replaced through another surgical procedure. Instead of focusing on developing the time the battery can last, many researchers have begun to focus on a continual energy source, the heart itself. Researchers are looking to power pacemakers through harvesting the vibrational energy given off by the heart during each heartbeat. One way this is being done is through introducing piezoelectric devices onto the pacemaker leads [14]. Another method is through electrostatic generators that can generate up to 36 μW [15]. Some energy harvesting devices can even generate up to 100 μW. In some cases, these energy harvesting techniques can completely power the pacemaker on their own, and can lead to further reduction in the size of pacemakers due to the removal of a battery from the pulse generator [14].

As we have seen, pacemakers are very important medical devices that fundamentally operate using various sensing technologies. Pacemakers have positively affected the lives of millions of patients and will continue to be a widely used medical device to restore proper heart rhythm.

7.3 Implantable Electrodes

In this section we will discuss systems that are categorized as electrical implantable sensors and devices. The key element in electrical sensing is the electrode that transduces signals between electronics and living tissues/cells. Electrodes for biomedical applications can be categorized into non-invasive and invasive electrodes [16]. Non-invasive electrodes are surface electrodes applied to the skin of the subject. They are used for electrocardiography (ECG), electromyography (EMG), or electroencephalography (EEG), which are techniques for recording and evaluating the electrical activities of the heart, skeletal muscles and neurons of the brain, respectively, from the surface of the skin. Invasive electrodes are implanted adjacent, around or within nerves or muscles. Here we mainly describe implantable electrodes. An electrode passively records extracellular signals from electrically active tissues including central nervous system, peripheral nerve neurons, and skeletal muscle cells. It can also stimulate cells with electrical signals or measure the impedance of nearby tissues to infer cell physiology. We start by describing theories, design, and fabrication of microelectrodes for neural measurement and excitation, followed by an introduction of practical applications for neural implants, retinal implants, hearing implants or hearing aids, and cochlear implants.

7.3.1 Microelectrode

Electrically active cells such as neural cells or muscle cells generate small currents in short spikes, which are called action potentials. These currents are actually due to ions (Na^+, K^+) moving across the cell membrane through protein channels. Extracellular recording from close proximity to cells can pick up these signals which ranges $100-1000\,\mu V$.

Many electrode designs have been devised and used for sensing/stimulating electrodes. For clinical applications, relatively large and robust electrodes fabricated by conventional manufacturing methods are mainly used. MEMS-based large arrays of microelectrodes are studied for better spatial resolution and selectivity. Figure 7.4 shows a typical structure of a MEMS-based microelectrode. The structure is relatively simple and easy to fabricate. The three essential elements are the substrate, electrodes, and top passivation layer. If the substrate is made of an insulating material such as glass or polyimide, the insulator layer beneath the electrode is not necessary.

The photolithography technology can potentially provide patterning of electrical connections much finer than practically required for recording of cellular activities. As far as planar electrodes are concerned, nearly arbitrary electrode sizes and geometries are possible, and multiple electrode materials can be used on a single device. Metals considered biocompatible and often used for electrodes include gold, platinum, titanium, and iridium.

One difficult requirement in practical measurement is that the sensing site should be arranged in three dimensional structures. Needle-like substrates are desirable for

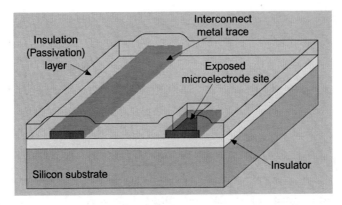

Figure 7.4: Schematic of a Typical Microelectrode.
From [17].

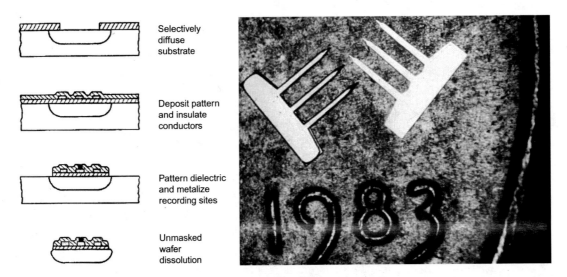

Figure 7.5: Silicon Microelectrode Probes.
From [19].

microelectrodes to penetrate into animal bodies. An arrayed structure is also often required to enable simultaneous parallel access to many neurons. In the late 1960s, early efforts on fabrication of micromachined probes were made. Wise et al. created microprobes utilizing chemical wet etching of silicon [18]. The important technique introduced by Najafi et al. in 1985 [19] is the very heavy boron diffusion that controls the shape of the probe substrate. The EDP (ethylene-diamine and pyrocatecol) etch stops when the boron concentration is higher than a level of about 5×10^{19} cm^{-3}. The fabrication procedure and a photograph of probes are shown in Fig. 7.5.

7.3.2 Neural Implants

Neural implants aim to create a neurotechnical interface to a human being. They look to achieve a bidirectional flow of information to either record neural activities or perform nerve or muscle excitation via an electrical stimulation. Electrical nerve stimulation has been clinically used for more than 100 years, while recording of electrical activity of intact nerves with implanted electrodes is a relatively new challenge [20]. Electrodes for neural implant include extraneural electrodes that provide simultaneous interface with many axons in the nerve, and penetrating electrode which may contact small groups of axons within a nerve fascicle [16]. Neural implants must secure reliability for both patient safety and acceptable functionality as a medical device. Furthermore, any design must be able to ensure that neither the neural implant nor the body's environment has any effect on the other [21].

Microprobes are effectively used in recording activity in the brain. While this method does not directly treat ailments, it is especially useful to doctors in order to learn more about the nervous system and develop valuable treatment for various disorders. It may also be used to assist the disabled. Possible functions include mentally controlling the motion of a machine such as a robot, a wheelchair, or a neuroprosthetic limb [22,23]. Wireless implantable microsystems are under development to achieve recordings of neural spikes via hard thresholding that reveal valuable information about the nervous system. In a particular research experiment, a neural processor capable of recording activity on 64 simultaneous channels was created [24]. The system is wireless and can transmit activity to an external observer. A simplified diagram of the workings of the system is shown in Fig. 7.6.

Silicon microprobes sense activities in the brain and send signals to an electronics package. A bidirectional telemetry module retrieves the data through the forward telemetry link and

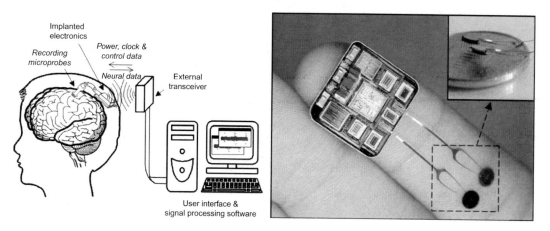

Figure 7.6: Wireless Implantable Microsystem.
From [24].

generates power for the entire system. The reverse telemetry link encodes, modulates, and transmits the information to outside. All the neural channels can be scanned for neural spikes [25].

For practical clinical applications, where robustness and safety are in high priority, smaller numbers of relatively large electrodes are still more commonly used. An interesting result regarding utilization of implantable electrodes was reported in 2011 [26]. It was reported that repeated periods of stimulation of the spinal cord and training increased control ability in a case of spinal cord injury. The patient was a 23-year-old man who lost voluntary motor function and partial preservation of sensation below the T1 cord segment as a result of a car accident. A 16-electrode array was surgically implanted on the dura (L1−S1 cord segments) for chronic electrical stimulation. Epidural stimulation enabled the man to stand for 4.25 min with assistance only for balance. Additionally, 7 months after implantation, the patient also recovered control of some leg movements, although this is only during epidural stimulation (Fig. 7.7).

7.3.3 Cochlear Implants

A cochlear implant is an implanted device that provides a sense of sound to patients who are profoundly deaf or severely hard of hearing. Cochlear implants are a very widely used form of implantable device. According to the US Food and Drug Administration (FDA), as of December 2010 approximately 219 000 people worldwide have received implants (NIH

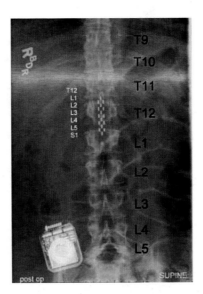

Figure 7.7: Implanted Electrodes for Epidural Stimulation.
From [26, Supplementary webappendix]

Publication No. 11-4798). Unlike a hearing aid which simply amplifies sounds, cochlear implants directly stimulate the auditory nerve, bypassing damaged portions of the ear. The implant gives a patient a representation of sounds that help him or her understand speech. Widely used coding strategies [27,28] are not simple presentation of analogue waveforms, since improved sampling strategies showed better speech recognition with cochlear implants [29].

The implant basically consists of four main parts as shown in Fig. 7.8.

- A microphone picks up sound from the environment.
- A processor processes sound signals and selects information to be sent to the electrode.
- A transmitter and receiver are used to transmit signals to the stimulator which generates electric impulses.
- An implanted array of electrodes stimulates different regions of the auditory nerve.

The diameter of the circular part of a cochlear is 5−8 mm, and insertion depth is typically 17−19 mm [30]. Figure 7.9(a)−(d) shows ideal insertion path of an electrode array [31].

Spatial density or number of electrodes is not considered a critical issue in cochlear devices. Modern cochlear implants have only 22 electrodes to replace the ∼16 000 hair cells used for normal hearing. It is reported that vowel and consonant recognition for cochlear-implant

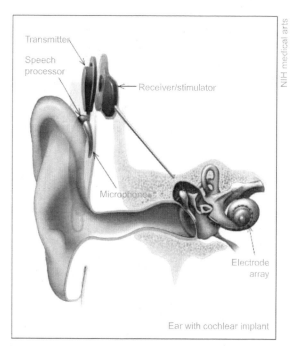

Figure 7.8: Cochlear Implant.
Image from http://www.nidcd.nih.gov/health/hearing/pages/coch.aspx

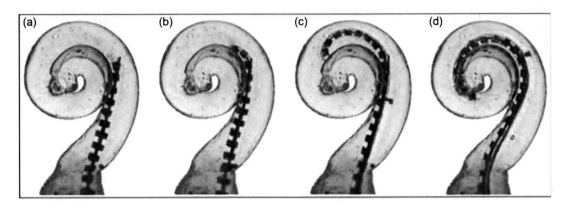

Figure 7.9: Insertion of Electrode.
(a) Intended insertion of spiral molded electrode beginning of first cochlear turn. (b)–(d) Ideal insertion path of an electrode array. *From [31].*

listeners did not improve with more than seven electrodes. Word and sentence recognition showed marginally significant increase with 22 electrodes from seven to ten electrodes [27].

7.3.4 Retinal Implants

One area where microfabricated high density electrode arrays play a critical role is retinal implants. Increase in the number of functional electrodes leads to the enhancement of the quality of images that can be presented to the patient. Several studies have been made in an attempt to create artificial vision systems based on the principle of electrical activation.

The retina is made of layers of neurons, part of which are called photoreceptors and are sensitive to light. Many patients lose their vision because their photoreceptors have degraded due to age-related macular degeneration (AMD) or retinitis pigmentosa (RP). However, other neurons in the retina that receive electrical signals from the photoreceptors may still be activated by electrical pulses. In this way, an implanted electrode array can produce the sensation of light in a blind person.

Compared to cochlear implants, which have been commonly used worldwide, retinal implants are still at a developmental stage. An early study of a chronic retinal implantation of 4×4 electrodes enabled a patient to detect motion and recognize simple shapes [32]. The improved system Argus II, which uses 60 stimulating microelectrodes [33] from the same group was given FDA approval in 2012. The Argus II system consists of an image sensor, image processor, pulse generating integrated circuits and the implantable electrodes (Fig. 7.10). Several important components such as eyeglass-mounted cameras, video processing unit and data communication hardware are installed external to the body. The power and the signals are wirelessly transferred to the implanted device.

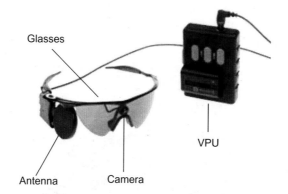

Figure 7.10: Camera Based Artificial Vision System.
Copyright © 2013 Second Sight Medical Products, Inc. (http://2-sight.eu).

One direction of improvement is the integration of denser array of microelectrodes. Zrenner et al. developed an implantable micro photodiode array that is positioned under the retina (Fig. 7.11). The array has 1500 active photodiodes, each of which is connected to an amplifier and a local stimulation electrode [34]. The system also includes 16 electrodes that can be used for light-independent electrical stimulation. Patients were able to read letters and combine them to words.

Another direction is development of flexible electrode arrays capable of stimulating the inner retina [35]. Rodger et al. developed a flexible parylene-based electrode array and used a heat molding process to conform it to the curvature of canine retinas (Fig. 7.12). The arrays were implanted in two canines for 6 months, and follow up observation showed normal vessel perfusion under the arrays.

7.3.5 Heart Blood Volume Sensor

Real time, in vivo measurements of pressure and volume within the left ventricle provide a framework for understanding cardiac mechanics. For example, researchers use pressure/volume data from active animals that have been administered a drug. Pressure and volume measurement is usually plotted as a PV loop, in which left ventricular (LV) pressure is plotted against LV volume at multiple time points during a single cardiac cycle.

There are two methods for volume measurements, classical conductance based PV loops and admittance based PV loops. The conductance method is performed with a conductance catheter with excitation and recording electrodes (Fig. 7.13). The excitation electrodes generate a field inside the heart by applying an alternating current between the electrodes. The resulting voltage is measured by the recording electrodes, and the conductance G is

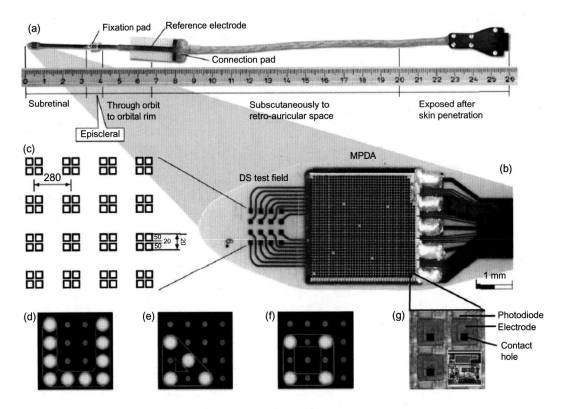

Figure 7.11: Electrode Array for Retinal Implant Image.
From[34]. Copyright © 2013 Second Sight Medical Products, Inc.

found based on the relationship $G = 1/R$, where R is the resistance. The conductance G is then related to the blood volume using Baan's Equation [36]:

$$V_t = 1/\alpha(\rho L^2)(G_{blood} - G_{||}), \tag{7.1}$$

where

V_t time-varying changes in volume
ρ resistivity of blood
L length between electrodes
α constant dependent on the stroke volume
G_{blood} conductance of blood
$G_{||}$ parallel conductance of muscle.

The admittance-based method is similar to the conductance method, but instead of just taking the resistance of the blood, conductive and capacitive properties of the blood and the muscles are taken into account. While blood can be considered as a resistive element,

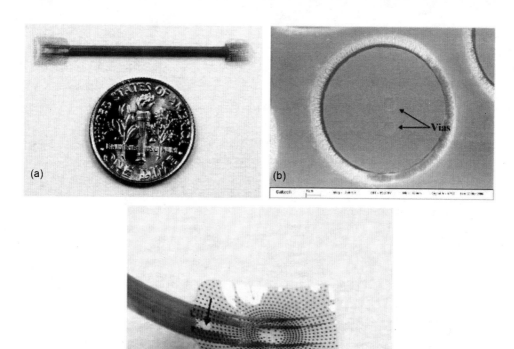

Figure 7.12: Flexible Electrode Array for Retinal Implant.
From [35].

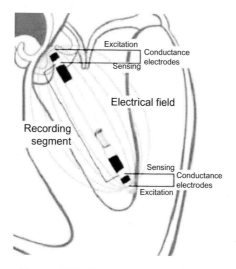

Figure 7.13: Conductance Catheter.
http://www.scisense.com/education/cv-theory-conductance.html

muscles have both a capacitive and resistive element. This method is more accurate because once the portions of the electrical signal due to the muscle can be removed and the only signal remaining is then due to the blood volume [37].

7.3.6 Subcutaneous Electrodes for Continuous Glucose Monitoring

Self-monitoring of glucose levels is a fundamental part of diabetes management. It is typically performed by intermittent measurements of capillary blood glucose. Patients with type 1 diabetes typically test their blood sugar five to ten times a day (see Chapter 4 Section 4.2.2 for more details). One problem of such tests is that isolated measurements do not reflect real-time variations of glucose levels which occur throughout the day and night. Testing at a higher frequency is desirable for the care of diabetes. It has been reported that the more frequent measurements of blood glucose levels, or glycemia, for individuals with type 1 diabetes showed the better glycated hemoglobin (HbA1c) concentration [38].

Continuous glucose monitoring (CGM) allows determination of how the glucose level reacts to food intake, insulin dosing, physical activities, and other factors. The measured data can be used to determine the insulin infusion rates for correction of hyperglycemia. Shichiri et al. in 1982 proposed the use of a needle-type subcutaneous sensor for the monitoring of glucose levels in subcutaneous tissue [39]. They developed a closed-loop glycemic control system that incorporates the glucose sensor and infusion pumps, and tested it with pancreatectomized dogs [39]. Although improvements have been made by several groups since then, the handmade subcutaneous sensor reported in [40] is a good example to illustrate the basic concept of such needle type electrodes. Figure 7.14 shows the schematic of the sensor. The sensing electrode was made of a Teflon-coated platinum-iridium wire with a diameter of 0.25 mm. The exposed part of the electrode was coated with cellulose acetate, on which glucose oxidase is immobilized. The reference Ag/AgCl electrode is a wrapped silver wire treated with HCl to form AgCl on surface.

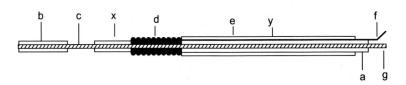

Figure 7.14: Schematic Diagram of a Handmade Subcutaneous Sensor.
(a) Teflon coated Pt-Ir wire; (b) Teflon tip; (c) sensing site (1 mm in length); (d) Ag/AgCl reference electrode; (e) heat-shrinkable tubing; (f) reference electrode terminal; (g) working electrode terminal. *From [40]. http://www.scisense.com/education/cv-theory-conductance.html Courtesy of Transonic Scisense.*

Now, several subcutaneous sensors are commercially available for CGM [41]. The first FDA approved device was the system from Medtronic, which also uses a subcutaneous needle-type amperometric enzyme electrode based on glucose oxidase. Other companies which provide CGM devices include Dexcom Inc. and Abbot Inc. Examples of the CGM devices are shown in Fig. 7.15. These systems are typically composed of a glucose sensor with an external electrical connector and a glucose monitor. A needle-type glucose-oxidase sensor is inserted subcutaneously. The earlier models relied on a physical connection between the sensor and the monitor. The newer prototypes utilize a wireless emitter attached to the sensor that transmits information to the monitor. Typically, the glucose levels are measured every 1 to 10 min, and the sensor has to be changed every 3 to 7 days [41,42].

The largest problem with the current technology is significant sensor drift. The causes of the drift are thought to include proteins trapped in the sensor, which impede the diffusion of glucose to the glucose oxidase layer, changes in tissue oxygen pressure (pO_2), altered local blood flow due to tissue interferents, and wound response [43]. As a result of the drift, CGM devices require calibration based on fingerstick measurements multiple times a day, which reduces the advantage of such systems over traditional devices.

7.4 Implantable Mechanical Sensors

7.4.1 Implantable Pressure Sensors

Blood pressure is an important form of vital sign that are mechanically measured from a human body. Implantable pressure sensors have been a topic of interest since the 1950s [44,45]. The history of catheter-based blood pressure sensors dates back to 1969, when Scheinman, Abbot, and Rapaport first measured blood pressure in the pulmonary artery using a sensor incorporated with a nylon catheter. The next year, Swan and

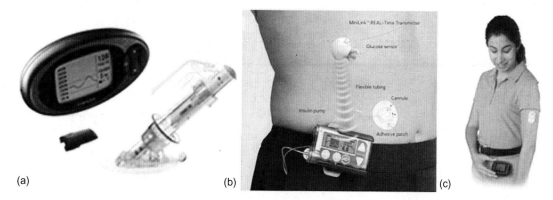

(a)   (b)   (c)

Figure 7.15: Examples of CGM Devices from (a) Dexcom, (b) Medtronic, and (c) Abbot.
From [41]. James A. Hunt, (2007) "Level sensing of liquids and solids — a review of the technologies",
Sensor Review, Vol. 27 Iss: 3, pp.200—206.

Ganz introduced a flow directed, balloon-tipped catheter [46]. Quickly becoming the standard for coronary care, their catheter was popular for its ease of use and ability to be used without fluoroscopy. These catheters, now known as Swan–Ganz or pulmonary artery catheters, utilize an inflated balloon wedged in a small pulmonary vessel to provide an indirect measurement of left atrium pressure. Subsequent techniques and alterations to the device have been created over the years for a variety of functions: the thermodilution technique for cardiac output measurement, addition of multiple lumens to allow injection of drugs through the catheter, incorporation of fiberoptic SvO_2 probes to obtain instantaneous measurements of oxygen saturation in the vessel, etc. [47].

Similarly, new procedures and devices are being constantly created for real-time, in vivo blood pressure measurements. Primarily motivated by ICU surgery and trauma care, new applications for intravascular blood pressure measurements include delivering real-time, direct measurement of pressure; offering information about potential occlusions and perforations; providing controlled catheter balloon inflation; and guiding the catheter through the vascular network are constantly being researched.

Pressure sensors can vary drastically in technology, design, performance, application suitability and cost. With the proliferation of catheters with multiple lumens, there are three main technologies that have emerged on the market for use as blood-pressure sensors based on piezoresistive, inductive–capacitive, and optical transduction.

Since the discovery of piezoresistivity in silicon in the mid 1950s, silicon-based pressure sensors have been widely produced. Microfabrication technology has greatly benefitted from the techniques and advancements in integrated circuit industry, taking advantage of materials, processes, and toolsets. Because of this, microelectromechanical systems (MEMS)-based devices have seen huge explosion and are now poised to catalyze the development of new markets.

The common feature of all of the pressures sensors is that they convert pressure to deflection of a mechanical element. Many devices are based on diaphragm deformation. Strain gauges were commonly used on diaphragm-based devices. For most part of the measurable range, amount of deflection is directly proportional to the applied pressure. In the case of a diaphragm with large built-in stress or large deflections, however, this direct proportionality does not hold true. A linear response in the deflection measurement is often desirable, since such systems are simple to calibrate and measure.

7.4.2 Implantable Piezoresistive Pressure Sensors

In 1954, Smith reported the piezoresistive effect of silicon and germanium, stating that the material undergoes a change of resistance with applied stress [48]. This discovery enabled production of semiconductor-based piezoresistive sensors. Commonly used for in vivo monitoring of blood pressure are silicon membrane-based piezoresistive pressure sensors positioned on the tip of an arterial catheter [49]. The catheter is a tube that can be inserted into a blood vessel or body cavity. Catheters allow drainage, injection of fluids, or access by surgical instruments, in addition to several functionalities that can be mounted on it. The process of inserting a catheter is catheterization. Piezoresistive-based pressure sensors have piezoresistors mounted on or in a diaphragm (see Section 2.4.1 for the fabrication procedure). For thin diaphragms and small deflections, the resistance change is linear with the pressure.

The evolution of the piezoresistive pressure sensor technology is illustrated in Fig. 7.16. Metal diaphragms attached with metal strain gauges were used initially. Metal diaphragms were quickly replaced with single-crystal silicon diaphragms with diffused piezoresistors (Fig. 7.16b) thereby eliminating the problems of hysteresis and creep associated with metal diaphragms. At room temperatures, silicon is perfectly elastic and will not plastically

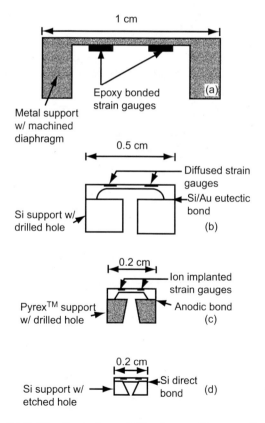

Figure 7.16: The Evolution of Diaphragm Pressure Sensors.
From [50].

deform [50]. Silicon obeys Hooke's law up to 1% strain, a tenfold increase over common metal alloys [51]. This made silicon-based diaphragms an immediate popular choice over metal diaphragms.

Some of the first silicon diaphragms were created by mechanical milling spark machining followed by wet chemical isotropic etching, to create a cup shape [52]. These diaphragms were bonded to silicon supports by a gold—silicon eutectic ($T = 370$ °C) [52].

Development of anodic bonding and the ability of device to withstand $500-1500$ V and $400-600$ °C enabled fabrication of silicon diaphragms bonded to pyrex glass supports [53]. Use of anodic bonding amounted to significant cost reduction in the sensor fabrication process. However, this did not bring about the required miniaturization of the piezoresistive based sensors for biomedical applications.

Introduction of silicon on insulator (SOI) technology in the 1980s offered a number of benefits to MEMS based pressure sensors, primarily because of the buried insulator that can act as an etch stop allowing precise control of the diaphragm thickness [54]. Since then, several miniaturized pressure sensors have been reported, which have silicon nitride [55–57] or polysilicon [58–60] diaphragms.

For catheter-based pressure sensors, silicon piezoresistor-based sensing has been considered standard since Millar Instruments introduced a miniature catheter tip pressure transducer in 1973, which comprised two silicon piezoresistors in a half Wheatstone bridge [61]. At the present time, commercially available devices from Millar instruments include an FDA-approved single-use pressure catheter Mikro-Cath™ (Fig. 7.17). It is intended for the use as a minimally invasive device with limited body contact less than 24 hours. The sensor is incorporated onto a medium-sized 3.5 French catheter, or 1.1 mm in diameter, and can

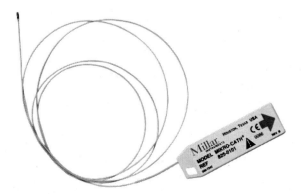

Figure 7.17: Mikro-Cath Pressure Catheter.
Image courtesy of Millar Instruments.

sense pressures from −50 mmHg to +300 mmHg. While the Millar Instruments sensor is mechanically stable, it is disadvantaged by its large power consumption, temperature dependence, instability in dynamic field conditions, and high stiffness. A similar type of catheter tip pressure sensor is commercially available from several companies such as Sentron and SciSense.

Study of chronically implantable blood pressure sensors is an area of ongoing research, and several interesting attempts have been reported. Mills et al. developed an implantable device for chronic measurement of pulsatile blood pressure (BP) from conscious, freely moving laboratory mice [62]. Figure 7.18 shows the structure of the device. A 5-cm-long fluid-filled catheter transfers intra-arterial pressure to a semiconductor strain gauge sensor which sizes measuring 1.5 × 1.5 mm. Blood pressure and heart rate were recorded from 16 chronically (30−150 days) implanted mice.

Millar Instruments now provides commercially available implantable telemetric pressure sensor systems. Reports have been made on pressure measurements in small animal models [63,64]. The system also allows measurements of biopotentials such as ECG (electrocardiogram), EMG (electromyogram), and EEG (electroencephalogram).

7.4.3 Capacitive Sensors for Implantable Pressure Sensors

Capacitive sensors employ parallel metalized plates as capacitors and measure the change in capacitance. A typical bulk-micromachined capacitive pressure sensor is shown in Fig. 7.19. The capacitance of a parallel plate can be found from

$$C = \frac{\varepsilon A}{d}$$

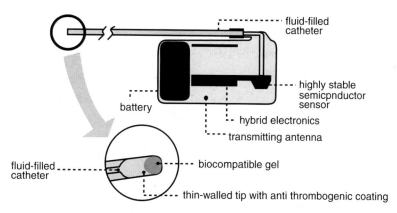

Figure 7.18: Chronically Implantable Pressure Sensor.
From [62].

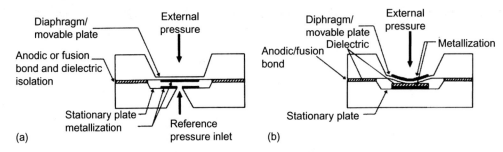

Figure 7.19: Cross-section Schematic of a Capacitive Pressure Sensor. (a) Non-contact Mode and (b) Contact Mode.
From [50].

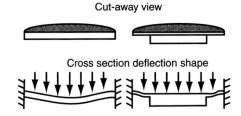

Figure 7.20: A Comparison of Deflection Shapes for Uniform-thickness (left) and Bossed (Right) Diaphragms.
From [50].

where, ε, A, and d are the permittivity of the gap, the area of the plates, and the separation of the plates, respectively (details of capacitors used as electrostatic actuators are described in Section 2.4.2, and capacitive sensing with microcantilevers is described in Section 6.2.4).

Since the deflection of a circular diaphragm is not linear, a capacitive sensor can be operated in contact mode to increase linearity. In contact mode, the electrode is on the other side of dielectric diaphragm and the capacitance becomes nearly proportional to the area of contact (see Fig 7.19(b)), and therefore exhibits good linearity with respect to applied pressures [50].

An alternative known method to achieve a linear response is to use bossed diaphragms. The thicker center portion (or boss) is much stiffer than the thinner portion of the membrane on the outside. The center boss contributes most of the capacitance of the structure and its shape does not distort appreciably under applied load (Fig. 7.20). Hence the capacitance-pressure characteristics are more linear [50].

The main advantages of using capacitive based sensors over piezoresistive pressure sensors include higher pressure sensitivity and decreased temperature sensitivity [65,66]. However, capacitive pressure sensors suffer from excessive signal loss due to parasitic capacitance, which hindered the development of miniaturized capacitive sensors until on-chip circuitry could be fabricated [44,67]. Historically, capacitive sensors have benefited from the same advantages in diaphragm etching and wafer bonding that piezoresistive sensors have. However, the piezoresistive approach generally has a complex transducer with simple circuit requirements, while the opposite is true of the capacitive approach [50].

An early study of an implantable capacitive pressure sensor has been reported by Schnakenberg [68] et al. They developed an intravascular pressure monitoring system, which consists of an implantable silicone capsule (Fig. 7.21). It contains a single transponder chip with a monolithically integrated capacitive pressure sensor and an antenna. The chip operates at a power supply voltage of 3 V with the power dissipation less than 450 μW.

As we have seen in previous sections, many implantable devices are telemetric. One advantage of using a capacitive pressure sensor is that it can be easily integrated into a wireless resonant circuit. Georgia Tech's MEMS research group and CardioMEMS developed a flexible micromachined passive pressure sensor [69]. Figure 7.22 shows an example of a passive pressure sensor. The sensor consists of a cavity sandwiched by two capacitor plates which are resonantly connected with an inductance. When one of the two plates is made of a deflectable diaphragm, the value of capacitance changes with external pressure, which induces a shift in the resonant frequency of the LC circuit. An equivalent circuit model is also shown in Fig. 7.22.

An implantable device from CardioMEMS has been implanted in 550 patients from 64 centers in the United States [70]. Patients were divided into a treatment group and a control group. The treatment group took medication instructions from clinicians who used daily measurement of pulmonary artery pressures in addition to standard of care. Over the six

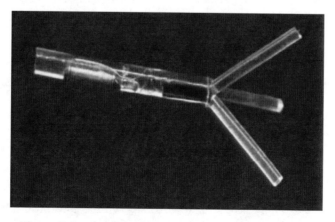

Figure 7.21: Implantable Capacitive Pressure Sensor.
From [68].

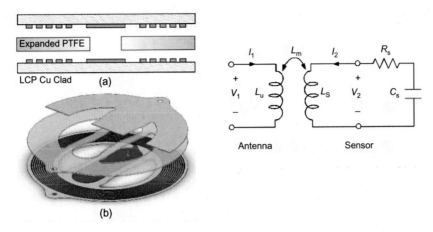

Figure 7.22: Capacitive Pressure Sensor.
From [69].

Figure 7.23: Implantable Bladder Pressure Monitoring System.
From [71].

months, 84 heart-failure-related hospitalizations were reported in the treatment group
($n = 270$) compared with 120 in the control group [70]. No device-related or system-related
complications were reported during the period. Coosemans and colleagues developed an
implantable bladder pressure monitoring system utilizing a capacitive pressure sensor
(Fig. 7.23). The system is intended for use of the diagnosis of urinary incontinence. The
system can be inductively powered inside a bladder, and is capable of wireless bidirectional communication. It demonstrated a pressure resolution of 0.04 kPa at a sampling
rate of 10 Hz. The system may be useful for patients with loss of proper bladder function
by damage to nervous system. The system may provide stimulation signals to the
contraction muscles to prevent incontinence [71].

Discussions on piezoresistive and capacitive sensing

Advancements in SOI-based piezoresistive pressure sensors did decrease the required die size and simplified integration with electronics, but at the cost of reduced sensitivity and reproducibility of mechanical properties. Piezoresistive sensors were more commonly used in catheter-based pressure sensing. They are better suited for periodic pressure measurement rather than for dynamic telemetric applications as the power consumption is high [50]. Since the resistivity of a semiconductor is temperature dependent (see Chapter 4, Section 4.3.1), a piezoresistive sensor could produce a large temperature error, which should be compensated. Capacitive sensors are more suited to implantable devices because of their lower power consumption, which is essential for wireless applications as power is not freely available. For this reason research in the area of implantable pressure sensors tends to focus on the use of capacitive MEMS devices [72]. Drawbacks of a capacitive sensor for implantable applications include the need for active implanted electronics and its instability over various environmental conditions.

7.4.4 Optical Transduction-Based Pressure Sensors

Several diaphragm-based optical sensors have been reported based on the Mach–Zehnder interferometry [73–75] and Fabry–Perot interferometry [76] principle, which measures pressure induced deflections (Fig. 7.24). In brief, a mechanical deflection changes the path length of the reflected light wave, which alters the interference of the reflected light with a

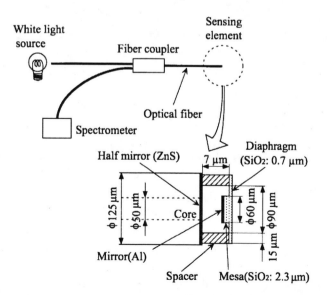

Figure 7.24: Schematic of a Diaphragm-type 125 μm Diameter Fiber-optic Pressure Sensor.
From [77].

reference light (see problems 5.7 and 5.8 for examples). The deflection derived from these devices varies linearly with pressure.

Optical sensors can be quite accurate, but often suffer from temperature sensitivity problems [50]. Moreover, aligning the optics and calibrating the sensors can be challenging and expensive. Furthermore, bending of the fiber can result in undesirable light intensity change, which results in output noise. A readout method using white-light interferometry was developed that succeeded in improving sensor performance [77,78] to a certain extent.

The Samba Sensors are pressure sensor products using fiberoptics. A micromachined silicon chip with an outside diameter (OD) of 0.43 mm is attached to an optical fiber with an OD of 0.25 mm. This sensor utilizes a reflective element at the end of the sensor tip which is compressed when a pressure wave passes over. Based on the principle of Fabry–Perot interferometry, the amount of compression translates to a change in the intensity of reflected light, which is correlated to a change in pressure [79]. This technology makes for very accurate sensors that are also very small. The sensor, however, is disadvantaged by its fragility and change in path lengths due to bending of the fiber, when used for in-vivo applications. Figure 7.25 shows a typical measurement setup.

7.4.5 Piezoelectric Pressure Sensors

Piezoelectric effect is the induction of an electric charge in response to an applied mechanical strain. It is different from piezoresistive effect, where a strain induces a change in resistivity. Details of piezoelectric effect are described in Chapter 6, Section 6.3.

For implantable applications, polymer-based piezoelectric devices are commonly used because of their mechanical flexibility. Polyvinylidene fluoride (PVDF) exhibits the highest piezoelectricity among all piezoelectric polymers. Along with the high electromechanical coupling coefficient and its semicrystalline nature, makes PVDF a popular choice for fabrication of robust and sensitive sensors (Fig. 7.26). PVDF is an ideal biomaterial because it is nontoxic, inert, resistant to water absorption (absorbs <0.04% by weight), biocompatible [80] and clean-room friendly.

A diaphragm made of PVDF is usually used as the sensing element. Compression or stress in the membrane produces charge, which is recorded using the metalized electrodes coated on either side of the diaphragm. M. Robert et al. [82] reported a process that spin coat the copolymer film directly onto a curved substrate in order to make uniform surface using press focusing and lapping over a curved surface. However, this approach also required multi-processing steps and its result was not satisfactory. In contrast, C. Li et al. [81] reported the film to be spin coated on a flat substrate and transferred onto the rounded catheter backing (Fig. 7.26). However, this extra step required the use of an adhesive layer and a proper fit to the final substrate. Such

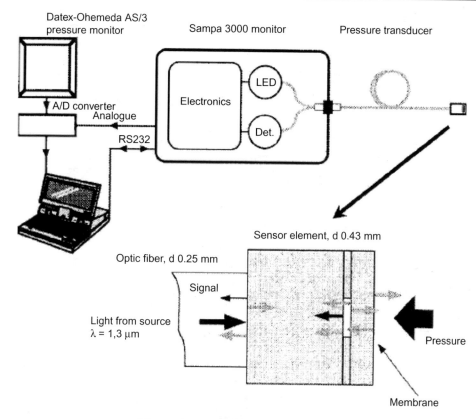

Figure 7.25: Samba Pressure Sensor.
From [79].

transferring the film from its original substrate to the curved backing of the sensors can result in damage to the fragile piezoelectric film. Another problem is that the adhesion layer can become a source of another interface, which may be a major concern at high frequencies where the thickness of the glue layer may approach the dimension of a wavelength of the acoustic signal.

While use of PVDF as membrane for ultrasound transducers has been well known, application of PVDF for pressure sensing application is a recent trend. In another study, Li et al. [83] reported the dual-mode operation of polyvinylidene fluoride trifluoroethylene (PVDF-TrFE) piezoelectric polymer diaphragm, capacitive, or resonant mode, as flexible intracranial pressure sensors (Fig. 7.27). The dual-mode capability of the sensor had advantages of (a) high linearity in small pressure ranges of 0–50 mmHg and insensitivity to environmental temperature variations in the capacitive mode; and (b) high sensitivity and resonant frequency as the output in the resonant mode which allows easy adaption for wireless application. In addition, this approach provided two detection methodologies on a

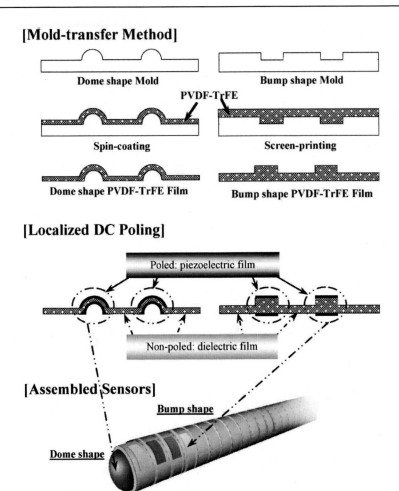

[Mold-transfer Method]

[Localized DC Poling]

[Assembled Sensors]

Figure 7.26: Dome and Bump shaped PVDF-TrFE Based Piezoelectric Tactile Sensor Modules for Smart Catheters.
From [81].

single device, thus allowing the independent measurement of pressure signals, as well as generating two sets of data for comparison and error checking. However the sensor was bulky (in thickness) and use of air-cavity can prove to be lethal when employed for in-vivo applications.

These problems were overcome by Sharma et al. [84,85] where a similar sensor was fabricated directly on a flexible catheter and piezoelectric mode was used for pressure measurement. Use of PVDF in piezoelectric mode yields the highest sensitivity. Using PVDF, they reported four times higher sensitivity over existing commercial piezoresistive pressure sensor. Such an increase was also possible due to the increase in crystallinity

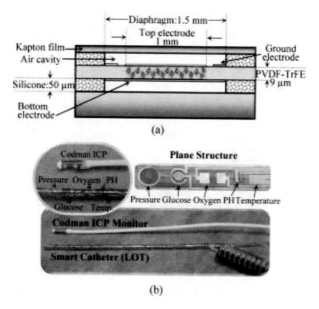

Figure 7.27: (a) Schematics of the Intracranial Pressure Sensor and (b) Photograph of the Developed Intracranial Pressure Sensor on Smart Multimodal Catheter.
From [83].

observed in thinner PVDF films. Increase in PVDF crystallinity in nanostructures has revived interest in the polymer and its applications. Sharma et al. also simulated relevant physiological conditions for testing of sensors (Fig. 7.28). A common disadvantage is that while piezoelectric materials are popular for dynamic monitoring they fail at static pressure measurements. However, by combining the piezoelectric mode of sensing along with other sensing modes (capacitive or resonant), it is possible to make highly sensitive pressure sensors. Such a sensor would address all the concerns that limit other current technologies from clinical adoption.

7.5 Drug Delivery Devices

Patient dosing and treatment techniques often require multiple painful injections in the form of inefficient bolus doses or inconvenient and invasive methods [86]. Current techniques still have much room for improvement in terms of safety, efficacy, pain, and convenience. A drug delivery or therapy producing device may be used to control the drug or treatment release to deliver on-demand pulsatile or adjustable continuous responses based on measurements from an in-body sensor or an inputted user setting. One of the important design concepts for drug delivery devices is whether the delivery is based on passive or active systems in order to release the prescribed drug dosages into the patient. Active systems allow for control of drug release after the device is implanted [87]. Currently, some

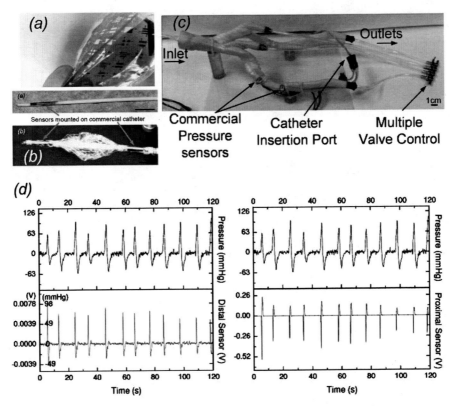

Figure 7.28: Photographs Showing (a) Flexible Thin-film Pressure Sensors Fabricated, (b) Mounted on a Catheter, (c) Tested Inside Vascular Testing Model; and (d) Results Showing High Sensitivity From Proximal and Distal Sensors Compared to Commercial Pressure Sensors.
From [84].

implantable MEMS-based drug delivery systems are being researched and put through trials. The use of these drug delivery systems will allow for controlled and accurate delivery of critically prescribed drug doses.

7.5.1 Working Principle

Drug delivery systems are designed with an array of microreservoirs etched into silicon. Each of these microreservoirs is then covered with individual gold membranes, and can be subsequently filled with either solid or liquid drug dosages. One method of dispersing the drug is to electrochemically dissolve the cathodic gold membranes using a chloride-containing solution when a voltage is applied (Figs 7.29 and 7.30).

For use as a drug delivery device, chemical sensors must be able to interface with the chemistry in the human body. The link between the body and the sensor would allow

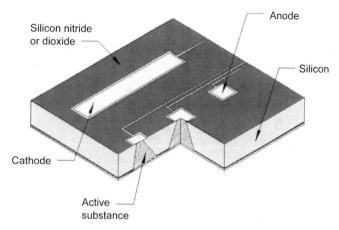

Figure 7.29: Prototype Microchip for Controlled Release Showing the Shape of a Single reservoir.
From [88].

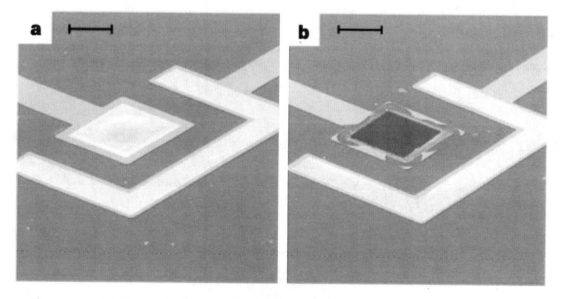

Figure 7.30: (a) Removal of an Anode Membrane to Initiate Release from a Reservoir and (b) After Application of Voltage.
From [88].

critical information to be passed from the body to the chip. This link would allow bridge diagnosis and therapeutics, which is the ultimate goal of a drug delivery system on a chip. The sensors would be used to measure some of the following signals: pH, analytes, and pressure of the tissues and body fluids. These drug delivery sensors could also eventually be transformed for use as glucose sensors [1].

7.5.2 Implementation Method

An effective drug delivery device should be made up of the following elements in order to achieve triggered responses: a microreservoir drug depot, micropump, valves, and sensors [1]. Due to the modular composition of many drug devices, various fabrication methods are applied depending upon the component. Microreservoirs have been approached from well-developed techniques of silicon microfabrication. Some research groups have used micromolded poly(dimethylsiloxane) (PDMS).

The IRD^3 is a device developed at MIT [87]. They claim to have devised "the first rapid drug delivery platform suitable for implementation as a smart micro-implant". Essentially, the device looks to release a bolus dose of drugs into the patient when certain targeted symptoms are detected. The underlying principle is that heating of a liquid increases pressure and cause membranes to burst. This quick heating releases drugs as the membranes are not selectively activated but rather quickly and in bulk. The actuation layer of the device is important to achieve the desired rate of drug release. Elman et al. used three resistive elements in their design in order to optimize for heat transfer into a localized volume [87]. The membranes shatter into micron and submicron fragments when high pressures are applied. The membranes are square due to simplicity in implementation.

The IRD^3 device has a modular composition, as shown in Fig. 7.31. Each modular section is constructed through a different fabrication method. The membrane layer is fabricated using micromachining and is composed of a silicon nitride membrane array combined with gold fuses designed to act as sensors. The actuation layer is comprised of micromachined 100 mm SCS substrates. The reservoir layer is composed of 2.25 mm thick Pyrex 7740 wafers drilled to define the reservoir then diced. The IRD^3 device is for use in ambulatory emergency care. Tests were conducted using vasopressin, and the results showed no

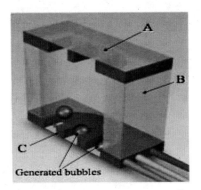

Figure 7.31: Three Layers of Modular Architecture of IRD^3: (A) Membrane Layer, (B) Reservoir Layer, and (C) Actuation Layer [87].

significant loss or degradation of the drug from use. The article states that the best use would be any application that requires a bolus shot.

Elman et al. argue that the proposed system is applicable to a number of modalities including the following: subcutaneous, intraperitoneal, intramuscular, and transdermal [87]. In addition, the IRD^3 has potential applications to high-risk populations such as individuals suffering from cardiac arrest, angina, strokes, and epilepsy. The design of the IRD^3 allows for attachment to current cardiac devices, specifically pacemakers and defibrillators.

The MicroCHIPS drug delivery device is a product from a spinoff company from the MIT group. It has multiple reservoirs placed on a microchip. The device is implanted into an individual and can then be remotely controlled to release drugs. The device could be programmed to open certain wells over varying periods of frequency. The programmable nature allows dosage and concentration to be controlled externally (Fig. 7.32).

The MicroCHIPS device uses standard microfabrication processes [89]. The primary components of the device are its reservoirs that can be implemented up to 100, drug formulations or biosensors, metal membranes to cap the reservoirs, and metal traces to help conduct current to the membranes when drug release is needed. The MicroCHIPS device has undergone clinical trials and is awaiting approval by the FDA. Currently there has been a First-In-Man (FIM) trial 1 [90].

7.6 Energy Harvesting Implantable Devices

Electric energy for pacemakers is supplied by batteries. Batteries with limited capacities will eventually drain long before the service life of the system. Their continuous replacement is an economical method. But the trend toward lower power consumption stimulates the investigation for alternative power sources to traditional batteries. Table 7.2 [91] summarizes requirements of different implantable medical devices for electricity.

Internal batteries have the disadvantage of duration of power supply because it is often limited by the capacity of the battery, that is, it is limited by the quantity of the available reactants in the system [92]. The energy generated by the battery is based on electrochemical reactions. Once the potential and current levels in the device has reduces the battery is unusable and replacement of batteries becomes necessary. Figure 7.33 shows a comparison of the energy densities of different battery types used for various applications presently. In order to maximize the lifespan and minimize the total volume of implantable devices, batteries with high energy density are sought. Presently, lithium ion-based batteries provide the maximum advantage for implantable applications, as it has the highest energy density.

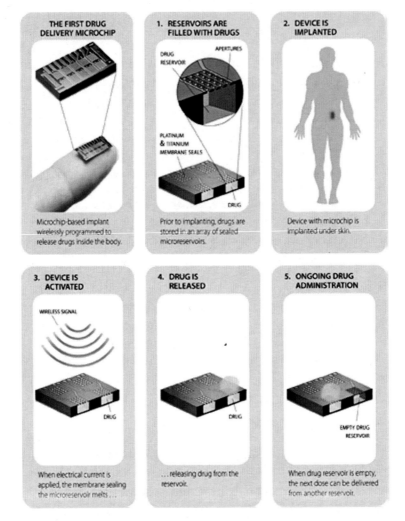

Figure 7.32: MicroCHIPS Drug Delivery Device.
MicroCHIPS(http://www.mchips.com/images/technology_infographic.jpg).

Replacing batteries [94] in implantable systems can pose a significant issue. Surgical interventions would be the only option to replace the batteries that have failed or that reach the end of its lifetime of operation. These operations could have detrimental effects such as scarring, infection or they could prove to be really expensive.

Alternate sources to batteries that could be implanted into biomedical devices were nuclear generators. This idea was conceived in 1960 and brought to the clinical environment in 1970. Numec Corporation developed this prototype where alpha particles emitted by a tiny slug of radioactive plutonium 238 bombarded with the walls of its container producing heat

Table 7.2: Requirement of Different Implantable Medical Devices for Electricity [91]

Implanted Device	Typical Power Requirement
Pacemaker	30–100 µW
Cardiac defibrillator	30–100 µW
Neurological stimulator	30 µW to several mW
Drug Pump	100 µW to 2 mW
Cochlear Implant	10 mW

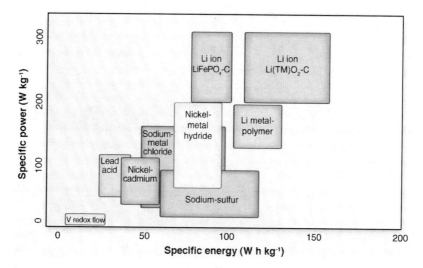

Figure 7.33: Power and Energy Density for Different Battery Types for Energy Storage.
From [93].

that was converted to electrical energy [95–97]. Although the plutonium is completely enclosed and radiation is considered to be small enough, there were concerns about the toxicity of it.

A number of biomedical based implantable medical devices would benefit from miniaturized MEMS-based implantable power supplies. One of the most important applications are cardiac pacemakers that currently use lithium ion batteries that power these devices with an operating power of 1 µW [98,99]. These batteries have an energy density of 1 W/mL. Biofuel cells created thus far have energy density of 10 mW/mL but this value could be improved to that of conventional lithium ion batteries. The advantages of biofuel systems are that it can provide constant energy density that is derived from ambient sugars present in the body.

Radiofrequency induction [100,101], ultrasound [102], and optical powering [103] have been studied as possible methods for delivering power transcutaneously to implanted medical devices or act as rechargeable batteries. These energy transmission methods do not require wire leads that penetrate the skin. RF transmissions have shown high efficiency but have also been prone to unwanted heating and tissue damage [104]. Ultrasound and optical methods have the disadvantage of large attenuation of energy passing through the skin. Optical methods using infrared transmission are sometimes associated with unwanted heating of the skin [105].

In a chemical fuel cell, electrical energy is generated by the electrochemical reaction of fuel and oxidant at two spatially separated electrodes [106]. Electrons, released upon the electro-oxidation of the fuel, flow from the anode through an external load circuit to the cathode, where the terminal electron acceptor, usually oxygen, is reduced. A number of groups such as those of Aston and Turner [107], Akiba [108], Katz [109,110], and Barton [111] have worked on bioelectrochemistry that use compounds such as sugars found in vivo in the system that convert to usable energy. Glucose, being the most abundant molecule, is also the most commonly targeted preferred fuel for an implantable fuel cell. The driving force of the electron flow is the difference in electrochemical potential of electrolyte near the anode and the cathode. Glucose and oxygen are typically present in the cells of human bodies. It is possible to utilize body resources to generate energy that can power number of implantable systems.

Pieozoelectric materials convert kinetic or mechanical energy into electrical energy. Material characteristics of different piezoelectric materials, namely PZT-5A, PZT-5H, and PVDF, for energy harvesting in bioMEMS applications are described in [112]. There have been approaches to develop piezoelectricity-based power generators for implantable medical device applications [113,114]. Recently, Platt et al. [115] developed a PZT-based in-vivo piezoelectric system that can be used to power knee replacement units in the human system. These were capable of sensing abnormally high forces exerted to the joints, misalignment and degradation of the system. These devices were capable of harvesting energy and providing sufficient amount to power and run the sensing circuitry in the device.

Wang et al. [116] developed a prototype in the nanomaterial-based piezoelectric generator that can potentially produce continuous direct current by harvesting energy from the environment such as ultrasound, vibrations and blood flow. Zinc oxide nanowires generate electricity when they are bent. This piezoelectric device is attractive towards biomedical applications for a fact that zinc oxide is non toxic and biocompatible. However, ZnO wires are inflexible and brittle with a large modulus of 21 GPa and fracture strain of only several percent [117]. The low fracture strain of ZnO wires means that breakage can occur when small deformations are applied. Even if many wires are incorporated into the device to increase power, fracture of a few would lead to catastrophic failure of many other wires since the same mechanical load would be distributed amongst fewer wires. This fact could compromise the efficacy of the power generation capabilities.

Piezoelectric polymers such as poly(vinylidene difluoride) (PVDF) have several significant advantages over ZnO or any other ceramic. PVDF has a fracture strain of many tens of percent [118] and modulus that is an order of magnitude lower than ZnO. In general, the mechanical properties of PVDF are more close to current implantable lead materials. PVDF devices will be more robust, with less impact on lead mechanical properties. PVDF is also an ideal biomaterial because it is nontoxic, inert, resistant to water absorption (absorbs <0.04% by weight), biocompatible [119] and compatible microfabrication techniques. In contrast, ZnO is difficult to pattern, assemble into devices, stack in multiple layers and soluble in even dilute (<1%) saline solutions [120,121] making encapsulation challenging, especially given that the lead sheath is not a perfect barrier to penetration of body fluids over the life of the implant [122].

Interesting power recovery schemes from body motion have been demonstrated using PVDF-based devices designed to fit within shoes [123] and backpack straps [124]. In such applications, transformation of up to 1% of the total mechanical energy into electrical energy was shown. Assuming similar efficiencies for PVDF devices attached on implantable leads, harvesting the one to five watts of mechanical power from the heart [125] could generate as much as 10–50 mW, which is available for the lifetime of the patient. Given that a typical pacemaker uses roughly 5 mW of power for the 4–5 years of battery lifetime, this analysis establishes the high potential of PVDF energy generators for recharging pacemaker batteries. There are no known demonstrations of PVDF-based devices designed to harvest energy from the mechanical action of the heart. However, in the late 1980s, 9 μm thick PVDF films (i.e., a bulk film with properties identical to macroscopic specimens) were incorporated into devices designed to harvest energy from respiration and it was determined that 1 mW of electrical power could be generated from respiration processes with ~1 W of total mechanical energy [126–128]. The devices were made by rolling thick PVDF films onto a flexible tube and tested by implantation into a dog. However, there are little known advances made in this area beyond these two initial demonstrations. It is important to note that in 1984 the understanding of piezoelectricity of bulk PVDF was just emerging [129] and it has since developed substantially. Only recently has the prospective of using PVDF nanofibers for energy generation gained attention [130]. Furthermore, there was little appreciation for how the properties of PVDF and other crystallizing polymers could be enhanced by confining the polymer to 10–100 nm layers till recently.

7.7 FDA Regulation of Medical Devices

With a large variety of items that interact with the human body and continuously more being created every year, it is important to ensure that these devices are safe for human consumption and use. The US Food and Drug Administration (FDA; http://www.fda.gov), is responsible for protecting the public health through guaranteeing the safety and effectiveness of food, drugs, medical devices, and biological products, among other things. FDA approval for medical devices is critical for use in patients.

In this section, we will introduce the FDA's definition of medical device, methods for classifying medical devices, and various regulatory procedures to gain approval. We will also discuss FDA regulation of nano and microfabricated silicon sensors.

7.7.1 Definition of Medical Device

In this section, we will focus on FDA regulations over medical devices. A medical device, as defined by the FDA, is "an instrument, apparatus, implement, machine, contrivance, implant, in vitro reagent, or other similar or related article, including a component part, or accessory" that fulfills some requirements including being "intended for use in the diagnosis of disease or other conditions, or in the cure, mitigation, treatment, or prevention of disease, in man or other animals" or for being "intended to affect the structure or any function of the body of man or other animals" without implementing chemical action or metabolism by the body. This definition includes items from toothbrushes to diagnostic products and sensors to nanotechnology for use in medical devices.

7.7.2 Classification of Medical Devices

The FDA has classified over 1700 generic types of medical devices. Each of these types is then grouped into one of 16 medical specialties (cardiovascular devices, dental, etc.), called panels. Each generic type of devices is also classified into one of three regulatory classes: Class I, Class II, or Class III. Each class varies in regards to regulatory procedures to demonstrate the safety and effectiveness of the device. Class I devices are not meant to help support or sustain life and usually do not present a high risk of patient injury through use. Examples of Class I devices are examination gloves and elastic bandages. Class II devices are meant to be used without causing injury to the patient or user. Examples include many diagnostic sensors and powered wheelchairs. Finally, Class III devices are those that support or sustain human life or potentially have a high risk of illness or injury through use. In addition, Class III devices include those types of devices that did not previously commercially exist prior to May 1976. The Medical Device Amendment of 1976 was passed to create these three regulatory classes for medical devices. To assure the safety and effectiveness of new types of devices, they are automatically classified as Class III devices. Examples of Class III devices are implantable pacemakers and automated external defibrillators.

7.7.3 Regulatory Procedures for Medical Devices

Regulatory control increases from Class I to Class III. All classes are subject to "General Controls", but Class II and Class III devices require additional regulation. Class II devices require "Special Controls", while Class III devices require "Pre-Market Approval". Many Class I and a few Class II devices may be exempt from certain regulations, but Class III almost always requires a full Pre-Market Approval. Pre-market notification is a method for the FDA to determine if the device is equivalent to a device already approved and placed in a class category. This is required if a manufacturer is looking to introduce a device into commercial distribution for the first time. By showing equivalency to another approved device, which was not subjected to requiring "Pre-Market Approval", the regulatory path becomes much simpler and required controls are limited.

General Controls

General controls cover a broad range of topics and apply to almost all medical devices. It includes provisions that relate to: adulteration; misbranding; device registration and listing; premarket notification; banned devices; notification, including repair, replacement, or refund; records and reports; restricted devices; and good manufacturing practices.

Special Controls

Special controls are usually device specific for Class II devices. These are meant to provide additional measures to prove the safety and effectiveness of the device, beyond what the general controls could provide. Special controls could potentially include: performance standards, post-market surveillance, patient registries, special labeling requirements, premarket data requirements, and guidelines.

Pre-Market Approval (PMA)

The PMA includes the toughest regulations. Approval is based on if the PMA provides sufficient scientific data to prove the safety and effectiveness of the device. A PMA includes technical sections containing data and information, non-clinical laboratory studies on topics such as immunology and biocompatibility, and clinical investigations showing device reactions and patient information.

Examples of FDA Regulation Pathways

Tables 7.3 and 7.4 are two examples of medical devices that have recently gone through the FDA approval process. Table 7.3 is for a Class II medical device that had a regulatory path of "Pre-Market Notification", or a 510(k), while the device in Table 7.4 went through the "Pre-Market Notification" regulatory path of a Class III medical device.

Table 7.3: The Regulatory Path of a Typical Class II Device

Type of Regulatory Path:	Pre-Market Notification (510 K)
Company name	Nihon Kohden Corporation
Device name	Nihon Kohden TG-970P CO2 Sensor Kit w/new Airway Adaptor
Classification name	Anesthesiology device
Device class	Class II
Predicate device	TG-970P CO, sensor kit
Intended use	To measure CO_2 concentration in a gas mixture to aid in determining the patient's ventilatory status
Standards compliance	IEC 60601-1, IEC 60601-1 Amendment 1, IEC 60601-1 Amendment 2, ISO 10993-1, ISO 14971

http://www.accessdata.fda.gov/cdrh_docs/pdf12/K122214.pdf

Table 7.4: The Regulatory Path of a Typical Class III Medical Device

Type of Regulatory Path:	Pre-Market Approval
Company name:	Dexcom, Inc.
Device name:	Dexcom G4 PLATINUM Continuous Glucose Monitoring System
Device class:	Class III
Intended use:	Glucose monitoring device for detecting trends and tracking patterns in persons with diabetes
Sensor information:	Sensor uses the enzyme glucose oxidase to convert glucose in the interstitial fluid into an electrical current proportional to concentration
Laboratory studies:	Bench performance testing, biocompatibility, sterility assurance, electromagnetic compatibility and interference, shelf-life and storage stability, packaging integrity/shipping testing, software validation, and human factors and usability testing
Animal studies:	No animal or additional studies were conducted
Clinical studies:	72 patients wore device for period of one week and participated in in-clinic and home use studies. Device was shown to be effective and safe for use.

http://www.accessdata.fda.gov/cdrh_docs/pdf12/P120005b.pdf

The Nihon Kohden TG-970P is a CO_2 sensor based on absorption spectroscopy (see Chapter 5, Section 5.5 for details). The sensor composes the main components of an infrared light emitting diode of specified wavelength and a photodetector.

The Dexcom G4 PLATINUM Continuous Glucose Monitoring System utilizes a subcutaneous sensor for glucose monitoring. Details of subcutaneous sensors are described in Section 7.3.6.

7.7.4 FDA Regulation of Devices Based on Silicon Nano/Microfabrication

Silicon nano/microfabrication has been playing an increasingly important role in biomedical applications in the past decades. Numerous commercial and trial devices, as introduced in this book, have been utilizing silicon to interface with the biological system. In the commercialization of implantable devices, the testing results of biocompatibility evaluation must be always included in submissions to the FDA and other medical regulatory bodies. The biocompatibility test must be also performed upon the final production device that has been processed and sterilized.

Because systems based on silicon nano/microfabrication are still relatively new to the FDA, extensive testing for various materials and techniques is yet to be made for sensors based on such technologies. In addition, it would be difficult for FDA to constitute policies that cover the wide array of all the existing technologies. One obvious reason is that countless numbers of new technologies emerge every day, and other difficulties include the fact that different industries (e.g. electronics, pharmaceuticals, and materials) have markedly different corporate cultures and intellectual property protection issues and strategies, and many of innovative sensing systems are synergetic integration of technologies from such different industries.

Material Biocompatibility

An early study of basic biocompatibility testing based on an internationally recognized test matrix (ISO 10993) has been performed for several silicon materials including silicon, silicon oxide, silicon nitride, titanium, and SU-8 [131]. Table 7.5 shows the results of the 1- and 12-week rabbit muscle implantation tests. Rabbits were implanted with four test and four negative sterile implant samples, for periods of 1 and 12 weeks. Negative controls were pieces of polyethylene. Each or test and negative sample size was approximately 1×10 mm. Implant site were observed both microscopically and macroscopically to assess tissue responses.

7.8 Conclusion

Implantable sensors and devices offer a promising and effective way to monitor or provide treatment in real time. The ease of transportation along with operations that are not reliant on user manipulation make them ideal for medical applications. There is a great need and interest in implantable sensors and devices, and as such there should be much desirable to implement these areas in new and exciting forms. Due to the many stages of testing and evaluation, implantable sensors are not able to get adopted as quickly as the potential fabrication technology advances. The devices that have been put on the market recently were tested for many years as any failure of the devices could potentially lead to patient death or other forms of disease.

With more testing and better technology, future implantable sensors should show great promise for helping to reform and revolutionize the medical industry. The use of these methods may set up a system of telemetric measuring of vital signs, which could help prevent

Table 7.5: Evaluation of MEMS Materials for Implantation

One- and 12-week Rabbit Muscle Implantation Results				
MEMS Material	Irritant Ranking Score		Δ between Test Sample and Control[a]	Non-irritant (0—2.9)
				Slight Irritant (3.0—8.9) Moderate Irritant (9.0—15.0) Severe Irritant (> 15.0)
	Test Sample	Control		
1-week implantation				
Silicon	6.0	6.7	0.0	Non-irritant
Thermal oxide	6.0	5.0	1.0	Non-irritant
N-doped poly	7.3	6.0	1.3	Non-irritant
Silicon nitride	6.3	6.7	0.0	Non-irritant
Titanium	7.3	8.0	0.0	Non-irritant
SU-8	8.3	7.3	1.0	Non-irritant
Silicon carbide	10.0	8.0	2.0	Non-irritant
12-week implantation				
Silicon	2.0	4.7	0.0	Non-irritant
Thermal oxide	2.0	4.7	0.0	Non-irritant
N-doped poly	2.3	4.0	0.0	Non-irritant
Silicon nitride	3.0	2.7	0.3	Non-irritant
Titanium	2.0	3.3	0.0	Non-irritant
SU-8	4.3	2.0	2.3	Non-irritant
Silicon carbide	1.7	2.0	0.0	Non-irritant

[a]Negative difference is recorded as zero.
From [131].

hospital misuse and overcrowding. On the patient side, these technologies may allow for the same quality of care found in a hospital setting from the comfort of one's own home.

The field of implantable sensors is not new, but there are still abundant new opportunities worth of investigation. Implantable sensors could be applied in more areas, patient monitoring in particular may be greatly improved with efficient implantable systems. Once the technology show promises, there will be more patients willing to consent to monitoring, especially when needles or other painful methods of treatment are eliminated. Overall, implantable sensors whether they be mechanical, electrical, chemical, or electrical, present great promise in revolutionizing the medical industry.

Problems

P7.1 Pacemaker

How many people in the US are currently using pacemakers?

P7.2 Pacemaker

Name three types of pacemakers. Explain which types should be used for which types of disease.

P7.3 Blood Pressure

What is the typical blood pressure range of healthy people? How much temporal and pressure resolutions are needed for blood pressure measurements?

P7.4 Implantable Sensor

Let us consider development of an implantable glucose sensor that can be integrated into a pacemaker. Will it be beneficial enough to invest time and money on? Discuss advantages and disadvantages. Also, discuss expected technical difficulties and how we can solve them.

P7.5 Data Acquisition

Design a system that transmits signals from electrodes of a pacemaker to an external computer.

1. What is the required sampling rate? How large is the data if we need chronic data recording?
2. What is the principle of data transmission? Is it advantageous compared to other possible candidates?
3. What is the power supply?

P7.6 Batteries

Based on Fig. 7.33, and Table 7.2, estimate typical battery lives for pacemaker, cardiac defibrillator, neurological stimulator, drug pump, and cochlear implant.

P7.7 FDA Regulation

1. Name three examples of molecular sensors and their applications that do not require FDA approval for commercialization.
2. Name three examples of molecular sensors and their applications that will be categorized as Class II devices by FDA.
3. Name three examples of molecular sensors and their applications that will be categorized as Class III devices by FDA.

References

[1] Richards Grayson AC, Scheidt Shawgo R, Li Y, Cima MJ. Electronic MEMS for triggered delivery. Advanced Drug Delivery Reviews 2004;56:173−84.

[2] Tomita Y, Imoto Y, Tominaga R, Yasui H. Successful implantation of a bipolar epicardial lead and an autocapture pacemaker in a low-body-weight infant with congenital atrioventricular block: report of a case. Surgery Today 2000;30:555−7.

[3] Greenspon AJ, Patel JD, Lau E, Ochoa JA, Frisch DR, Ho RT, et al. Trends in permanent pacemaker implantation in the United States from 1993 to 2009 increasing complexity of patients and procedures. Journal of the American College of Cardiology 2012;60:1540−5.

[4] Lau CP, Tai YT, Fong PC, Li JPS, Leung SK, Chung FLW, et al. Clinical experience with an activity sensing DDDR pacemaker using an accelerometer sensor. Pacing and Clinical Electrophysiology 1992;15:334−43.

[5] Dell'Orto S, Valli P, Greco EM. Sensors for rate responsive pacing. Indian Pacing and Electrophysiology Journal 2004;4:137.

[6] Hayes DL, Asirvatham SJ, Friedman PA. Cardiac Pacing, Defibrillation and Resynchronization: A Clinical Approach. Oxford: Wiley-Blackwell; 2012.

[7] Coats A, Adamopoulos S, Radaelli A, McCance A, Meyer T, Bernardi L, et al. Controlled trial of physical training in chronic heart failure. Exercise performance, hemodynamics, ventilation, and autonomic function. Circulation 1992;85:2119−31.

[8] Leung SK, Lau CP, Tang MO, Leung Z, Yakimow K. An integrated dual sensor system automatically optimized by target rate histogram. Pacing and Clinical Electrophysiology 1998;21:1559−66.

[9] Webster JG. Design of Cardiac Pacemakers: IEEE, Piscataway, 1995.

[10] Israel CW, Hohnloser SH. Current status of dual-sensor pacemaker systems for correction of chronotropic incompetence. American Journal of Cardiology 2000;86:K86−94.

[11] Bax JJ, Marwick TH, Molhoek SG, Bleeker GB, Van Erven L, Boersma E, et al. Left ventricular dyssynchrony predicts benefit of cardiac resynchronization therapy in patients with end-stage heart failure before pacemaker implantation. American Journal of Cardiology 2003;92:1238−40.

[12] Barold SS, Mugica J. The Fifth Decade of Cardiac Pacing. Oxford: Wiley-Blackwell; 2008.

[13] Rod Gimbel J, Bello D, Schmitt M, Merkely B, Schwitter J, Hayes DL, et al. Randomized trial of pacemaker and lead system for safe scanning at 1.5 Tesla. Heart Rhythm 2013;10(5):685−91.

[14] Karami MA, Inman DJ. Powering pacemakers from heartbeat vibrations using linear and nonlinear energy harvesters. Applied Physics Letters 2012;100:042901-4

[15] Tashiro R, Kabei N, Katayama K, Tsuboi E, Tsuchiya K. Development of an electrostatic generator for a cardiac pacemaker that harnesses the ventricular wall motion. Journal of Artificial Organs 2002;5:0239−45.

[16] Navarro X, Krueger TB, Lago N, Micera S, Stieglitz T, Dario P. A critical review of interfaces with the peripheral nervous system for the control of neuroprostheses and hybrid bionic systems. Journal of the Peripheral Nervous System 2005;10:229−58.

[17] Kovacs GT. Micromachined Transducers Sourcebook. New York, NY: WCB/McGraw-Hill; 1998.

[18] Wise KD, Angell JB, Starr A. An integrated-circuit approach to extracellular microelectrodes. IEEE Transactions on Biomedical Engineering 1970;238−47.

[19] Najafi K, Wise K, Mochizuki T. A high-yield IC-compatible multichannel recording array. IEEE Transactions on Electron Devices 1985;32:1206−11.

[20] Hoffer J, Loeb G. Implantable electrical and mechanical interfaces with nerve and muscle. Annals of Biomedical Engineering 1980;8:351−60.

[21] Stieglitz T. Manufacturing, assembling and packaging of miniaturized neural implants. Microsystem Technologies 2010;16:723−34.

[22] Taylor DM, Tillery SIH, Schwartz AB. Information conveyed through brain-control: cursor versus robot. Neural Systems and Rehabilitation Engineering, IEEE Transactions on [see also IEEE Trans. on Rehabilitation Engineering] 2003;11:195–9.

[23] Schwartz AB. Cortical neural prosthetics. Annual Review of Neuroscience 2004;27:487–507.

[24] Sodagar AM, Perlin GE, Yao Y, Najafi K, Wise KD. An implantable 64-channel wireless microsystem for single-unit neural recording. *IEEE Journal of Solid-State Circuits* 2009;44:2591–604.

[25] Sodagar AM, Wise KD, Najafi K. A Fully Integrated Mixed-Signal Neural Processor for Implantable Multichannel Cortical Recording, IEEE Biomedical Circuits and Systems Conference 2006 November, 29–December 1, 2006, British Library London, UK (BioCAS'06).

[26] Harkema S, Gerasimenko Y, Hodes J, Burdick J, Angeli C, Chen Y, et al. Effect of epidural stimulation of the lumbosacral spinal cord on voluntary movement, standing, and assisted stepping after motor complete paraplegia: a case study. The Lancet 2011;377:1938–47.

[27] Friesen LM, Shannon RV, Baskent D, Wang X. Speech recognition in noise as a function of the number of spectral channels: comparison of acoustic hearing and cochlear implants. The Journal of the Acoustical Society of America 2001;110:1150.

[28] Skinner MW, Clark GM, Whitford LA, Seligman PM, Staller SJ, Shipp DB, et al. Evaluation of a new spectral peak coding strategy for the nucleus 22 channel cochlear implant system. The American Journal of Otology 1994;15:15.

[29] Wilson BS, Finley CC, Lawson DT, Wolford RD, Eddington DK, Rabinowitz WM. Better speech recognition with cochlear implants. Nature 1991;352:236–8.

[30] Escudé B, James C, Deguine O, Cochard N, Eter E, Fraysse B. The size of the cochlea and predictions of insertion depth angles for cochlear implant electrodes. Audiology and Neurotology 2006;11:27–33.

[31] Rebscher SJ, Hetherington A, Bonham B, Wardrop P, Whinney D, Leake PA. Considerations for the design of future cochlear implant electrode arrays: electrode array stiffness, size and depth of insertion. Journal of Rehabilitation Research and Development 2008;45:731.

[32] Humayun MS, Weiland JD, Fujii GY, Greenberg R, Williamson R, Little J, et al. Visual perception in a blind subject with a chronic microelectronic retinal prosthesis. Vision Research 2003;43:2573–81.

[33] Ahuja A, Dorn J, Caspi A, McMahon M, Dagnelie G, Stanga P, et al. Blind subjects implanted with the Argus II retinal prosthesis are able to improve performance in a spatial-motor task. British Journal of Ophthalmology 2011;95:539–43.

[34] Zrenner E, Bartz-Schmidt KU, Benav H, Besch D, Bruckmann A, Gabel V-P, et al. Subretinal electronic chips allow blind patients to read letters and combine them to words. Proceedings of the Royal Society B: Biological Sciences 2011;278:1489–97.

[35] Rodger DC, Fong AJ, Li W, Ameri H, Ahuja AK, Gutierrez C, et al. Flexible parylene-based multielectrode array technology for high-density neural stimulation and recording. Sensors and Actuators B: Chemical 2008;132:449–60.

[36] Baan J, Van Der Velde ET, De Bruin HG, Smeenk GJ, Koops J, Van Dijk AD, et al. Continuous measurement of left ventricular volume in animals and humans by conductance catheter. Circulation 1984;70:812–23.

[37] Porterfield JE, Kottam AT, Raghavan K, Escobedo D, Jenkins JT, Larson ER, et al. Dynamic correction for parallel conductance, GP, and gain factor, α, in invasive murine left ventricular volume measurements. Journal of Applied Physiology 2009;107:1693–703.

[38] Karter AJ, Ackerson LM, Darbinian JA, D'Agostino Jr RB, Ferrara A, Liu J, et al. Self-monitoring of blood glucose levels and glycemic control: the Northern California Kaiser Permanente Diabetes registry. American Journal of Medicine 2001;111:1–9.

[39] Shichiri M, Yamasaki Y, Kawamori R, Hakui N, Abe H. Wearable artificial endocrine pancreas with needle-type glucose sensor. The Lancet 1982;320:1129–31.

[40] Bindra DS, Zhang Y, Wilson GS, Sternberg R, Thevenot DR, Moatti D, et al. Design and in vitro studies of a needle-type glucose sensor for subcutaneous monitoring. Analytical Chemistry 1991;63:1692–6.

[41] Girardin CM, Huot C, Gonthier M, Delvin E. Continuous glucose monitoring: a review of biochemical perspectives and clinical use in type 1 diabetes. Clinical Biochemistry 2009;42:136–42.

[42] Klonoff DC. Continuous glucose monitoring roadmap for 21st century diabetes therapy. Diabetes Care 2005;28:1231–9.

[43] Oliver N, Toumazou C, Cass A, Johnston D. Glucose sensors: a review of current and emerging technology. Diabetic Medicine 2009;26:197–210.

[44] Hin-Leung C, Wise KD. An ultraminiature solid-state pressure sensor for a cardiovascular catheter. IEEE Transactions on Electron Devices 1988;35:2355–62.

[45] Parsonnet V, Driller J, Cook D, Rizvi SA. Thirty-one years of clinical experience with nuclear-powered pacemakers. Pacing and Clinical Electrophysiology 2006;29:195–200.

[46] Swan H, Ganz W, Forrester J, Marcus H, Diamond G, Chonette D. Catheterization of the heart in man with use of a flow-directed balloon-tipped catheter. New England Journal of Medicine 1970;283: 447–51.

[47] Robin ED. The cult of the Swan-Ganz catheter. Overuse and abuse of pulmonary flow catheters. Annals of Internal Medicine 1985;103:445.

[48] Smith CS. Piezoresistance effect in germanium and silicon. Physical Review 1954;94:42.

[49] Chau H-L, Wise KD. An ultraminiature solid-state pressure sensor for a cardiovascular catheter. IEEE Transactions on, Electron Devices 1988;35:2355–62.

[50] Eaton WP, Smith JH. Micromachined pressure sensors: review and recent developments. Smart Materials and Structures 1999;6:530.

[51] Gieles A, Somers G. Miniature pressure transducers with a silicon diaphragm. Philips Technical Review 1973;33:14–20.

[52] Peake E, Zias A, Egan J. Solid-state digital pressure transducer. Electron Devices, IEEE Transactions on 1969;16:870–6.

[53] Wallis G, Pomerantz DI. Field assisted glass-metal sealing. Journal of Applied Physics 1969;40:3946–9.

[54] Shaikh M, Kodad S, Jinaga B. Performance analysis of piezoresistive MEMS for pressure measurement, 2005.

[55] Shimaoka K, Tabata O, Kimura M, Sugiyama S. Micro-diaphragm pressure sensor using polysilicon sacrificial layer etch-stop technique, in Technical Digest: 7th International Conference on Solid-State Sensors and Actuators (Transducers' 93), Yokohama, Japan, June 7–10, 1993, pp. 632–635.

[56] Sugiyama S, Suzuki T, Kawahata K, Shimaoka K, Takigawa M, Igarashi I. Micro-diaphragm pressure sensor, in 1986 International Electron Devices Meeting, Los Angeles, CA, December 7–10, 1986, pp. 184–187.

[57] Sugiyama S, Shimaoka K, Tabata O. Surface micromachined micro-diaphragm pressure sensors, in Solid-State Sensors and Actuators, 1991. Digest of Technical Papers, 1991 International Conference on Solid-State Sensors and Actuators (TRANSDUCERS'91), June 24–27, 1991, San Francisco, CA, pp. 188–191.

[58] Guckel H, Burns D. Fabrication techniques for integrated sensor microstructures, in 1986 International, Electron Devices Meeting, 1986, pp. 176–179.

[59] Guckel H. Surface micromachined pressure transducers. Sensors and Actuators A: Physical 1991;28:133–46.

[60] Burns D, Zook J, Horning R, Herb W, Guckel H. Sealed-cavity resonant microbeam pressure sensor. Sensors and Actuators A: Physical 1995;48:179–86.

[61] Millar H, Baker L. A stable ultraminiature catheter-tip pressure transducer. Medical and Biological Engineering and Computing 1973;11:86–9.

[62] Mills PA, Huetteman DA, Brockway BP, Zwiers LM, Gelsema AM, Schwartz RS, et al. A new method for measurement of blood pressure, heart rate, and activity in the mouse by radiotelemetry. Journal of Applied Physiology 2000;88:1537–44.

[63] Ye Z-Y, Li D-P, Li L, Pan H-L. Protein kinase CK2 increases glutamatergic input in the hypothalamus and sympathetic vasomotor tone in hypertension. The Journal of Neuroscience 2011;31:8271–9.

[64] Li D-P, Byan HS, Pan H-L. Switch to glutamate receptor 2-lacking AMPA receptors increases neuronal excitability in hypothalamus and sympathetic drive in hypertension. The Journal of Neuroscience 2012;32:372–80.

[65] Wise KD, Angell JB. An IC piezoresistive pressure sensor for biomedical instrumentation. Biomedical Engineering, IEEE Transactions on 1973;:101−9.

[66] Samaun S, Wise K, Nielsen E, Angell J. An IC piezoresistive pressure sensor for biomedical instrumentation, in Solid-State Circuits Conference. Digest of Technical Papers. 1971 IEEE International, Philadelphia, PA, 17−19 February 1971, pp. 104−105.

[67] Goustouridis D, Normand P, Tsoukalas D. Ultraminiature silicon capacitive pressure-sensing elements obtained by silicon fusion bonding. Sensors and Actuators A: Physical 1998;68:269−74.

[68] Schnakenberg U, Krüger C, Pfeffer J-G, Mokwa W, vom Bögel G, Günther R, et al. Intravascular pressure monitoring system. Sensors and Actuators A: Physical 2004;110:61−7.

[69] Fonseca MA, Allen MG, Kroh J, White J. Flexible wireless passive pressure sensors for biomedical applications, in Technical Digest Solid-State Sensor, Actuator, and Microsystems Workshop, Hilton Head, 2006, 2006.

[70] Abraham WT, Adamson PB, Bourge RC, Aaron MF, Costanzo MR, Stevenson LW, et al. Wireless pulmonary artery haemodynamic monitoring in chronic heart failure: a randomised controlled trial. The Lancet 2011;377:658−66.

[71] Coosemans J, Puers R. An autonomous bladder pressure monitoring system. Sensors and Actuators A: Physical 2005;123:155−61.

[72] Jiang G. Design challenges of implantable pressure monitoring system. Frontiers in Neuroscience 2010;4:2.

[73] Wagner C, Frankenberger J, Deimel PP. Optical pressure sensor based on a Mach-Zehnder interferometer integrated with a lateral a-Si: H pin photodiode. Photonics Technology Letters, IEEE 1993;5:1257−9.

[74] Dziuban J, Gorecka-Drzazga A, Lipowicz U. Silicon optical pressure sensor. Sensors and Actuators A: Physical 1992;32:628−31.

[75] Hoppe K, Andersen L, Bouwstra S. Integrated Mach-Zehnder interferometer pressure transducer, in Technical Digest of The 8th International Conference on Solid-State Sensors and Actuators, 1995 and Eurosensors IX. (Transducers' 95.), Stockholm, Sweden, 25−29 June, 1995, pp. 590−591.

[76] Chan M, Collins S, Smith R. A micromachined pressure sensor with fiber-optic interferometric readout. Sensors and Actuators A: Physical 1994;43:196−201.

[77] Katsumata T, Haga Y, Minami K, Esashi M. Micromachined 125 μm diameter ultra miniature fiber-optic pressure sensor for catheter. The Transactions of the Institute of Electrical Engineers of Japan. A publication of Sensors and Micromachines Society 2000;120:58−63.

[78] Wolthuis RA, Mitchell GL, Saaski E, Hartl JC, Afromowitz MA. Development of medical pressure and temperature sensors employing optical spectrum modulation. IEEE Transactions on Biomedical Engineering 1991;38:974−81.

[79] Sondergaard S, Karason S, Hanson A, Nilsson K, Hojer S, Lundin S, et al. Direct measurement of intratracheal pressure in pediatric respiratory monitoring. Pediatric Research 2002;51:339−45.

[80] Wang H-Y, Bernarda A, Huang C-Y, Lee D-J, Chang J-S. Micro-sized microbial fuel cell: a mini-review. Bioresource Technology 2011;102:235−43.

[81] Li C, Wu PM, Lee S, Gorton A, Schulz MJ, Ahn CH. Flexible dome and bump shape piezoelectric tactile sensors using PVDF-TrFE copolymer. Journal of Microelectromechanical Systems 2008;17:334−41.

[82] Robert M, Molingou G, Snook K, Cannata J, Shung KK. Fabrication of focused poly (vinylidene fluoride-trifluoroethylene) P (VDF-TrFE) copolymer 40−50 MHz ultrasound transducers on curved surfaces. Journal of Applied Physics 2004;96:252−6.

[83] Li C, Wu PM, Shutter LA, Narayan RK. Dual-mode operation of flexible piezoelectric polymer diaphragm for intracranial pressure measurement. Applied Physics Letters 2010;96:. pp. 053502-053502-3

[84] Sharma T, Aroom K, Naik S, Gill B, Zhang JX. Flexible thin-film PVDF-TrFE based pressure sensor for smart catheter applications. Annals of Biomedical Engineering 2012;1−8.

[85] Sharma T, Je S-S, Gill B, Zhang JX. Patterning piezoelectric thin film PVDF−TrFE based pressure sensor for catheter application. Sensors and Actuators A: Physical 2012;177:87−92.

[86] LaVan DA, McGuire T, Langer R. Small-scale systems for in vivo drug delivery. Nature Biotechnology 2003;21:1184−91.

[87] Elman N, Duc HH, Cima M. An implantable MEMS drug delivery device for rapid delivery in ambulatory emergency care. Biomedical Microdevices 2009;11:625−31.

[88] Santini JT, Cima MJ, Langer R. A controlled-release microchip. Nature 1999;397:335−8.

[89] Maloney JM, Uhland SA, Polito BF, Sheppard Jr NF, Pelta CM, Santini Jr JT. Electrothermally activated microchips for implantable drug delivery and biosensing. Journal of Controlled Release 2005;109:244−55.

[90] Farra R, Sheppard NF, McCabe L, Neer RM, Anderson JM, Santini JT, et al. First-in-human testing of a wirelessly controlled drug delivery microchip. Science Translational Medicine 2012;4: pp. 122ra21-122ra21

[91] Jia D, Liu J. Human power-based energy harvesting strategies for mobile electronic devices. Frontiers of Energy and Power Engineering in China 2009;3:27−46.

[92] Lanmüller H, Sauermann S, Unger E, Schnetz G, Mayr W, Bijak M, et al. Battery-powered implantable nerve stimulator for chronic activation of two skeletal muscles using multichannel techniques. Artificial Organs 1999;23:399−402.

[93] Dunn B, Kamath H, Tarascon J-M. Electrical energy storage for the grid: a battery of choices. Science 2011;334:928−35.

[94] Shill HA, Shetter AG. Reliability in deep brain stimulation. IEEE Transactions on Device and Materials Reliability 2005;5:445−8.

[95] Parsonnet V, Manhardt M. Permanent pacing of the heart: 1952 to 1976. American Journal of Cardiology 1977;39:250−6.

[96] Laurens P. Nuclear-powered pacemakers: an eight-year clinical experience. Pacing and Clinical Electrophysiology 1979;2:356−60.

[97] Parsonnet V, Berstein AD, Perry GY. The nuclear pacemaker: is renewed interest warranted? American Journal of Cardiology 1990;66:837−42.

[98] Holmes CF. Electrochemical Power sources and the treatment of Human Illness. The Electro Chemical Society Interface 2003;Fall.

[99] Linden D, Reddy TB. Handbook of Batteries. third ed. New York: McGraw-Hill; 2002.

[100] Schuder JC. Powering an artificial heart: birth of the inductively coupled-radio frequency system in 1960. Artificial Organs 2002;26:909−15.

[101] Smith B, Tang Z, Johnson MW, Pourmehdi S, Gazdik MM, Buckett JR, et al. An externally powered, multichannel, implantable stimulator-telemeter for control of paralyzed muscle. IEEE Transactions on Biomedical Engineering 1998;45:463−75.

[102] Suzuki S-N, Katane T, Saito O. Fundamental study of an electric power transmission system for implanted medical devices using magnetic and ultrasonic energy. Journal of Artificial Organs 2003;6:145−8.

[103] Goto K, Nakagawa T, Nakamura O, Kawata S. An implantable power supply with an optically rechargeable lithium battery. IEEE Transactions on Biomedical Engineering 2001;48:830−3.

[104] Foster KR, Adair ER. Modeling thermal responses in human subjects following extended exposure to radiofrequency energy. Biomedical Engineering Online 2004;3.

[105] Tang Q, Tummala N, Gupta SK, Schwiebert L. Communication scheduling to minimize thermal effects of implanted biosensor networks in homogeneous tissue. IEEE Transactions on Biomedical Engineering 2005;52:1285−94.

[106] Kreysa G, Sell D, Kramer P. Bioelectrochemical fuel-cells. Berichte Der Bunsen-Gesellschaft-Physical Chemistry Chemical Physics 1990;94:1042−5.

[107] Turner APF, Aston WJ, Higgins IJ, Davis G, Hill HAO. Applied aspects of bioelectrochemistry − fuel-cells, sensors, and bioorganic synthesis. Biotechnology and Bioengineering 1982;:401−12.

[108] Akiba T, Bennetto HP, Stirling JL, Tanaka K. Electricity production from alkalophilic organisms. Biotechnology Letters 1987;9:611−6.

[109] Katz E, Buckmann AF, Willner I. Self-powered enzyme-based biosensors. Journal of the American Chemical Society 2001;123:10752−3.

[110] Katz E, Willner I. A biofuel cell with electrochemically switchable and tunable power output. Journal of the American Chemical Society 2003;125:6803−13.

[111] Barton SC, Atanassov P. Enzymatic biofuel cells for implantable and micro-scale devices. Abstracts of Papers of the American Chemical Society 2004;228:. pp. 004-FUEL

[112] Ramsay MJ, Clark WW. Piezoelectric energy harvesting for bio-MEMS applications, in SPIE's 8th Annual International Symposium on Smart Structures and Materials, 2001, pp. 429−438.

[113] Lewin G, Myers G, Parsonnet V, Raman KV. An improved biological power source for cardiac pacemakers. ASAIO Journal 1968;14:215−9.

[114] Starek PJ, White DL, Lillehei CW. Intracardiac pressure changes utilized to energize a piezoelectric powered cardiac pacemaker. ASAIO Journal 1970;16:180−3.

[115] Platt SR, Farritor S, Haider H. On low-frequency electric power generation with PZT ceramics. IEEE/ASME Transactions on Mechatronics 2005;10:240−52.

[116] Wang ZL, Song J. Piezoelectric nanogenerators based on zinc oxide nanowire arrays. Science 2006;312:242−6.

[117] Desai AV, Haque MA. Mechanical properties of ZnO nanowires. Sensors and Actuators a-Physical Feb 2007;134:169−76.

[118] Fang F, Zhang MZ, Yang W. Strain rate mediated microstructure evolution for extruded poly (vinylidene fluoride) polymer films under uniaxial tension. Journal of Applied Polymer Science 2007;103:1786−90.

[119] Nalwa HS, editor. Ferroelectric Polymers: Chemistry, Physics and Applications. New York: Marcel Dekker; 1995.

[120] Agren MS. Influence of 2 vehicles for zinc-oxide on zinc-absorption through intact skin and wounds. Acta Dermato-Venereologica 1991;71:153−6.

[121] Snyders H. On the corrosion of Zinc by water and saline solutions. Minutes of Proceedings of the Institution of Civil Engineers 1878;27.

[122] Bruck SD, Mueller EP. Materials aspects of implantable cardiac-pacemaker leads. Medical Progress through Technology 1988;13:149−60.

[123] Shenck NS, Paradiso JA. Energy scavenging with shoe-mounted piezoelectrics. IEEE Micro 2001;21:30−42.

[124] Granstrom J, Feenstra J, Sodano HA, Farinholt K. Energy harvesting from a backpack instrumented with piezoelectric shoulder straps. Smart Materials & Structures 2007;16:1810−20.

[125] Avraham R. The Circulatory System. Philadelphia, PA: Chelsea House Publishers; 2000.

[126] Hausler E, Stein L. Implantable physiological power supply with PVDF film. Ferroelectrics 1984;60:277−82.

[127] Hausler E, Stein L. Hydromechanical and physiological mechanical-to-electrical power converter with pvdf film. Ferroelectrics 1987;75:363−9.

[128] Starner T. Human-powered wearable computing. IBM Systems Journal 1996;35:618−29.

[129] Lovinger AJ. Ferroelectric polymers. Science 1983;220:1115−21.

[130] Chang CE, Tran VH, Wang JB, Fuh YK, Lin LW. Direct-write piezoelectric polymeric nanogenerator with high energy conversion efficiency. Nano Letters 2010;10:726−31.

[131] Kotzar G, Freas M, Abel P, Fleischman A, Roy S, Zorman C, et al. Evaluation of MEMS materials of construction for implantable medical devices. Biomaterials 2002;23:2737−50.

Appendix 5A: Fresnel Diffraction and Fraunhofer Diffraction

Here we describe two types of optical diffraction: Fresnel diffraction and Fraunhofer diffraction. Comparison of the two types of diffraction in photolithographic applications is discussed in Chapter 2, Section 2.3.2 Photolithography.

5A.1 Fresnel Diffraction

Consider the diffraction of the electric field passing through an aperture (Fig. 5A.1) $t(x', y', 0)$. The diffraction pattern at a point (x, y, z) can be expressed by:

$$E(x, y, z) = \frac{z}{j\lambda} \iint t(x', y', 0) \frac{e^{-jkr}}{r^2} \, dx' dy' \tag{5A.1}$$

$$t(x', y') = \begin{cases} 1 & \text{aperture} \\ 0 & \text{elsewhere} \end{cases} \tag{5A.2}$$

where $k = 2\pi/\lambda$ is the wavenumber, and r is the distance between $(x', y', 0)$ and (x, y, z).

5A.2 Fresnel Condition

The distance r can be

$$r = \sqrt{(x-x')^2 + (y'-y)^2} = \sqrt{a^2 + z^2} = z\sqrt{1 + \frac{a^2}{z^2}} \tag{5A.3}$$

where

$$a = \sqrt{(x-x')^2 + (y'-y)^2} \tag{5A.4}$$

Here a is related to the minimum pattern size that can be imaged. To simplify (A.3) in the integral, we can use Taylor series expansion:

$$\sqrt{1+u} = (1+u)^{1/2} = 1 + \frac{u}{2} - \frac{u^2}{8} \cdots \tag{5A.5}$$

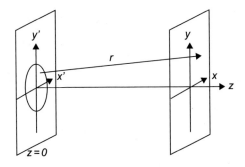

Figure 5A.1: Fresnel Diffraction of an Aperture.

Using Eq. 5A.5, Eq. 5A.3 can be rewritten as

$$r = z + \frac{a^2}{2z} - \frac{a^4}{8z^3} + \cdots \tag{5A.6}$$

Multiplying $k = 2\pi/\lambda$ on both sides,

$$kr = 2\pi \left(\frac{z}{\lambda} + \frac{a^2}{2z\lambda} - \frac{a^4}{8z^3 \lambda} + \ldots \right) \tag{5A.7}$$

The Fresnel condition is a case where we can neglect the third term, but not the second term. It may be represented as a condition where the wavelength λ is smaller than the related dimensions, and the distance is larger than the minimum feature size, namely

$$\lambda << a \left(= \sqrt{(x-x')^2 + (y'-y)^2} \right) << z. \tag{5A.8}$$

A popular form of the Fresnel condition is:

$$F = \frac{a}{\lambda d} \gtrsim 1 \tag{5A.9}$$

where F is the **Fresnel number**, a is the aperture size and d is the distance from the aperture to the object along the z direction. It should be noted that Eq. 5A.9 does not necessarily provide a sufficient condition that allow us to neglect the third term, although it is useful in many practical cases.

Based on the Fresnel condition, the minimum aperture size α is related the aperture size a and calculated as:

$$\alpha_{min} = \sqrt{\lambda d}. \tag{5A.10}$$

Under the Fresnel condition, we can take the first two terms of Eq. 5A.6, and kr can be approximated as:

$$kr = 2\pi \left(\frac{z}{\lambda} + \frac{a^2}{2z\lambda} \right) \tag{5A.11}$$

Then the integral (Eq. 5A.1) becomes:

$$E(x,y,z) = \frac{e^{-jkz}}{j\lambda z} e^{-j\frac{k}{2z}(x^2+y^2)} \iint t(x',y',0) e^{-j\frac{k}{2z}(x'^2+y'^2)} e^{j\frac{k}{z}(x'x+y'y)} dx'dy' \tag{5A.12}$$

Note that the denominator r^2 is approximated as z^2 assuming $r \approx z$. To simplify the description, we can use C instead of the full expression $\frac{e^{-jkz}}{j\lambda z} e^{-j\frac{k}{2z}(x^2+y^2)}$.

$$E(x,y,z) = C \iint t(x',y',0) e^{-j\frac{k}{2z}(x'^2+y'^2)} e^{j\frac{k}{z}(x'x+y'y)} dx'dy' \tag{5A.13}$$

which is called the Fresnel diffraction integral. The integral provides analytical solution of the diffraction on the object screen only if the given condition is satisfied.

5A.3 Fraunhofer Diffraction

The Fraunhofer diffraction is considered when the system fulfills the far-field condition:

$$F = \frac{a^2}{\lambda d} << 1 \tag{5A.14}$$

which means the scale of the distance between the aperture and the screen is much larger than that of the aperture size. Usually $F < 0.01$ is recognized as far-field. In this case, the propagating wavefront is assumed to be planar. The expression for the Fraunhofer diffraction is similar to the Fresnel diffraction. The far-field condition enables to ignore the exponential first term of Eq. 5A.12, since $\frac{k}{2z}(x'^2+y'^2) << \frac{k}{z}(x'x+y'y)$. Thus the equation for the Fraunhofer diffraction becomes:

$$E(x,y,z) = C \iint t(x',y') e^{j\frac{k}{z}(x'x+y'y)} dx'dy' \tag{5A.15}$$

The Eq. 5A.15 is identical to its Fourier transform expression:

$$E(u,v) = C \iint t(x',y') e^{j2\pi(ux'x+vy')} dx'dy' \tag{5A.16}$$

where $u = x/\lambda z$ and $v = y/\lambda z$. Fraunhofer diffraction pattern is given as the Fourier transform of the aperture.

5A.3.1 *Fraunhofer Diffraction with a Slit*

Consider a two-dimensional slit aperture of infinite depth:

$$t(x', y') = \begin{cases} 1 & \text{for} \quad -b/2 \leq y' \leq b/2 \\ 0 & \text{elsewhere} \end{cases} \tag{5A.17}$$

Converting the aperture by the Fourier transform given in Eq. 5A.16 results in:

$$E(u, v) = C \cdot b \cdot \delta(u) \cdot \sin c(\pi v b) \tag{5A.18}$$

where

$$\text{sinc} \, x = \frac{\sin x}{x} \tag{5A.19}$$

The intensity distribution at the object field is:

$$I(v) = I_0 \cdot \text{sinc}^2(\pi v b), \tag{5A.20}$$

A plot of $I(v)$ is shown in Fig. 5A.2.

5A.3.2 *Fraunhofer Diffraction with a Lens*

Consider a lens that has features shown in Fig. 5A.3. The radii of curvature on the right and left surfaces are given as r_1 and r_2, respectively. It is possible to assume:

$$x_1 \approx x_2 = x, y_1 \approx y_2 = y, \tag{5A.21}$$

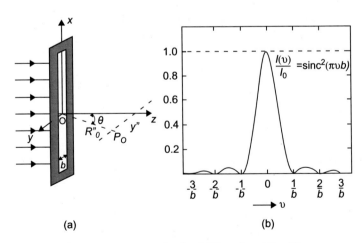

(a)   (b)

Figure 5A.2: Fraunhofer Diffraction of a Slit.

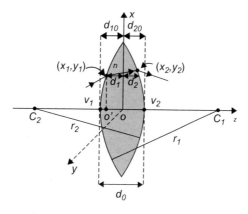

Figure 5A.3: Model of a Thin Lens.

since the transverse displacement of a ray in a thins lens is negligible. Also the intensity of the wave does not change, while the phase varies. Therefore, the system equation of the lens can be expressed as:

$$E_t(x, y) = T(x, y)E_i(x, y),$$ (5A.22)

where

$$T(x, y) = e^{-j\Delta\Phi(x,y)}.$$ (5A.23)

$E_t(x,y)$ and $E_i(x,y)$ are the transmitted wave and the incident wave, respectively. The phase retardation at (x,y) through the lens is:

$$\Delta\Phi(x, y) = k[1 - \{d_0 - d(x, y) + nd(x, y)\}],$$ (5A.24)

where

$$\begin{aligned} d(x, y) &= d_1(x, y) + d_2(x, y) \\ &\approx d_{10} - \frac{x^2 + y^2}{2r_1} + d_{20} - \frac{x^2 + y^2}{2(-r_2)} \\ &= d_0 - \frac{1}{2}\left(\frac{1}{r_1} - \frac{1}{r_2}\right)(x^2 + y^2) \end{aligned}$$ (5A.25)

Substituting Eq. 5A.25 into 5A.24 yields:

$$\Delta\Phi(x, y) = k\left[nd_0 - \frac{x^2 + y^2}{2f}\right]$$ (5A.26)

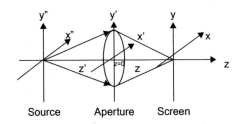

Figure 5A.4: Image of a Point Light Source in a Lens System.

where n is the refractive index of the lens and f is the focal length of the lens in the air:

$$\frac{1}{f} = (n-1)\left(\frac{1}{r_1} - \frac{1}{r_2}\right) \tag{5A.27}$$

Thus, the system equation of the lens can be rewritten as:

$$T(x,y) \approx e^{jk(x^2+y^2)/2f}. \tag{5A.28}$$

Note that the constant term knd_0 can be neglected if the lens thickness is relatively thin.

Assuming we have a point source as shown in Fig. 5A.4, the diffraction from the point source to the lens and aperture combination is:

$$U_p(x',y') = C_1 e^{\left[-j\frac{k}{2z'}(x'^2+y'^2)\right]}. \tag{5A.29}$$

where z' is the coordinate distance from the point source and the lens. In terms of the point source and the lens, the new transmission function of the aperture $t'(x',y')$ becomes:

$$t'(x',y') = U_p(x',y') \cdot T(x,y) \cdot t(x',y'). \tag{5A.30}$$

The substitution of Eq. 5A.30 into 5A.15 results in:

$$
\begin{aligned}
E_s(x,y) &= C \iint t'(x',y') e^{-j\frac{k}{2z}(x'^2+y'^2)} e^{j\frac{k}{z}(x'x+y'y)} \, dx'dy' \\
&= C \iint C_1 e^{\left[-\frac{jk}{2z'}(x'^2+y'^2)\right]} e^{jk\frac{(x'^2+y'^2)}{2f}} \cdot t(x',y') \cdot e^{-j\frac{k}{2z}(x'^2+y'^2)} e^{j\frac{k}{z}(x'x+y'y)} \, dx'dy' \\
&= C \iint C_1 e^{\left[-\frac{jk}{2}(x'^2+y'^2)\left(\frac{1}{z'}+\frac{1}{z}-\frac{1}{f}\right)\right]} \cdot t(x',y') \cdot e^{j\frac{k}{2z}(x'x+y'y)} \, dx'dy'
\end{aligned}
\tag{5A.31}
$$

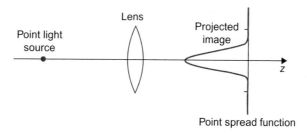

Figure 5A.5: Point Spread Function of a Lens System.

If the source plane and the object plane are conjugated in terms of the lens:

$$\frac{1}{z'} + \frac{1}{z} - \frac{1}{f} = 0,\tag{5A.32}$$

then the quadratic exponential term becomes one and can be neglected. Under this condition, the equation (5A.31) reduces to

$$E_s(x, y) = C \iint C_1 t(x', y') \cdot e^{j\frac{k}{2z}(x'x + y'y)} \, dx'dy',\tag{5A.33}$$

which is identical to Eq. 5A.15. When the source and the object plane lie in the conjugate planes of the lens, the Fraunhofer diffraction pattern of a point light source imaged by a lens is found by the Fourier transform. This pattern is called the point spread function (PSF, see Fig. 5A.5), and it defines the theoretical limit of the resolution of an optical system, such as a UV exposure system or an optical microscope (see the discussion in Section 2.3).

Figures in this appendix are redrawn from: Optics: Principles and Applications by Sharma, KK.

Appendix 5B: Coupling Between the Evanescent Wave and the Surface Plasmon

The surface plasmon resonance sensor is based on the coupling between the evanescent wave and the surface plasmon. A general sensor system using this plasmon coupling mechanism is called attenuated total reflection (ATR) method.

Figure 5B.1 shows a typical configuration using a cylindrical prism. In this configuration, the metal film layer is in-between the prism and the dielectric (Kretschmann geometry, see Section 5.xx). If a light transmits into the prism with an incident angle larger than the critical angle, the evanescent wave propagates along the prism-metal interface. If the metal film is thin enough, the evanescent wave can reach to the outer boundary of the metal film so that the penetrated evanescent wave can be coupled into surface plasmon when their momentums are well matches to each other.

The propagation constant of the surface plasmon can be described as:

$$k_{SP} = k_{SP0} + \Delta k = \frac{\omega}{c} \sqrt{\frac{\varepsilon_m \varepsilon_d}{\varepsilon_m + \varepsilon_d}} + \Delta k \tag{5B.1}$$

where:

$$\Delta k = r_{pm} e^{2jk_{xm}D} 2 \frac{\omega}{c} \left(\frac{\varepsilon_m \varepsilon_d}{\varepsilon_m + \varepsilon_d} \right)^{3/2} \frac{1}{\varepsilon_d - \varepsilon_m} \tag{5B.2}$$

Here k_{SP0} is the metal-dielectric propagation constant (Eq. 5.58) of the surface plasmon when the prism does not exist and Δk is for the presence of the prism and the finite thickness (D) of the metal film. The prism-metal reflection coefficient r_{pm}^p can be calculated by the p-polarized reflection coefficient equation:

$$r_{pm}^p = \left(\frac{k_{zp}}{\varepsilon_p} - \frac{k_{zm}}{\varepsilon_m} \right) \Big/ \left(\frac{k_{zp}}{\varepsilon_p} + \frac{k_{zm}}{\varepsilon_m} \right), \tag{5B.3}$$

where k_{zp} and k_{zm} can be derived from Eq. 5B.4:

$$|k_i|^2 = k_{zi}^2 + k_{xi}^2 = \left(\frac{\omega}{c} \right)^2 \varepsilon_i, \quad i = p \ or \ m. \tag{5B.4}$$

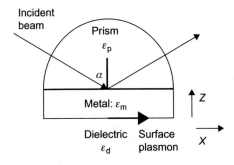

Figure 5B.1: Prism Based Coupling for SPR Sensor.

Note that $k_{xp} = k_{xm} = \frac{2\pi}{\lambda} n_p \sin \alpha$ and $\omega/c = 2\pi/\lambda$, where n_p is the refractive index of the prism. This shows the coefficient r^p_{pm} and the exponential term $e^{2jk_{xm}D}$ in Δk are related to the incident angle.

The propagation constant of the penetrated evanescent:

$$k_x^{EW} = \frac{2\pi}{\lambda} n_p \sin \alpha \qquad (5B.5)$$

The coupling between evanescent wave and the surface plasmon requires $k_{SP} = k_x^{EW}$. From Equations 5B.1, 5B.2, and 5B.5, assuming Δk is relatively small, we have:

$$n_{ef} = n_p \sin \alpha_{coup} = \text{Re}\left\{ \sqrt{\frac{\varepsilon_m \varepsilon_d}{\varepsilon_m + \varepsilon_d}} \right\} + \frac{\Delta k \lambda}{2\pi} \qquad (5B.6)$$

Note that α_{coup} denotes the coupling angle.

Finally, we will consider the reflectivity of the system which consists of the prism, the metal film and the dielectric medium. The reflectivity R for p-polarized light, or TM mode, of the three-layer system is given below:

$$R = \left| \frac{E_r^p}{E_i^p} \right|^2 = \left| \frac{r^p_{pm} + r^p_{md}\exp(2jk_{zm}D)}{1 + r^p_{pm}r^p_{md}\exp(2jk_{zm}D)} \right|^2 \qquad (5B.7)$$

Assuming ε_m satisfies with the specific condition $|\varepsilon'_m| \gg 1$ and $|\varepsilon'_m| \gg |\varepsilon''_m|$, the equation above can be approximated to:

$$R = 1 - \frac{4\Gamma_t \Gamma_{rad}}{(n_p \sin \alpha - n_{ef})^2 + (\Gamma_t + \Gamma_{rad})^2} \qquad (5B.8)$$

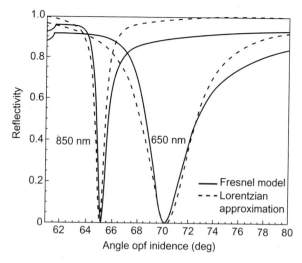

Figure 5B.2: Angular Dependence of Reflectivity Based on Fresnel Model and Lorentzian Approximation.

From Homola J, Surface Plasmon Resonance Based Sensors, Springer Series on Chemical Sensors and Biosensors, Volume 4, 2006, Springer, New York, p 32.

where:

$$\Gamma_t = \mathrm{Im}(k_{SP0})\lambda/2\pi \tag{5B.9}$$

$$\Gamma_{rad} = \mathrm{Im}(\Delta k)\lambda/2\pi. \tag{5B.10}$$

Here Γ denotes the attenuation coefficient. This approximation is called Lorentzian approximation. Figure 5B.2 shows the comparison between the rigorous approach and the Lorentzian approximation. The optimal coupling happens when $\Gamma_t = \Gamma_{rad} = \Gamma$ is satisfied. This condition is fulfilled only for a single thickness of the metal film. The optimal coupling metal thickness is related to the wavelength and materials. If the optimum condition $\Gamma_t = \Gamma_{rad} = \Gamma$ is satisfied, the angular width of the dip at $R = 0.5$ can be found from Eq. 5B.8 as:

$$\Delta\alpha_{1/2} = \frac{4\Gamma}{n_p \cos \alpha_{coup}} \tag{5B.11}$$

Equation 5B.11 indicates the decrease of the surface plasmon is proportional to the angular width of the dip while the factor $\cos \alpha_{coup}$ also varies but slowly.

In terms of the spectral reflectivity, we can calculate the optimum coupling spectral half-width of the dip $\Delta\lambda_{1/2}$:

$$\Delta\lambda_{1/2} = \frac{4\Gamma}{\left| \frac{dn_p}{d\lambda} \sin \alpha - \frac{dn_{ef}}{d\lambda} \right|} \tag{5B.12}$$

where $dn_p/d\lambda$ is the prism dispersion and $dn_{ef}/d\lambda$ is the effective index dispersion of the surface plasmon. Increasing the wavelength will decrease the attenuation coefficient Γ_i and the difference in dispersions of effective indices of the evanescent wave and that of the surface plasmon. Thus, these effects may compensate each other.

Appendix 6A: Analysis of Beam Bending

As we described in Chapter 6.xx, bending of a beam is a result of the internal strain distribution induced by the external force. The internal strain creates moment $M(x)$ that counters the rotational moment M_f from the external force. The actual expression of M_f has to be found for each condition. Moment $M(x)$ $(= -M_f)$ is then related to the curvature of the beam y'' in the following equation:

$$y'' = -\frac{M}{EI} \tag{6A.1}$$

Solving beam deflection thus comes down to the two steps of finding: (1) Moment $M(x)$ from the equilibrium equation, and (2) appropriate boundary conditions for y, y', and y'' to solve the differential equation (6A.1).

Table 6A.1 shows deflection of beams with different load distributions and boundary conditions. Finding M_f for each condition is good practice in learning about analysis of beam bending.

Table 6A.1: Beam Deflection

Type of Beam	Deflection	Curvature
Cantilever with a concentrated force F Example: Force sensor, AFM cantilever, nanowire resonator	$y = \frac{Fx^2(3L-x)}{6EI}$	$y'' = \frac{F(L-x)}{EI}$
Cantilever with an external moment M_f at the end (or cantilever with an uniform surface stress on top surface) Example: Bilayer cantilever, surface functionalized cantilever	$y = \frac{M_f \cdot x^2}{2EI}$	$y'' = \frac{M_f}{EI}$
Cantilever with an uniformly distributed force ω (N/m) Example: Pressure sensor	$y = \frac{\omega x^2(x^2 - 4Lx + 6L^2)}{24EI}$	$y'' = \frac{\omega \cdot (L-x)^2}{2EI}$

(Continued)

Table 6A.1: (Continued)

Type of Beam	Deflection	Curvature
Beam fixed at ends with a concentrated force F at the center	$y = \frac{Fx^2(3L-4x)}{48EI}$	$y'' = \frac{F(L-4x)}{8EI}$
	for $0 < x < \frac{L}{2}$	for $0 < x < \frac{L}{2}$
Example: Nanowire resonator, electrostatic actuator		
Beam fixed at ends with an uniformly distributed force ω (N/m)	$y = \frac{\omega(L^2 - 6Lx + 6x^2)}{12EI \cdot L}$	$y'' = \frac{\omega x^2(L-x)^2}{24EI \cdot L}$
	for $0 < x < \frac{L}{2}$	for $0 < x < \frac{L}{2}$
Example: Pressure sensor		
Beam simply supported at ends, with a concentrated force F at the center	$y = \frac{Fx}{12EI}\left(\frac{3}{4}L^2 - x^2\right)$	$y'' = = \frac{Fx}{2EI}$
	for $0 < x < \frac{L}{2}$	for $0 < x < \frac{L}{2}$
Beam simply supported at ends, with an uniformly distributed force ω (N/m)	$y = \frac{\omega x(L^3 - 2Lx^2 + x^3)}{24EI}$	$y'' = = \frac{\omega x(x-L)}{2EI}$
	for $0 < x < \frac{L}{2}$	for $0 < x < \frac{L}{2}$

Index

Edwards Brothers Malloy
Thorofare, NJ USA
September 22, 2015